FORECASTING AND MANAGEMENT OF TECHNOLOGY

FORECASTING AND MANAGEMENT OF TECHNOLOGY

ALAN L. PORTER
Georgia Institute of Technology, Atlanta, Georgia

A. THOMAS ROPER and THOMAS W. MASON
Rose-Hulman Institute of Technology, Terre Haute, Indiana

FREDERICK A. ROSSINI
George Mason University, Fairfax, Virginia

JERRY BANKS
Georgia Institute of Technology, Atlanta, Georgia

with software and documentation by

BRADLEY J. WIEDERHOLT
Institute for Software Innovation, Alpharetta, Georgia

A WILEY-INTERSCIENCE PUBLICATION
JOHN WILEY & SONS, INC.
NEW YORK   CHICHESTER   BRISBANE   TORONTO   SINGAPORE

Copyright © 1991 by John Wiley & Sons, Inc.

Library of Congress Cataloging in Publication Data:

Forecasting and management of technology / by Alan L. Porter . . . [et al.].
        p. cm. — (Wiley series in engineering and technology management)
  "A Wiley-Interscience publication."
  Includes index.
  ISBN 0-471-51223-0
  1. Technological forecasting. I. Porter, Alan L. II. Series.
T174.F67 1991
601'.12—dc20   90-45269
                                        CIP

Printed in the United States of America

10 9 8 7 6 5

PREFACE

This book calls for using visions of the future to guide present actions (Chapter 19). We were guided by a vision of this book as a *basic textbook on technology forecasting for business and engineering students*. Technology forecasting and assessment of the potential effects of technological change (impact assessment) must be used by those who wish to manage technology.

This is a "methods" book. It aims to show the reader "how to" apply a variety of methods at a basic, practical level. Part I gives a framework for the methods presented in Part II; Part III introduces the assessment and evaluation of technological change. Chapter 1 (Section 1.3) gives an overview of the three parts and their chapters. Although basic, the book's balanced treatment of these subjects will enable readers to apply the methods to significant issues in managing technological development.

The methods of technology forecasting are straightforward to learn. This book requires only basic mathematics to understand its presentation of forecasting methods. Many of the methods are qualitative or involve simple arithmetic. Linear regression is introduced (Chapter 9), beginning with basics, and it is developed to the level needed for technology trend extrapolations (Chapter 10). Matrices are essential to such techniques as cross-impact simulation (Chapter 12) and the analytic hierarchy process (Chapter 18), but they are developed with minimal algebra.

We have done our best to write an "Ex"-rated book: explicit cases illustrate application of the methods; many numbered Exhibits appear throughout; and Exercises are provided at the end of each chapter to develop skill in using the methods. We ask for your feedback on possible improvements and suggestions for additional case materials that could be included in future editions.

We thank many friends and colleagues for their help. In particular, our heartfelt thanks to:

- Our families, for their forebearance
- Scott Cunningham, for his discussion of the Lotka-Volterra models (Chapter 10)
- Will Fey, for his presentation of system dynamics (Chapter 12)
- Preston White, for multiple objective methods (Chapter 18)
- Mary Anne Pierce, Billie Holechko, and Brenda Downs, for typing, editing, and putting the pieces back together—over and over
- Chris Dede, Hal Linstone, Joe Martino, Jim Swain, and three anonymous reviewers, for helpful criticism, advice, and specific guidance on various aspects
- Dundar Kocaoglu, for his encouragement and drive
- Frank Cerra and his colleagues at John Wiley, for making this come together as a book

Without further ado, let's dive into a most fascinating but frustrating, momentous but doable subject—technology forecasting.

ALAN L. PORTER

November 1990

CONTENTS

PART I: OVERVIEW AND PRINCIPLES

1 Management, Technology, and the Future   **3**

1.1 Introduction, 3

1.2 Technology and Competitive Advantage, 4

    1.2.1 Technology and National Competitive Advantage, 4

    1.2.2 Technology and Competitive Advantage within the Firm, 9

1.3 The Book, 13

2 Principles of Sociotechnical Change   **17**

2.1 Theories of Social Change, 17

2.2 Developing a Framework for Sociotechnical Change, 20

    2.2.1 Technology Forecasting, 21

    2.2.2 Impact Assessment, 22

    2.2.3 Technology Management, 22

2.3 The Technology Delivery System, 22

2.4 Innovation Processes, 24

2.5 Data for Managing and Studying Sociotechnical Change, 26

2.6 Conclusion, 27

3 Technology Planning   **30**

3.1 Perspective on Planning, 30

    3.1.1 Planning, 30

    3.1.2 General Planning versus Technology Planning, 33

3.2 Technology Planning Methods, 34

3.2.1 Planning Processes, 34

3.2.2 Impact Assessment Principles, 36

3.2.3 A Seven-Step Planning Approach, 39

3.2.4 Some Illustrations of Goal Setting and Mission Statements for Planning, 41

3.3 Tools for Technology Planning, 46

4   Forecasting   **47**

4.1 Introduction, 47

4.1.1 The Future Is Uncertain, 49

4.1.2 Forecasting Targets, 51

4.1.3 Criteria for Good Forecasts, 52

4.1.4 Common Errors in Forecasting, 56

4.2 Technology Forecasting, 57

4.2.1 Models of Technology Growth and Diffusion, 58

4.2.2 Technology Forecasting in Context, 62

4.2.3 Technology Forecasting Methods, 63

4.2.4 Forecasts Impact Decision Making, 66

4.3 Social Forecasting, 67

4.4 Conclusion, 69

5   Managing the Forecasting Project   **71**

5.1 Introduction, 71

5.2 Information Requirements, 72

5.3 Project Scheduling, 75

5.3.1 PERT, 76

5.3.2 Gantt Chart, 78

5.3.3 Project Accountability Chart, 78

5.4 Management and Communications, 80

5.4.1 Managing the Forecasting Project, 80

5.4.2 Communications Patterns, 84

5.5 Conclusion, 85

PART II: METHODS

6   General Methodological Issues   **89**

6.1 Inquiring Systems, 89

6.2 Scientific Inquiry and Forecasting, 92

6.3 Strategies and Techniques for Forecasting, 93

6.3.1 Strategy for Technique Selection and Application, 93

6.3.2 Strategy for Communicating Forecasts to Users, 97

6.4 Conclusion, 98

7   The Stimulation of Creativity   **100**

7.1 Five Elements of Creativity, 100

7.2 Individual Creativity, 101

7.2.1 Lateral Thinking, 101

7.2.2 Metaphors and Analogies, 104

7.2.3 Checklists, 104

7.2.4 Morphological Analysis, 105

7.2.5 Random Words, 106

7.3 Group Creativity, 108

7.3.1 Brainstorming, 108

7.3.2 Crawford Slip Writing, 110

7.3.3 Other Group Techniques, 110

7.4 Conclusion, 111

8   Monitoring   **114**

8.1 Introduction, 114

8.2 Principles, 117

8.2.1 Assumptions, 117

8.2.2 Types of Monitoring, 118

8.3 Monitoring Steps, 125

8.3.1 Monitoring Objectives and Focus, 125

8.3.2 Technology Description and Contextual Mapping, 126

8.3.3 Monitoring Strategy, 126

8.3.4 Interpretation and Communication, 129

8.4 Implementation, 130

8.4.1 Who Should Conduct Monitoring? 130

8.4.2 How Should Monitoring Be Conducted? 133

8.4.3 Integrating Monitoring Results with Other
Forecasting Techniques, 134

8.5 Recommended Sources, 134

9   Basic Tools for Quantitative Trend Extrapolation   **138**

9.1 Introduction, 138

9.2 Selecting the Variables to Extrapolate, 140

9.3 Naive Models, 141

9.4 Adaptive Weighting, 145

9.5 Linear Regression, 147

    9.5.1 Regression Model Validity, 148

    9.5.2 Fitting the Regression Equation, 148

    9.5.3 Using Causal Relationships, 157

    9.5.4 Evaluating the Regression Model, 160

9.6 Transformations, 163

9.7 Regression and Forecasting—Some Extensions, 165

10   Extrapolating Technological Trends   169

10.1 Trend Analysis in Technology Forecasting, 169

10.2 Steps in Trend Analysis, 169

10.3 The S-Shaped Curves, 175

    10.3.1 The Fisher-Pry Model, 176

    10.3.2 The Gompertz Model, 182

    10.3.3 Choosing Fisher-Pry or Gompertz, 185

    10.3.4 Selecting an Upper Bound for the Forecast, 187

10.4 Toward a General Model of Technological Change: The Lotka-Volterra Equations, 187

    10.4.1 Theory, 187

    10.4.2 Functional Equivalences of the Model, 190

    10.4.3 Issues, 197

10.5 Conclusion, 197

11   Expert Opinion   200

11.1 Introduction, 200

11.2 Selecting Experts, 201

11.3 Selecting the Technique, 204

11.4 Techniques for Gathering Expert Opinion, 205

    11.4.1 Individual Input, 206

    11.4.2 Committees, Seminars, and Conferences, 207

    11.4.3 The Nominal Group Process, 209

    11.4.4 Surveys, 212

    11.4.5 The Delphi Technique, 214

    11.4.6 Delphi Variants, 218

    11.4.7 POSTURE, 219

11.5 Conclusion, 220

12   Simulation   222

12.1 Introduction, 222

12.2 Cross-Impact Analysis, 223

    12.2.1 Traditional Cross-Impact, 223

12.2.2 Time-Dependent CI, 229

12.2.3 Other Time-Dependent CI Schemes—XimpacT, 238

12.2.4 Closure, 239

12.3 KSIM, 241

12.4 Forecasting with System Dynamics, 246

12.4.1 The System Dynamics Philosophy of Human
Systems, 247

12.4.2 System Dynamics Methodology, 249

12.4.3 Technology Forecasting with System Dynamics, 250

12.5 Gaming Simulation, 254

12.6 Conclusion, 257

13   Scenarios   **259**

13.1 Introduction—What are Scenarios? 259

13.2 Types of Scenarios, 260

13.3 How Are Scenarios Used? 261

13.4 Constructing Scenarios, 263

13.4.1 The Topical Dimensions, 263

13.4.2 The User, 264

13.4.3 The Time Frame, 264

13.4.4 The Assumptions, 264

13.4.5 The Dimensions, 265

13.4.6 The Number and Emphasis of Scenarios, 266

13.4.7 Building and Presenting Scenarios, 267

13.4.8 Examples of Scenarios, 267

13.5 Critiquing Scenarios, 269

13.6 Conclusion, 270

14   Economic Forecasting and Analysis   **272**

14.1 Introduction, 272

14.2 Technology and the Economy, 273

14.3 Markets and Innovation, 274

14.4 Forecasting the Economy, 276

14.5 Input-Output Analysis, 279

14.6 Conclusion, 283

PART III: ASSESSMENT TO MANAGE TECHNOLOGICAL CHANGE

**15   Impact Assessment: General Issues and the Identification
of the Impacts of Technology**   **289**

15.1 Introduction, 289

15.1.1 General Considerations, 290

15.1.2 Some Distinctions, 291

15.2 General Issues in Impact Assessment, 292

    15.2.1 Assessment Steps, 292

    15.2.2 Issues in Assessment Quality, 293

    15.2.3 Impact Assessment at Different Scales of Effort, 294

    15.2.4 Problem Formulation and Bounding, 296

15.3 Impact Identification, 298

    15.3.1 Scanning Techniques, 298

    15.3.2 Tracing Techniques, 299

    15.3.3 Narrowing the Impact Set and Estimation of Effects, 300

15.4 Conclusion, 300

16   Analysis of the Impacts of Technologies   **303**

16.1 Introduction, 303

16.2 Impacts on Technology, 304

    16.2.1 Vertical Impacts, 304

    16.2.2 Horizontal Impacts, 305

    16.2.3 Integrative Impacts, 307

16.3 Institutional/Organizational Impacts, 308

    16.3.1 Internal Organizational Changes, 308

    16.3.2 External Institutional Changes, 308

16.4 Social Impacts, 309

    16.4.1 Social Impact Assessment, 309

    16.4.2 Socioeconomic Impact Assessment, 310

16.5 Cultural and Behavioral Impacts, 313

    16.5.1 Impacts of and on Values, 313

    16.5.2 Impacts on Behavior, 315

16.6 Political/Legal Impacts, 315

    16.6.1 Political Impacts, 315

    16.6.2 Legal Analysis, 316

16.7 International Impacts, 317

16.8 Environmental Impacts, 318

    16.8.1 Environmental Impact Assessment Methods, 318

    16.8.2 Topical Areas, 319

16.9 Health Impacts, 321

16.10 Conclusions, 322

17   Benefit/Cost and Risk Analysis   **324**

17.1 Introduction: Opportunity Costs and Choices—Again, 324

17.2 Benefit/Cost Analysis—Within the Firm, 325

17.3 Accounting for Risk—Within the Firm, 329

17.4 Benefit/Cost Analysis—Society's Stake and the Manager's Response, 335

17.5 Accounting for Risk in Social Decisions—Risk Assessment, 340

    17.5.1 A Four-Part Framework for Health Risk Assessment, 343

    17.5.2 Perceived Risk, 345

    17.5.3 Risky Decisions, 347

    17.5.4 Assessing Risks: Monetary and Basic Values, 350

17.6 Conclusion, 351

18 Evaluation of Technologies and Their Impacts   **354**

18.1 Introduction, 354

18.2 Criteria, 356

18.3 Alternatives, 358

18.4 Measures, 359

    18.4.1 Types and Levels of Measures, 359

    18.4.2 Measurement Inputs and Combinations, 361

18.5 The Analytic Hierarchy Process (AHP), 363

18.6 Multiple Objective Methods, 369

18.7 Participation and Mediation, 374

    18.7.1 Participation, 374

    18.7.2 Mediation, 375

18.8 Conclusion, 376

19 Managing the Present from the Future   **380**

19.1 Alternative Temporal Perspectives, 380

19.2 Looking Back: Historical Lessons on Technological Change, 381

19.3 Looking Back: Previous Forecasts and Assessments, 383

19.4 Looking Back: From the Future, 386

19.5 Visions, 387

Appendix A   **The Technology Forecasting TOOLKIT**   **391**

A.1 Introduction, 391

A.2 Hardware Systems Required to Use the Technology Forecasting TOOLKIT, 392

A.3 Installing and Running the Technology Forecasting TOOLKIT, 393

    A.3.1 Installing the Technology Forecasting TOOLKIT, 393

    A.3.2 Running the Technology Forecasting TOOLKIT, 393

A.4 A Quick Tutorial, 393

A.5 Individual Tools Available in the Technology Forecasting TOOLKIT, 397

A.5.1 Analytic Hierarchy Process, 401

A.5.2 Creativity Stimulation, 401

A.5.3 Cross-Impact Analysis, 402

A.5.4 KSIM—Kane Simulation, 403

A.5.5 Project Scheduling, 403

A.5.6 Trend Extrapolation, 404

Appendix B.1 F Statistic Values   **406**

Appendix B.2 t Statistic Values   **417**

Appendix C Sensitivity Analysis   **419**

Bibliography   **421**

Index   **441**

Enclosure: Technology Forecasting TOOLKIT Disk (for the IBM PC and compatibles)

FORECASTING AND MANAGEMENT OF TECHNOLOGY

PART I
OVERVIEW AND PRINCIPLES

___1
MANAGEMENT, TECHNOLOGY, AND THE FUTURE

OVERVIEW

Two themes confront the modern manager: technology as a key to productivity and change as a fact of life. These themes imply that today's technology manager must be able to forecast and assess technological change to obtain competitive advantage. This chapter links the "management of technology" to the future and explains the organization of the book.

1.1 INTRODUCTION

According to a classic definition, industrial management deals with "the four Ms" — men (and women), materials, money, and machines. A manager's emphasis, of course, can vary according to the context in which he or she operates — production, service, government, or even education (with some reservations about associating the notion of management with either faculty or students). Achievement in any of these contexts in the 1990s demands attention to technology as a critical force. For instance,

- *Production*: An estimated half of Hewlett-Packard's 1990 *products* were developed in the previous three years; improved *processes*, such as X-ray lithography to etch fine lines in semiconductors, determine which companies (and countries) remain competitive.
- *Service*: The way in which firms interact with customers is changing with the introduction of automatic teller machines (ATMs), electronic mail (E-mail),

facsimile transmission (FAX), home services through computerized networks such as Prodigy™, and other innovations.

- *Government*: Defense depends heavily on superior technology as the nations of the world strive to balance smart missiles, cruise missiles, and the Strategic Defense Initiative (SDI); now even the Internal Revenue Service (IRS) is using electronic income tax filing.
- *Education*: Senator Kennedy introduced another bill in 1990 to remedy America's coming shortfall of people with critical technical skills; computers have invaded schools, and hypermedia techniques bid to replace the traditional lecture/recitation format.

Although not as alliterative as the four Ms, a more succinct description of the focus of today's manager in any of these contexts would be *technology* and *people*.

This book stresses technology, especially technological change, but it does so with a purpose in mind—to support the *management* of technology. That management requires full attention to the "people" issues associated with technological change. Those issues involve both people outside the organization who care how a technology will affect them and act to influence its development, and people within the organization who must adapt their skills and alter their working relationships as new technology is introduced. Technological change improves economic productivity more than any other factor (see Section 14.2); consequently, it is a major concern of governments. Likewise, the firm must manage technological innovation in the context of its markets, its structure, and its competition (Section 14.3). Section 1.2 relates technology to competitive advantage.

1.2 TECHNOLOGY AND COMPETITIVE ADVANTAGE

During the 1980s the United States went from the world's greatest trade surplus to the world's greatest trade deficit, running well over $100 billion per year through the late 1980s. What happened? Attempts to resolve unfair trading practices and devaluation of the dollar did not restore balance. The prime culprit appeared to be a loss in American competitive advantage—an advantage formerly secured by a lead in technology. Section 1.2.1 explores the relation between technology and competitive advantage at a national level; Section 1.2.2, at the level of the individual firm.

1.2.1 Technology and National Competitive Advantage

Markets are becoming increasingly global as a result of the internationalization of customer desires, advances in communications, decreases in transportation costs, a lowering of trade barriers, and the growing prominence of multinational companies—which account for more than two-thirds of U.S. industrial output (National Academy of Engineering, 1987)! Exhibit 1.1 demonstrates how complex multinationalism is becoming. Technology-based goods and services account for a growing proportion of the value of international trade (Roessner and Porter, 1990). Exhibit 1.2 tells the story of a recent technological success and predicts a coming one. International competitiveness in technology-based industries is the new metric of national economic achievement. A nation's competitiveness depends on the capacity of its industry to innovate (M. Porter, 1990). Countries use various means to build the capacity to create and market new technology, including the adaptation of technology developed in other countries. Analysts are trying to understand these means and their effectiveness. Segal (1986) notes the importance of informal mechanisms in the adoption of new technology; Wallender (1979) points to the importance of national support systems; Dahlman and Westphal (1981) focus on enhancing technological mastery; Cohen and associates (1984) and Scott and Lodge (1985) look for relative comparative advantage.

■ **Exhibit 1.1 The Multinational Automobile (Based on *Consumer Reports*, 1990).**

The Geo Prizm, a renamed Chevrolet Nova, comes off the same California assembly line that turns out Toyota Corollas. The Prizm is one of the Geo cars sold by Chevrolet dealers but made by Japanese companies. The Geo Metro is a subcompact made by Suzuki; the Geo Tracker is a clone of the Suzuki Sidekick sport-utility vehicle; and the Geo Storm is a sporty car made by Isuzu.

Ford, meanwhile, is to introduce two 1991 models based on the Mazda Protege: the Escort, made in Michigan, and the Mercury Tracer, made in Mexico. ■

Such factors can be brought together in a dynamic socio-technical context as a *technology delivery system* or *TDS* (Wenk and Kuehn, 1977), a concept discussed further in Chapter 2. This perspective on the development of technology emphasizes the need for an effective enterprise to combine the necessary inputs (for instance, technical know-how, raw materials, capital, and managerial abilities) to deliver marketable products. The TDS also maps the critical environmental influences on the success or failure of a technological development (see Figure 1.1). Four main factors appear critical to the performance of nations in the international market for technology-intensive products:

- *National Orientation*: A nation's use of directed action to enhance technology-oriented production
- *Socioeconomic Infrastructure*: The social and economic institutions that support and maintain the physical, human, organizational, and financial resources essential to the functioning of a modern, technology-based industrial nation
- *Technological Infrastructure*: The institutions and resources needed to apply technology-intensive manufacturing processes and to develop, manufacture, and market technology-intensive products
- *Productive Capacity*: The physical and human resources committed to manufacturing and the efficiency with which these are used

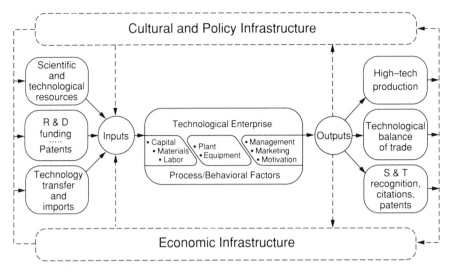

Figure 1.1 National capabilities to absorb and utilize technology. (*Source:* The American Society of Mechanical Engineers, *Manufacturing International '88*, Vol. 2. Reprinted by permission.)

■ **Exhibit 1.2   Two Tales of Technology: The PC and the Optical Cable (Based on Keyworth, 1990).**

In 1980 one million Americans used computers; in 1990 fifty million do so. Perhaps no technology has ever taken off so rapidly and changed society so profoundly.

The source of this change is the personal computer, the PC. Not many forecasted its importance—it ran counter to the trend of bigger and more powerful computers. Above all, it was not clear what useful functions it would serve. Three factors continue to drive the PC to success:

- *Technological*: Constantly increasing power at less cost "pushes" creation of new uses.
- *Psychological*: Individualistic Americans love the independence the PC provides.
- *Market "Pull"*: The exploding growth of the market attracts producers and allows economies of scale.

The PC market has been the most effective wealth-generating machine of the American economy in the 1980s and the fastest-growing source of new jobs. The PC turned the computer industry on its head; the "sons of PC" (microprocessors, workstations) will turn more industries on their heads in the 1990s.

The fiber optic network is coming. Japanese forecasters anticipate about 30 percent of the Japanese GNP in 25 years will come from new goods and services made possible by high-capacity (broadband) fiber optic networking that will go everywhere the telephone system is now connected. Fiber to the home will provide the physical infrastructure for computing and communications needs, just as roads provided the transportation infrastructure for the car. The digital network will create vast economic opportunities, such as digital FAX, digital photography, and multimedia displays that blend pictures, video, sound, and text.   ■

Consider the story in Exhibit 1.2 that describes the coming infrastructure for the Information Age—the fiber optic network; its success depends on national policy and economics to encourage laying that fiber to the home. In turn, the presence or absence of this network will affect industrial opportunities and therefore international competitiveness.

Companies and countries confront growing national and international competition that puts a premium on productivity, rapid product introduction, quality, and reliability. And, increasingly, technology is becoming a measure of national power and an implement of public policy. Figure 1.2 shows the rapid emergence of international competitors in what was once almost the sole property of the United States—high technology (that is, high-value-added products that tend to embody the current state of the art and have a large research and development—R&D—content). High technology *standing* combines the critical areas of aerospace; chemicals and plastics; computers; drugs; electronic components; engines, machines, and

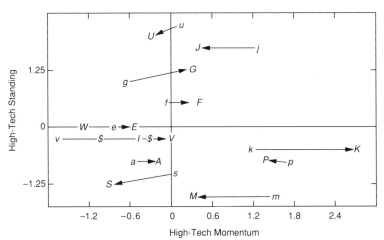

C: Canada   *U*: US   *J*: Japan   *K*: South Korea   *M*: Malaysia   *P*: Singapore   *F*: France
G: West Germany   *I*: Italy   *E*: UK   *N*: Netherlands   *V*: Switzerland   *S*: Spain   *A*: Australia
W: Sweden   *$*: multiple occurence

Note: $ (multiple occurence) would show the Netherlands increasing from the left $ to the right $;
Canada, decreasing from the right $ to the left.
The small letters show the indicators as of 1985 (with momentum from a 1981 base); the
large letters show the 1987 indicators (with momentum from 1985).
Changes from 1985 to 1987 for Sweden and Italy are too small to show.
All indicators are relative to the countries in this sample; both indicators are standardized
for the sample to have a mean of 0 and a standard deviation of 1.

Figure 1.2   Indicators of high-technology competitiveness among nations.

robotics; professional and scientific instruments; and radio, TV, and telecommunications equipment. It is a composite indicator based on actual exports and expert opinions about each country. High technology *momentum* gauges recent changes in capabilities, based on four component measures. Figure 1.2 and its supporting data show that

- The U.S. lead in high tech has shrunk; in 1987 the United States exported about $89 billion; Japan, $83 billion; and a combined Germany will challenge these levels. Also the United States imported almost as much in high tech products as it exported.
- The newly industrialized countries (NICs) of Asia are real tigers, competing aggressively, not just in generic types of products, but in high technology too (Xu, Porter, and Roessner, 1990).

For 20 years following World War II the United States assumed and retained world technological leadership and gloried in rapid growth in productivity and income. The next 20 years saw America lose its leadership position in most smokestack industries (those production processes reflecting moderate technological requirements and yielding relatively standard products) (Landau and Rosenberg, 1986). The data in Figure 1.2 show U.S. leadership now dissolving in high-technology production also.

In 1989 the MIT Commission on Industrial Productivity put forth recommendations on what needs to be done to improve America's competitive posture (Dertouzos, Lester, and Solow, 1989). These boil down to remedying six recurring patterns:

- Outdated strategies
- Short time horizons
- Technological weaknesses in development and production
- Neglect of human resources
- Failures of cooperation
- Work at cross-purposes by government and industry

The commission recommended that universities take a hard look at how engineering and management educational practices can be altered to produce students and faculty with

1. Knowledge of real problems with social, economic, and political dimensions
2. Ability to work effectively in teams to create new products, processes, and systems
3. Ability to work effectively across disciplines
4. Ability to integrate a sound understanding of science and technology with practical knowledge, an international orientation, and managerial skills

The first three recurring patterns in the MIT report point toward the need to monitor and forecast technological change and then to integrate this knowledge into governmental and corporate management. Reich (1989) hits the same chord:

> The first step in rapidly assimilating new technologies is to discover what they are. American firms are often slow to learn of a new technological insight. . . . Japanese firms regard global scanning for technological insights as an integral part of their business strategies.

The MIT educational recommendations emphasize linkage of technical knowledge with "people"—a theme echoed throughout this book, particularly in directing attention to the impacts of technological change (Chapters 15–18). Reich (1989) points out the need to create such links at many levels—between R&D and commercial production within a firm; among firms and government agencies in a collaborative effort to adopt industrywide standards early; and between academia and industry to train a sophisticated work force knowledgeable about both technology and business.

1.2.2 Technology and Competitive Advantage within the Firm

In asserting national competitiveness the firm is the center of action because in capitalist economies the firm is almost always the enterprise that must deliver the technology (Figure 1.1). Exceptions certainly exist; for instance, NASA (the National Aeronautics and Space Administration) must deliver space shuttles and space stations. Other nations long committed to central planning have recently moved abruptly toward a private economy, now convinced that the private firm is indeed more effective than the government ministry at delivering most products and services.

Engineers and managers must reduce cost, waste, overhead, and inventory if their firm is to be competitive. New electronic and computer-based technologies are revolutionizing design, manufacturing, and service functions, changing forever the ways in which organizations conduct routine business. At the same time technology managers face the research, development, design, and production of increasingly sophisticated products—which enjoy shorter and shorter market lives. The dazzling increases in power, reductions in costs, and kaleidoscopic parade of new computer products only exemplify the pace of technological change. Simultaneously, the manager must account for safety, environmental effects, and societal impacts. These involve both complex social issues and sophisticated science and engineering. On the other hand, engineers must deal with strategic as well as technical considerations as they strive to advance technology. The *management of technology* plays an ever-increasing role in management and in engineering.

Management of technology for competitive advantage begins with *R&D*. R&D efforts must be directed toward technologies that are most likely to impact a company's prime markets. R&D project selection and evaluation must keep customer desires and market dynamics in view far better than has traditionally been the case in the United States. The technological progress function (Chapter 10) can help assess the remaining potential to be mined in a given technological area.

New technologies emanating from R&D must work their way through design and manufacturing to emerge in the form of beneficial *innovations*. The United States is a leader in invention, but not in moving inventions through to innovation and diffusion. Too often the transition of new inventions from R&D into products or processes has been left to serendipity; it needs to be managed. The thrust toward concurrent engineering usefully prods firms into considering manufacturability, marketing, and maintenance—throughout the *design* process—by including representatives of those functions on the design team from the beginning. The project-dominant model of innovation presented in Chapter 2 addresses such concerns. Teece (1988) warns that innovators must manage technological change carefully or risk losing any potential competitive advantage to followers or imitators. Monitoring the competitive environment (Chapter 8) supports this managerial function. Technology managers need to link innovation processes closely to business strategy.

The increasing pace of technological change confronts technology managers with tough investment choices. Technology managers agree that there is too much emphasis in the United States on risk avoidance (Rouse and Rogers, 1990). On the one hand, the notion of the aggressive *entrepreneur* pursuing opportunity with a passion draws great favor. Large companies take special pains not to stifle entrepreneurial spirit within their bureaucracy. The notion of "intrapreneurship"—the innovative spirit within a company—enjoyed recent popularity leading to a variety of corporate structures, such as separate divisions (used to launch IBM's PC), spin-off companies (CDC's failed ETA supercomputer company), and anything else that might work. On the other hand, entrepreneurial risk requires careful assessment (see Dickson and Giglierano, 1986). The stakes involved in some innovations are huge—for instance, the estimated $1 billion cost for developing X-ray lithography to etch finer structures in silicon for computer chips, for an uncertain payoff. To make such decisions the manager must combine sound technology forecasts with economic analyses and risk assessment (Chapters 14 and 17).

Manufacturing conveys a whole alphabet soup of technology management issues: QC, JIT, FMS, CIMS, and the like. These popular ideas offer potential gains to the firm: quality control to provide quality products for highly competitive markets; just-in-time manufacturing to reduce inventories; flexible manufacturing systems to produce specialized products for niche markets; and computer-integrated manufacturing systems to provide common information bases to enhance process control. Whether such initiatives make sense for a given plant depends on how they affect the firm's competitive edge. Specifically, Miller and Vollmann (1985) warn companies not to focus too much on innovations to reduce direct operational costs. Rather, they point out the relatively greater importance of indirect costs, or overhead. Not since 1850 have direct labor costs contributed more than overhead to total value added in manufacturing; today direct labor amounts to less than 25 percent of value added! Furthermore, each of these process innovations implies change in organizational structure to take full advantage (for instance, cut the levels of middle management with CIMS). And each implies a need to adjust task and job designs. These "people" changes, in turn, affect autonomy, teaming, and other factors likely to have an impact on the morale of workers and managers (see Klein, 1989).

Services are also candidates for technology-based process innovations. Services account for about 70 percent of U.S. gross national product (GNP) (Quinn and Gagnon, 1986). The service sector is at least as capital-intensive as manufacturing, and many industries within it are highly technological. Some 80 percent of information technology equipment is sold to the service sector in the United States. This information technology can alter the bases of competition by enabling firms to cope with greater complexity, cross traditional market boundaries (as when banks sell stocks and bonds), and achieve economies of scale and scope (Quinn and Gagnon, 1986). Investment in information technology is not reflected in better organizational performance—unless the technology ties into competitive strategy (see Porter and Millar, 1985).

The competitive strategies of firms can be classified as *technology-driven* or *market-driven*. Technology-driven firms tend to be dominated by engineers, emphasize R&D to devise new products, and let the customers "beat a path to their better mousetraps." Market-driven firms let the marketing function dictate what the firm will develop and produce. Ansoff (1987) calls for a combined strategy that recognizes the heightened importance of technological change today (suggesting elevation of engineers to leadership positions in companies where they have been ignored) yet balances this with keen awareness of market opportunities (suggesting a reduction in the emphasis on technology for its own sake in certain high-tech companies). Business plans and technology plans should be tightly coupled (Rouse and Rogers, 1990).

Not to be ignored in the strategic balance is manufacturing. Wheelwright and Hayes (1985) distinguish four stages in manufacturing's strategic role, wherein top management

1. Considers manufacturing incapable of influencing competitive success; just keeps it from causing negative impacts
2. Considers manufacturing neutral; just keeps it the same as the competition (similar workforce agreements, economies of scale)
3. Expects manufacturing actively to strengthen the company's competitive position; screens manufacturing investments for consistency with the business strategy and considers longer-term manufacturing developments and trends systematically
4. Seeks manufacturing-based competitive advantage (anticipates the potential of new manufacturing technologies); involves manufacturing up front in major corporate decisions; and pursues long-range programs to establish capabilities in advance of needs

Exhibit 1.3 illustrates the competitive potential in stage 4.

■ **Exhibit 1.3   The General Electric Dishwasher (Based on Wheelwright and Hayes, 1985).**

In the late 1970s GE's dishwasher strategic business unit (SBU) analyzed its operations and saw an aging process—20-year-old product design; 10- to 20-year-old manufacturing process; and a strong traditional union. The SBU proposed a stage 3 commitment to update the product and the manufacturing process, even though they still held the leading position in the U.S. market, selling one-third of all units.

GE's senior management considered the long-term prospects for the business and turned down the requested investment! They told the SBU to up its thinking to a more strategic level (that is, to stage 4). Senior management saw potential to significantly improve operating product performance and make profitable changes in production processes—all with an eye to a market with real growth potential (only 55 percent of U.S. households owned dishwashers).

The final plan entailed more than double the initially proposed investment. It called for a complete redesign of the product around an innovation of a single-piece plastic tub and a single-piece plastic door. The manufacturing process was redesigned, with the factory workers participating heavily (this took over two years). This improved the factory's working environment through better communication and greater worker involvement. Quality standards were toughened internally, and external suppliers were required to reduce defects to one-twentieth of former levels. This netted pronounced improvements in unit costs, worker productivity, and product reliability. Following introduction of the new product, GE market share jumped within 12 months and *Consumer Reports* rated it as offering the best value among U.S. dishwashers.   ■

Regardless of what is managed, management implies concern about the future. Even ignoring the future involves the implicit assumption that tomorrow will not differ in any important way from today. All managers must make decisions today that affect the organization's future. Sales managers, for example, must project sales to help other managers identify staffing needs, establish production goals, or plan to meet financial requirements. Personnel managers help determine the talent needed to face tomorrow's problems. A professional basketball team manager must consider whom to draft or trade to build a winning team next year—or the year after—or the year after that. As Smits, Rossini, and Davis (1987) phrase it, the challenge lies in *managing the present from the future*.

Figure 1.3 illustrates how technology forecasting can position the manager in the future. This diagram suggests that the business core of the telephone-operating companies—providing exchange services to connect callers and manage the flow of information between them—is likely to be eaten away by alternative information routes in the future. This implies that telephone company management must initiate new business areas today or watch their companies decline tomorrow.

The future is not predestined. It is unknown and unknowable. The good news here is that the future will be shaped by today's decisions; the bad news, that those decisions must be based on extensions of today's knowledge and thus involve un-

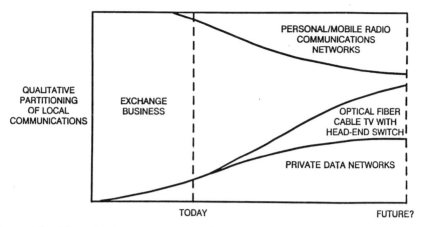

Figure 1.3 The technology squeeze on exchange business. (Reprinted by permission of Bell South.)

certainty. Therefore, the technology manager must develop the organizational and planning tools to deal with uncertainty—and the mental toughness to make decisions that will affect the well-being of his or her organization in that uncertain future. Tools used to generate the information on which those decisions will be based—whether they involve trends, modeling, or expert opinion—must illuminate the possible range of alternative futures. Planning and decision making must acknowledge the uncertainty of any future associated with a current choice. Perhaps the most important thing that can be learned about managing technology is the mindset that provides the flexibility to plan for and adapt to the unexpected.

Some decisions the manager must make affect the organization immediately. These usually require but modest extensions of existing knowledge and often can be made by using time-tested techniques that account mainly for factors internal to the firm and the technology involved. Other decisions will affect the positioning of the organization relative to broader market or social factors in the future. These require longer-term (and more uncertain) extensions of knowledge, consideration of a variety of external factors, and more complex forecasting and assessment techniques. All extensions of existing knowledge involve levels of uncertainty that increase rapidly with the time horizon of the decision. Regardless of their immediacy, however, all decisions will affect the future. The manager must base them on the clearest possible picture of the bounds of important future parameters if he or she is to act wisely. That is what this book is about.

1.3 THE BOOK

This book emphasizes two essential elements in the management of technology: forecasting technological change and assessing the impacts of such change. It aims to serve two groups who play major roles in managing technology: engineers and managers. The book provides principles and methods to extend present knowledge into the future and thus to support sound decisions about technology.

Technology works well only when it meshes smoothly with the people involved. Good forecasters and managers appreciate the importance of integrating technology into the social and organizational context. They can develop this appreciation by assessing the impacts of technology and considering the present and future social contexts affecting technology forecasts. At this point, the authors want to make clear that the book does not address all the factors that must be weighed in managing technology. These include substantive knowledge of specific technologies, specific governmental and corporate policies that affect technology, and organizational behavior with respect to technology development and implementation. Rather, it emphasizes the essential tools for forecasting and assessing technology in the context of technology management. Numerous examples illustrate the application of these tools in real contexts.

The book is organized into three parts: Part I (Chapters 1 through 5) provides an overview and principles; Part II presents a detailed discussion of methods in Chapters 6–14; Part III focuses on the analysis of technological changes and their effects, with a concluding perspective in Chapter 19.

Part I defines the framework. Chapter 2 links technology to society by providing principles of socio-technical change, observations on the process of technological innovation, and a model (called the *technology delivery system*) for the immediate social/economic context of a technology. Developing and applying new technology are critical elements in technology management. Chapter 3 introduces means to assess technological change and to plan strategically. Chapter 4 provides the basics that underlie forecasting. Chapter 5 contributes an approach to managing forecasting projects. Together, these chapters build the structure within which one can understand the specific methods considered in Part II.

Part II presents methods that can be used to create information about the future upon which to base managerial decisions. Emphasis is placed on understanding the basics of these methods and on choosing the method appropriate to a particular situation. Many examples are given. Some illustrate how to carry out the method in simple contexts; others sketch more complex applications. The methods presented are those the authors deem most widely usable based on their diverse professional perspectives. Chapter 6 develops general methodological considerations and identifies the core methods that constitute technology forecasting. Chapters 7–13 provide specific forecasting methods. Chapter 14 addresses economic forecasting.

Part III is based on the strong conviction that technological change must be considered together with its likely effects. This applies at two levels. First, to get the forecasts right the would-be forecaster must consider changes in socioeconomic context and people's likely reactions to new technologies. Second, the technology manager must estimate the economic benefits, determine the risks, and assess the full range of impacts before deciding whether to go ahead with development or adoption of a new technology.

Chapters 15 through 18 cover the impact assessment issues that arise in planning for the direct and indirect effects of technological changes. Chapter 17 addresses economic costs and benefits, and risk analysis, within this framework; it also relates to the treatment of economic forecasting in Chapter 14. Chapter 19 presents the considerations necessary for the manager to anticipate and evaluate change so as to make the decisions today that will lead to the desired future—to manage the present from the future.

Appendix A describes the accompanying software (for IBM PCs and compatibles) that supplements the text. The Technology Forecasting TOOLKIT provides the power to implement many of the methods described in the book and is thus an invaluable aid. Using the TOOLKIT, the reader can learn to apply the methods and develop sensitivity to parameter precision and initial conditions. The TOOLKIT software will also handle real problems with moderate data requirements.

Throughout, the book underscores the importance of the future in managing the technology of today. The chapters that follow explain and illustrate the basic concepts and methods required for this complex and uncertain task. For those who wish to pursue topics or approaches more deeply, extensive references are provided as guides to the literature.

EXERCISES

1.1 The U.S. Department of Commerce categorizes business activities into the following *sectors*:

- Mining
- Private households
- Farming
- Construction
- Transportation and utilities
- Finance, insurance, and real estate
- Government
- Manufacturing
- Trade
- Services

Here is a list of selected information-oriented *technologies* likely to advance technically and/or in extent of application during the 1990s.

- Animation/visualization
- Displays (such as helmet-mounted displays)
- Input devices (such as speech recognition)

- Multimedia (such as digital video interactive — DVI — able to combine full-motion video, graphics, sound, data, and text on compact optical disks for interactive play)
- Extensive networking of computers
- Robots with visual capabilities
- Expert systems with some capability to learn from their mistakes

(A) Pick one of these technologies (or some other) and identify possible applications in five different sectors during the next 10 years. *Hint*: Think in terms of products, services, communication mechanisms (internal or customer-related), training, integration of information, or other *functions* that could be changed by the technology.

(B) For each application identified in (A), indicate how it would affect the competitive advantage of a firm that produces the new application and of a firm that uses the new application.

(C) For one of your possible applications, identify various *management* issues posed. *Hint*: Choose an appropriate, specific perspective; for instance, assume one of the following roles:

- Federal regulator
- Sales manager of an established company developing a new product
- CEO of a start-up company trying to produce its first product or service
- Engineering project manager tasked with designing a new product
- Manufacturing engineer in charge of incorporating the new technology into an existing production process

1.2 Take the role of one professional, such as those listed in Exercise 1.1 (C), working for a specific organization concerned primarily with one of the sectors identified in Exercise 1.1. Identify a range of technologies pertinent to the effective performance of your work. For each technology, identify one possible major change that could affect your work in the next 10 years.

1.3 Monitor print media for one week and clip articles dealing with technological advances. Analyze these articles by describing the technology prior to the advance, the advance itself, and the technological capability after the advance. (*Hint*: You might use such print media as your local newspaper, the *New York Times*, *Wall Street Journal*, *USA Today*, *Time*, *Newsweek*, and *Business Week*, for example.)

___2
PRINCIPLES
OF SOCIOTECHNICAL
CHANGE

OVERVIEW

This chapter provides a coherent perspective for forecasting and managing sociotechnical change. Technology and society are interrelated, and both are changing rapidly. Changes in technology feed upon themselves, producing a stock of concepts that may be refined, developed, and used as the basis for further change—and the development and dissemination of technology create forces that can cause change in every aspect of society. On the other hand, changes in society produce conditions that cause technological change. We must understand these interrelationships to forecast and manage technology effectively. Unfortunately, a macrotheory of sociotechnical change has not yet been devised. Nevertheless, major social theories are presented as a backdrop for sketches of the significant aspects of change. The innovation-specific technology delivery system (TDS) maps the institutional linkages involved in a particular innovation. Finally, observations about data requirements are presented.

2.1 THEORIES OF SOCIAL CHANGE

This section briefly and selectively presents some of the principal perspectives that have shaped our concepts of social change. It is largely based on an analysis by Rossini and Porter (1982).

Theories of social change rest on descriptions of social systems, their dynamics and patterns of change. These theories are usually expressed in ordinary language, but they may also be expressed as mathematical or computer-based models. They seek to answer the questions "What is it?" and "How does it work?" Of and by itself a theory is a pure postulate. It may or may not be based on data or concrete observation. Traditional approaches to social change have generally been developed by thinkers who, operating at high levels of aggregation and abstraction, had minimal interest in relating their theories to systematic observations or social data. Theories of social change as understood today had their origins in the nineteenth century. Typically their scale is broad and they focus on global issues. Theories dealing with change in restricted environments, such as an organization, arose in the twentieth century.

Theories of social change are based on one or more of several major concepts: evolutionary, cyclic, functional, conflict, and critical/psychoanalytic. *Evolutionary theories* of social change emphasize the movement from lower to higher forms of society (unlike earlier views of society as decaying from a prehistoric golden age). Theorists proposed that, just as biological species evolve, so societies adapt through progressive change. Nineteenth-century thinkers such as Comte (1875) posited stages of development focused on the areas of their major interest. For instance, Comte's own interest was cognitive style, and his evolutionary trajectory for societies extended from the theological through the metaphysical to the positivistic or scientific. Maine, on the other hand, described progress from a patriarchal, status-oriented social order to a libertarian, contractual one (see, for example, Appelbaum, 1970).

One generic weakness of evolutionary theories is that they identify a desired end state and arbitrarily assume that it is the best state. This forecloses the issue of future changes that may be equally substantial. The viability of the mechanisms that drive the evolution and thus the dynamics of society must also be questioned. Yet societies do change, and long term-changes often occur in a consistent direction.

As contrasted to the linear progression envisioned by evolutionary theorists, *cyclic theories* posit that social development may be seen as a recurring pattern. The cyclic model of social or cultural change is an analogy to the growth, maturity, and decay of an organism as described by Spengler (1939) who even attached times to the phases of the cycle. Other examples of cyclic theories are Pareto's theory of the circulation of elites within a society (Pareto, 1935) and Sorokin's cyclic theory of types of cultures (Sorokin, 1962). Another famous cyclical theory is that of Kondratieff who suggested that the ebb and flow of technological developments caused capitalist societies to experience long waves of economic fluctuation with major depressions approximately every 60 years (see Drucker, 1985; Volland, 1987). One rather obvious difficulty with cyclic theories is that no society ever reenters precisely the same state. Thus it might be better to speak of spiral theories or, if no direction could be identified, oscillating theories. For indeed, society may exhibit oscillatory behavior. The theoretical frameworks that emphasize cyclic and oscillatory behavior do not readily support predictions.

Functionalist theories hypothesize equilibrium states for society. They emphasize stable social systems and the mechanisms that return them to equilibrium after they have been perturbed. Such mechanisms are like gyroscopic feedback that resists change. Although change obviously occurs, functionalist theories emphasize the persistence of the existing structure. Thus such theories can be seen as conservative, maintaining the status quo. Classic examples of functionalism include the work of Parsons in sociology (Parsons, 1951) and Malinowski in anthropology (Malinowski, 1944). Parsons, for example, distinguishes between processes that produce structural stability and those that produce structural change. However, he emphasizes the former, which effect "latent pattern maintenance."

Stabilizing mechanisms certainly play important roles in social change. Yet it is questionable that they return the social system to a preexisting state. It is more likely that they lead to a new state of equilibrium, producing the kind of metastability described in Ogburn's view of cultural lag (see, for example, Appelbaum, 1970). For Ogburn, culture is divided into material (most notably, technological) and nonmaterial (customs, beliefs, philosophies) elements. If one element, say technology, changes more rapidly than another, related element, a lag is created. Stabilizing feedback brings the system to a new equilibrium point. The magnitude of the maladjustment and the length of the lag may be important indicators.

Conflict theory focuses on change and is illustrated by the work of Marx (1967). He analyzed social dynamics on a broad scale and emphasized the role of the economic aspects of production. Although Marx's analysis had evolutionary elements, it rested on conflict between socioeconomic classes. He viewed this conflict as inevitable and analyzed it in depth. Functionalism and conflict theory each emphasize complementary processes that the other ignores. There is no doubt that conflict is a critical element in social change—look at the changes produced by revolutions in the past two centuries alone. Yet neither conflict nor stability provides the entire picture. The concepts of functionalism and conflict analysis seem to beg for synthesis.

Critical social theory (see, for instance, Habermas, 1971) is the social analog of individual psychoanalysis. Its goal is a self-understanding that society can use to transform itself. While the goal of psychoanalysis is to allow the individual to function effectively within an existing social order, not even this somewhat shaky norm exists for critical social theory. Indeed any norm would itself be subject to criticism as creating a normative view of society. At best this theory depicts a society that continually reestablishes its norms through an ongoing process of self-criticism. Obviously, this process is intimately involved with social change.

The classical theories of change are highly abstract and aggregated. Thus they are difficult to operationalize and to relate to commonly available data. Each theory adopts a single perspective and emphasizes a particular causal mechanism and direction. All offer valuable insights into social change, yet each is too impoverished alone to grasp the complex variety of social change. There is no theory that is broad enough to embrace all sociotechnical change. Technology managers must substitute ad hoc analyses involving techniques such as the technology delivery system (Section 2.3) and various forecasting methods. Yet this leaves the very real problem of establishing a conceptual framework that is robust enough to deal with the bewildering variety of complex situations that characterize sociotechnical change. Without this framework, there is no unified way of talking about commonalities in the diverse sociotechnical changes of concern in technology management and forecasting. For example, both the atomic bomb and the oral contraceptive arose from societal need, and both had profound behavioral and economic consequences. The next section strives to develop a framework robust enough to deal with such issues.

2.2 DEVELOPING A FRAMEWORK FOR SOCIOTECHNICAL CHANGE

"Society" and "technology" are abstractions for a wide range of entities with various levels of concreteness and aggregation. Therefore, a useful analysis must begin with a more specific definition of the sociotechnical system. The elements of the system can be either social or technical, although there will, of course, be gray areas. While any element can be a cause or an effect, there is no exclusively unidirectional causality. Thus the view that technologies cause societies to change by restructuring institutions and behaviors is as valid as the perspective that societies cause technologies to change by investment or by institutionalizing the processes of innovation.

The sociotechnical system is "open" in the sense that new elements may be either produced within the system or introduced from without. These elements include technological devices and principles, scientific knowledge, institutions, individuals, money, natural resources, and values. The system changes over time because of the interaction of its elements. External elements may enter the system or produce effects within it from the outside (see, for example, Exhibit 2.1). Likewise, elements within the system may leave it or may influence the world outside. The boundaries of the system are, in general, arbitrary. Processes involved in sociotechnical change include the development of scientific knowledge, technological principles, and prototypes; the production and diffusion of technological devices; changes in the characteristics and objectives of institutions; changes in the characteristics and beliefs of human beings; changes in wealth and resources; and changes in values.

■ **Exhibit 2.1   External Elements Sometimes Produce Effects (Based on Toffler, 1969).**

Toffler illustrates how an external technology, the match, altered sexual behavior in an African community. As described by the Dutch sociologist Egbert de Vries, this community believed that a new fire should be lit after each sexual act. Because this could be done only by borrowing fire from a neighbor's hut, adultery could be both embarrassing and dangerous. The introduction of the common match removed the need to "advertise" an individual's sexual adventures and so changed the community's behavior patterns. Toffler notes that the birth control pill may have played a similar role in industrial societies.   ■

None of the traditional theories discussed in Section 2.1 fits all the processes that constitute sociotechnical change, although some change processes may correspond to one or more of the types these theories describe. For example, the development of scientific knowledge is not cyclic; institutional change does not necessarily maintain equilibrium; changes in the state of resources are not necessarily the result of conflict; and so forth. Instead, sociotechnical change can be looked upon as a spiral process in which an open system changes under the mutual causal influences of its elements. Thus, although system change is not linear, neither is it truly cyclic in the sense that the system enters and reenters the same range of states.

A number of generic concomitants of change should be mentioned (Rossini and Porter, 1982). These include three possible outcomes: system stabilization, incremental change, or discontinuous change. Change may occur to achieve the goals of a system or any of the elements within it. It may occur to correct mismatches between a system and its environment and/or between elements of a system. Goal achievement and system maintenance are never fully synchronized, and the resulting tension may lead to changes. Conflicts between different goals/values/world views may lead to either incremental or discontinuous change.

The systems studied for the purpose of forecasting and managing technology are much smaller than the global ones treated by social change theorists. For example, the systems considered in this book may be a single organization or a single technological development. No matter how small the system, however, the first step must be to determine its boundaries, that is, to define which elements will be treated as internal and which external to the system. Both the elements and their linkages must be defined to establish a static initial structure. The potential change processes must also be identified. These may arise from the goals and values of elements of the system *that can act*, such as institutions and individuals. In a sense these change processes closely relate to the technology delivery system discussed in Section 2.3. However, the analysis in this section goes beyond the delivery system to hypothesize specific dynamics involved in change and thus to create a framework for understanding aspects of sociotechnical change.

It is important to realize that forecasting, management, and assessment of technology are not separate activities. However, because each involves distinct needs and concerns for the analyst or manager, it is convenient to treat them separately for clarity. Each is considered in the subsections that follow.

2.2.1 Technology Forecasting

Technology forecasting deals with causal elements of any sort—social, economic, or technological. However, the effects of interest are new technologies and incremental and/or discontinuous changes in existing technologies. The focus of the analysis is on a technology or family of technologies. Thus the end processes are the development of technological principles and/or prototypes and the diffusion of technological devices.

For starting points, consider the supply of scientific knowledge and technological principles, economic and societal demand, and the state of the technology itself. Intervening between the initial and final states is the process of using human, intellectual, material, and financial resources to develop and diffuse the technology. In addition there is the process of organizing institutions to manage, promote, and/or inhibit changes in the technology. Feedback may occur at any phase of the process. It may reinforce the direction of technological change, for instance, when intermediate results intensify demand. Or it may inhibit change because of disappointing results or changes in social preferences.

Although the specifics differ from forecast to forecast, the general framework of analysis is similar. The forecaster looks to theories of sociotechnical change to provide a perspective on the dynamics of technological change. Because no macrotheory exists, ad hoc techniques such as trend extrapolation, Delphi, and scenarios attempt merely to identify the outcomes and largely neglect the mechanisms that cause them.

2.2.2 Impact Assessment

Impact assessment begins where technology forecasting ends. The causal elements responsible for the impacts of technology are the development and diffusion of that technology. The outcomes are changes in any part of the sociotechnical system. In an impact assessment there may be many generations of effects arising from the initial state of the technology. The initial or first-order effects act as causes to produce the second-order effects and so on. These higher-order, usually unanticipated, effects may overwhelm the first-order effects, just as air pollution has been an enormous long-term and unanticipated effect of widespread automobile use in industrial societies. Assessment may be complicated by the fact that mechanisms other than those related to the technology may also be causing change.

Interlinked causes and effects that extend over long periods and lead to complex causal patterns are the most difficult aspects of impact assessment to handle. Lacking a macrotheory of sociotechnical change, the assessor must substitute judgmental techniques for detailed causal analysis. Indeed the two most common ways of classifying impacts are by impacted group and by class of impact (such as economic, social, or technological). In impact assessment, systems tend to be broader than in either forecasting or management because important technologies may penetrate all aspects of society. The impact assessment process may, however, be limited to certain ranges of consequences.

2.2.3 Technology Management

Technology management usually functions within the framework of a single organization. Its objectives are to benefit the organization in both its internal functioning and its external relationships. The management function is more immediate and ongoing than forecasting and assessment, which are more prospective. However, the realization of its objectives lies in both the present and the future. Its causal mechanisms are the goals, intellect, expertise, and management skills of the organization, along with its corporate momentum and the resources it can bring to bear. Whereas forecasting is the intellectual exercise of determining what could happen, management is the practical exercise of making it happen. Likewise, impact assessment deals with the effects of technology on society, while technology management deals with the impacts of technology on the organization.

2.3 THE TECHNOLOGY DELIVERY SYSTEM

The technology delivery system (TDS) addresses sociotechnical change in a limited environment. The TDS portrays the institutional linkages involved in developing a single technological innovation (Wenk and Kuehn, 1977). Figure 2.1 illustrates the system as a simple "boxes-and-arrows" model where direct information flows are

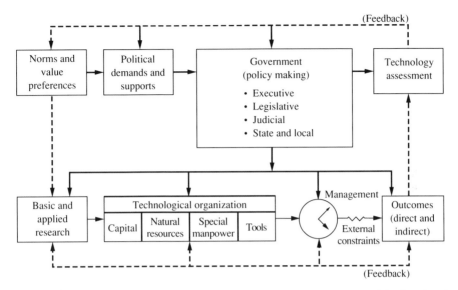

Figure 2.1 The technology delivery system. (*Source:* Wenk and Kuehn, 1977. Reprinted by permission.)

shown by solid lines and feedback loops by dashed ones. In this figure, institutions directly involved in developing the technology are displayed along the bottom. Displayed at the top are institutions in the larger society that are not directly involved in development but influence and are influenced by the technology. The TDS model has four basic elements:

1. *Inputs to the system*, including capital, natural resources, manpower, tools, knowledge from basic and applied research, and human values
2. *Institutions and organizations*, both public and private, that play roles in the operation of the TDS or that modify and control its output
3. *System processes* by which institutions interact through information linkages, market, political, legal, and social means
4. *System outcomes*, including both direct (intended) and indirect (unintended) effects on the social and physical environments

The TDS thus defines a limited sociotechnical system composed of institutions directly or indirectly involved in developing a single technology. All other portions of society are defined as external to the system. The formal links among these system elements are described by solid or dashed lines, although other, informal linkages no doubt exist and may sometimes be more important. Thus Figure 2.1 portrays the static initial structure of the system (Section 2.2). Change is represented by alterations over time in basic and applied research; the resources of the technological organization; the outcomes; the policies of government and its various implementations (such as laws, regulations, incentives); the level and composition of political demand and support; and the norms and value preferences of society. In other words, the TDS in effect represents a microtheory of sociotechnical change.

The TDS concept can be applied to any technology. However, the elements that appear in the model will change from application to application. Figure 2.2, for instance, represents a TDS constructed for microcomputer technology in developing countries (Shi, Porter, and Rossini, 1985). The TDS is very useful for conceptualizing the development of a single technology and the kinds of sociotechnical change that may result.

Such TDS models provide several critical inputs to the technology manager, including

- Mapping the key players, both institutions and individuals that can affect the development
- Depicting the essential thrusts of the technological development (note the intertwined hardware, software, and networking shown in Figure 2.2.)
- Identifying gaps in needed resources such as capital, skilled labor, and critical materials
- Emphasizing leverage points (Figure 2.2, for example, shows that development based upon imported technology requires scarce foreign exchange to finance import purchases.)
- Highlighting important technological enterprises (for instance, companies that can provide software in a country's native language)

2.4 INNOVATION PROCESSES

Organized technological creativity is one of the most exciting developments of the twentieth century. The process has been widely studied from economic, sociological, and geographic perspectives. From the perspective of the innovating organization, however, there are two major models for organizing the process: the phase-dominant model and the project-dominant model. Each provides a view of sociotechnical change constrained to a very limited system, the organization that develops the technological innovation. Everything else in society is taken as external to the process. Thus, in a sense, these are micro models for the internal functions of the developing organization in the TDS (Figure 2.1). To the extent that they help better understand those internal functions or manage them, they are important.

The *phase-dominant* (or linear) model illustrated in Figure 2.3 was developed by Kelly and colleagues (1978). The model portrays a linear sequence of functions, each of which is typically replicated by the different subunits of the innovating organization. The "handoffs" or interfaces between the various subunits, from basic research through engineering and manufacturing, across which the innovation passes are critical for an organization that adopts this model. This is because the mindsets, procedures, values, cultures, and goals of each subunit are typically quite different. Therefore, at each phase of the process the technology is altered to reflect subunit needs. For example, a product not designed to be easily manufactured may be partly or completely redesigned by the manufacturing unit so that it can be more easily or inexpensively built.

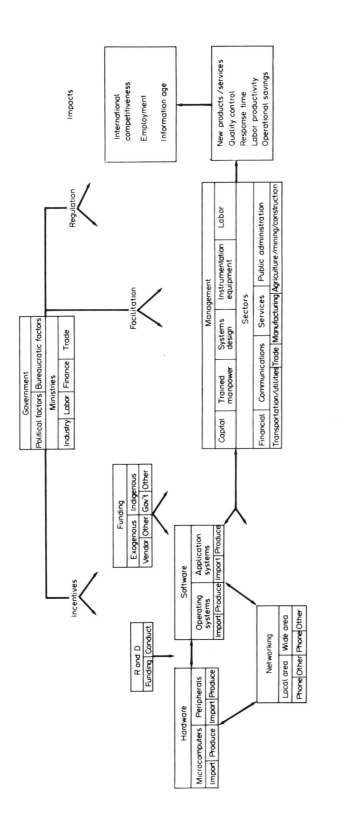

Figure 2.2 A technology delivery system for the application of microcomputers in developing countries. (*Source:* Shi, Porter, and Rossini, 1985. Reprinted with permission from the *International Journal of Applied Engineering Education*, pp. 321-327, © 1985, Pergamon Press.)

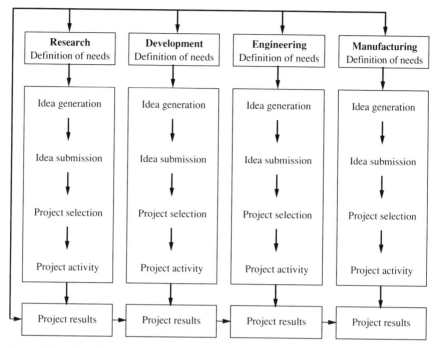

Figure 2.3   The phase-dominant model of the R&D process. (*Source:* Kelly et al., 1978. From P. Kelly and M. Kranzberg, Eds., *Technological Innovation*, © 1978, San Francisco Press, Inc., by permission.)

A second model is the *project-dominant* model (Kelly et al., 1978) shown in Figure 2.4. In this model, knowledge-based activities (such as idea generation or project activity) cut across the various phases of the innovation process (research, development, and so forth) instead of being replicated as they are in the phase-dominant model. Concurrent engineering reflects a project-dominant model in which a design team includes representatives from manufacturing, and perhaps sales and marketing, throughout the innovation process. It is arguable that the project-dominant model is more effective for major, discontinuous innovations and the phase dominant model is more appropriate for minor, incremental innovations.

Organizational characteristics and the characteristics of the innovation itself are important in determining how innovation is accomplished. They also help determine how an organization will deal with technological threats and opportunities. To the degree that these models help understand different approaches to developing innovative technology, they help forecast, manage, and assess technology and to anticipate and manage sociotechnical change.

2.5   DATA FOR MANAGING AND STUDYING SOCIOTECHNICAL CHANGE

Data are pieces of information about the present and past. Forecasts and impact assessments generate information about likely futures. To be useful in forecasting,

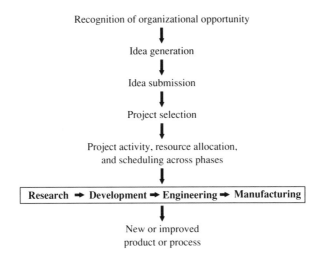

Recognition of organizational opportunity

↓

Idea generation

↓

Idea submission

↓

Project selection

↓

Project activity, resource allocation,
and scheduling across phases

↓

Research ➡ Development ➡ Engineering ➡ Manufacturing

↓

New or improved
product or process

Figure 2.4 The project-dominant model of the R&D process. (*Source:* Kelly et al., 1978. From P. Kelly and M. Kranzberg, Eds., *Technological Innovation*, © 1978, San Francisco Press, Inc., by permission.)

assessing, and managing sociotechnical change, data must relate to the elements that describe the system of interest. Thus the perspective and interests that define the system determine the data that are useful and the parameters that should be tracked and estimated to portray system change.

Although interesting work has been done on the indicators of technological change, it has not, in general, been related to the larger societal contexts in which change occurs. Likewise, the social indicators movement has attempted to develop indicators of the status and prospects of society analogous to commonly used economic indicators. However, this effort has not produced a generally accepted and useful set of social indicators. Thus there is no definitive set of free-standing social or technological data independent of the framework of a particular task.

Until the relation between data and theory is more robust or sets of valid independent indicators of sociotechnical change are developed, data needs remain context-driven. That is, in each situation data requirements will be specified by the characteristics of the system and the needs of the analysis.

2.6 CONCLUSION

Classical theories of macroscopic social change operate at high levels of abstraction and aggregation and generally view change from a single perspective. Although incomplete, they represent useful analyses that provide insights for analyzing and managing sociotechnical change. If not accepted dogmatically, they complement one another and lead to integrated perspectives that offer a descriptive framework for forecasting and managing change in sociotechnical systems.

For example, viewing sociotechnical change as a spiral in which some system elements move linearly while others move cyclically allows conceptual integration of the evolutionary and cyclic perspectives. Assuming that no system is unchanging or continuously changing its major characteristics, functionalism and conflict analysis address the interplay between mechanisms pulling the system apart and those restoring it. Critical theory adds the perspective that a social system of any sort is made up of individuals and institutions that can and do critically reflect on the meaning and value of the processes of stasis and change as well as their outcomes. The picture emerges of a system moving from one metastable state to the next. The manner in which these competing mechanisms achieve their long-term outcomes is therefore best described as a spiraling process.

The perspectives of forecasting, impact assessment, and management occupy different positions in the spiral of sociotechnical change. Impact assessment starts with a technology, which is where forecasting ends. Management uses organizational resources to move technology from forecast to realization. Thus it operates between forecast and impact assessment.

The major system elements of scientific knowledge, technological principles and devices, people, institutions, values, funds, and resources interact. These interactions lead to the development of new scientific knowledge, technological principles, and prototypes; the production and diffusion of technological devices; institutional change; changes in human knowledge, values, and beliefs; environmental change; and changes in wealth and resources. Although these interactions cannot be portrayed across the full scope of sociotechnical change by a single theory, they can be viewed in the limited scope provided by the technology delivery system. Further, analysis of the process of organizational innovation within an organizational context provides some understanding of very specific activities at the micro level.

Data for forecasting and technology management must be developed for each situation because there are no generally useful variables or indicators that cut across all aspects of sociotechnical change. Therefore, ad hoc techniques, especially ones that involve heavy use of human judgment such as the TDS, must be substituted for theory at an operational level to analyze change in a single innovation or within a single organizational context.

EXERCISES

2.1   The chapter distinguished five theories of social change. Predict American technological competitiveness in 15 years from the perspective of the first of these theories. Then do the same with each of the other four theories in turn. Comment on the extent to which this social perspective affects a forecast.

2.2   Divide the class into paired groups. Set up a "pro" and a "con" group to debate one or more of these topics. Each side should present a brief statement, citing specific examples, then field questions from other class members.

   (A)   Technological change is not necessarily progress.

   (B)   Technology strengthens the power of an elite minority within society.

   (C)   The more people "change," the more they remain the same. Principles of human behavior are timeless.

2.3 Consider the personal transportation system of a typical American city in 20 years (for instance, cars, buses, bicycles, personal jetpaks to fly to work).

  (A) Specify two social factors that could affect personal transportation and make explicit hypotheses as to how each will affect this development.

  (B) Specify two technological transportation advances and speculate on how these will or will not be integrated into these urban transportation systems.

  (C) Contrast your responses to (A) and (B). (A) considers social factors affecting the development of the technology; (B) considers technological factors affecting sociotechnical systems. Be able to distinguish these complementary relationships between social and technological factors.

2.4 Sketch a TDS for one of the technologies identified in Exercise 2.3(B).

  (A) Discuss the key social/economic factors within the technological enterprise.

  (B) Discuss important external social/political factors and the likelihood that these will affect development of the technology.

  (C) Identify three parties impacted by the technology and predict their attitude toward its development.

2.5 Exercises 2.3 and 2.4 considered possible technological improvements to urban transportation systems. Devise a "social" mechanism to improve the functioning of urban transportation over the next 20 years. How might one compare the merits of a technological and a social solution to a social need such as transportation?

2.6 Assume that you are a bright young engineer who has just devised an exciting new sensing technology. This portable device can detect both organic and inorganic trace elements in water within seconds, cheaply and conveniently. Nothing like it is on the market. You want to start your own company to build and market this device. Sketch a TDS to deliver this technology to market.

  (A) Identify the critical components required within the technological enterprise (your company) for success.

  (B) Identify key external elements that will greatly influence your chances for success.

  (C) Based on this information, would you go ahead with this venture?

2.7 Consider the technology of Exercise 2.6 again.

  (A) Assume you have decided to proceed with development. Argue for either a phase-dominant or a project-dominant innovation strategy to develop your product.

  (B) Assume instead that you have licensed your patent on this technology to a major multinational corporation in the field of scientific instrumentation. Argue whether it should use a phase-dominant or a project-dominant innovation strategy.

___3
TECHNOLOGY PLANNING

OVERVIEW

This chapter surveys the idea of planning and distinguishes technology planning from general planning. It introduces certain general planning approaches that apply well to technology planning and describes specific aids to goal-oriented planning. In addition, it details several qualitative and quantitative techniques that can facilitate strategic or project-level technology planning.

3.1 PERSPECTIVE ON PLANNING

Technology planning relates both to general planning and to special features of technology forecasting and management. This section provides a perspective on planning and examines differences between technology and general planning.

3.1.1 Planning

"Everybody plans" (Allio and Pennington, 1979). Personal plans tend to be short-range and casual, but the reader who is a student has, at least in this respect, planned activities with a somewhat distant goal in mind. Organizational plans may more frequently be longer-range and formal, but they are not as easily internalized and converted to action as personal plans. At an even higher level, nations plan governmental and, to a greater or lesser degree, societal activities. This chapter emphasizes planning at the intermediate level—in an organizational context. This is the level at which the technology managers carry out their responsibilities. Effective organizational planning links daily tasks to organizational objectives.

There are several models of planning. Marxist regimes epitomized commitment to planning with their centrally planned economies and five-year plans. The early efforts of the Soviet Union to operate its economy with central objectives and specific national goals were probably the first examples of planning as a focus of industrial decision makers. Recent events in that country and central Europe have raised serious questions about the possibilities of success for planned economies. At the other extreme, the United States has had an ideological commitment to a market economy and decentralized decision making. However, the large firms in this country—and certainly the multinational firms, which operate globally—have had to coordinate and plan in order to function in a rapidly changing world. Like their socialist counterparts, the bureaucrats who run these large enterprises have been criticized for the negative effects of central planning on innovation and effectiveness. All economic systems and the industrial enterprises within them are struggling to find techniques for planning that adequately gather and process information to generate guides for activity. These must be flexible and appropriate for a changing competitive environment and have the commitment of individuals at all levels of the organization.

Schmidtlein and Milton (1989) assert that a reasonable model of planning at the organizational level exhibits the following characteristics:

- Organizational goals exist and can be specified.
- Alternative courses of action can be identified and evaluated in relation to organizational goals.
- Logic and analytic procedures can help decide courses of action.
- Implementation of decisions made through planning activities is feasible and even likely.

This decision-centered view of planning guides this chapter.

Planning can address various domains, some of which are shown in Table 3.1. Both substantive knowledge and planning techniques have been accumulated within these domains, making for a rich but dispersed knowledge base. The focal area of this book—technology—overlaps most of these domains and can make use of the progress that has been made in them.

TABLE 3.1   Domains Addressed by Planning

Agriculture	Finance
Defense	Hazardous waste
Development	Health care
Economic concerns	Production
Energy	Regional/urban concerns
Environment	Research
Facilities	Transportation

Several organizational functions that relate closely to planning are

- *Budgeting*: Ideally, the shorter-term implementation of longer-term plans; in fact, often driven by quite different short-term motivations. Meaningful planning must link to budgeting to properly affect resource allocation.
- *Design*: Emphasizes the substantive implementation of change for which strategic planning sets the objectives and project planning addresses the process; design, as of new systems, is closely linked to technology planning (see Nadler, 1987, for cross-fertilization possibilities).
- *Issues Management or Monitoring*: Provides information from (also to) the external environment vital to planning (Heath, 1988).
- *Impact Assessment*: Weighs the effects of alternatives; a critical part of informed planning.
- *Policy Analysis*: Weighs the available options to implement plans.

Several forms of planning should be differentiated. Strategic planning helps an organization adapt to its environment (Cope, 1987). It aims to initiate changes for the organization's longer-term (typically more than one year) well-being. Some people distinguish strategic from long-range planning, with strategic planning focusing on one aspect at a time and long-range planning viewing the organization holistically (Thierauf, 1987, p. 120). By analogy, business strategy typically considers a given business area for the organization and corporate strategy compares the advantages to the organization of engaging in alternative business areas (Hax and Majluf, 1988).

Peter Drucker (1974) has stressed the role of planning in management and clarified the definitions of strategic, long-range, and short-range planning. He has said the following about *strategic planning*:

It is the continuous process of making present entrepreneurial *(risk-taking) decisions* systematically and with the greatest knowledge of their futurity; organizing systematically the *efforts* needed to carry out these decisions; and measuring the results of these decisions against the expectations through organized *systematic feedback*. (1974, p. 125)

Drucker stresses that planning in a business environment should be based neither on specific techniques nor on specific forecasts. In fact, he says that planning is all the more important because forecasts are so often incorrect. The process should be one of rigorous analysis in an effort to answer the following three questions:

What is the business?
What will the business be?
What should it be?

The answers to these questions need to be viewed in proper relation to one another. It is probably desirable to separate the first two conceptually because it is important to avoid simply extrapolating the present into the future. However, the last two must be related, and it clearly makes sense to view "what is" as a starting point for making the organization what it "should be." The critical issues, therefore, center on the impacts of managerial decisions needed to carry out a plan to reach this goal (Drucker, 1974, p. 122).

The nature of the planning process will differ depending upon what is considered, what approach is taken, and what actions are recommended. A problem-finding approach to planning identifies current and future problems and provides solutions (Thierauf, 1987). Contingency planning carries this a step further by preadapting the organization to possible future environmental conditions or competitors' actions (Sutherland, 1988). These largely defensive orientations should be complemented by seeking organizational opportunities. In fact, successful innovation in technology or management often involves turning the challenges resulting from change into opportunities. More stringent environmental regulations may constrain companies, but they also open markets for new products and processes. A new technology may make a firm's product line obsolete, but it does the same thing to competitors' product lines. The literature on planning also distinguishes forward planning—first taking stock of current situations and processes, then considering future alternatives—from backward planning—starting with desired outcomes (goals, objectives), then devising steps needed to reach them. (Saaty, 1980).

This discussion emphasizes the rich diversity possible in planning. Consolidation of these variations suggests three sorts of information basic to strategic planning: external information, internal information, and information on available options for action by the organization. It is the last type that really separates planning from forecasting. All the variations can be viewed in relation to types of forecasting that will be introduced in Chapter 4. For example, Saaty's (1980) notion of backward planning is very close to the notion of normative forecasting. However, it is important to remember that the results gained from using forecasting techniques are only one input to the planning process. They do not in themselves provide the guide for decisions. Such a guide must come from using all three sorts of information to develop comprehensive strategy and the short-term budgets and other business tactics to implement it.

3.1.2 General Planning Versus Technology Planning

Technology planning is a specific subset of activities that should be an integral part of the organization's overall strategic plan. This textbook is directed at the future technology manager whose decisions will primarily affect a particular firm. However, in a finite and interdependent world those decisions can have industrial, local, national, and even international effects. The five organizational functions mentioned (budgeting, design, and so on) relate directly to technology planning, but organizations seem to find it difficult to make the connections.

Technology pervades many of the public and private sector planning domains that were listed in Table 3.1. Because of its pervasiveness, technology planning is not treated as a separate, established planning domain. In fact, it is usually hard for managers to identify their organizations' technologies. In interviews with managers of 18 British corporations, only 5 perceived a clear, substantial technological dimension to their strategic plans (Clarke, Ford, and Saren, 1989). Instead, most of the plans emphasized financial, marketing, or manufacturing issues, which in turn involved many different technologies in many different process/product applications. Clarke, Ford, and Saren conclude that technologies are deeply embedded in organizations, only partially codified, and therefore difficult to analyze strategically. They suggest that analysts trying to examine the role of technology in corporate plans do the following:

- Consider *technology clusters* meaningful to the organization rather than individual technologies.
- Locate organizational technology interests along a chain that extends from primary technology supplier through intermediate users to final customers.

To illustrate how these two considerations operate in contingency planning, consider Whirlpool Corporation's investigation of the implications of a possible constriction in petroleum supply in the 1970s (Davis, 1973). Whirlpool's analysts identified two important technology clusters—energy supplies and lubricants. They then considered how corporate interests might be served by becoming an energy supplier on a modest scale and by using alternative process technologies (such as silicon lubricants). Using the approach suggested here, they could have extended their analysis of the implications of energy shortages and resultant price increases by considering Whirlpool customers. This would have revealed another possible technology planning alternative—development of energy-conserving appliances. This might have enabled them to anticipate the consumer interest and government regulatory actions of the next decade.

Technology planning relates the processes of technological change to the organization's strategic plan. The concepts of the innovation process (see Chapter 2) help provide the conceptual framework to plan for technological change and corresponding organizational actions. The following sections will review techniques to assist the technology manager in developing the guidelines for these actions.

3.2   TECHNOLOGY PLANNING METHODS

This section provides a range of methods to accomplish key technology planning functions.

3.2.1   Planning Processes

The process used in planning may be more important than the plan produced. Effective planning directs managers' and employees' attention to objectives and to possible courses of action to attain those objectives. A proper approach to the planning process considers all points of view in the organization. This builds commitment to the objectives and fosters team spirit. The planning process should be considered carefully, but it need not be formal. In fact, informal and understandable approaches can lead to a greater sense of employee ownership of the plan and hence more effective implementation.

Hax and Majluf (1988) weigh the pros and cons of various strategic processes, recommending that one balance a number of factors, favoring a highly participatory process. They suggest the following rules:

- Have top management manage the process, while encouraging wide participation.
- Adopt a synoptic (all-encompassing), disciplined process.
- Strive for consensus by using negotiation to set strategy.
- Combine past and future (evaluate past actions; formulate novel strategies).
- Communicate resulting strategic choices widely.

Several sources (such as Aaker, 1988) offer formal business planning systems that include workshops on international analyses and on strategy development, mechanisms to compare two or three alternative strategies, vehicles to deliver selected strategies, and programs to implement strategies. Aaker favors a highly focused endeavor, concentrating on a few key actions. Argenti (1988) proposes that planning teams include an experienced outsider, such as a senior member of the organization's accounting firm. All these business planning suggestions apply to technology planning, as do the methods of planning in other domains. Such discussions of planning techniques supplement those described here, as well as providing information on overall organization planning within which technology planning must be developed and implemented.

Like other types of organizational planning, technology planning needs to follow systematic, participative processes. Unfortunately, instances of unsatisfactory technology planning processes are too frequent. A few proponents may press to develop or adopt a technology without involving all organizational parties with an interest in it. Past technological changes are often not evaluated. Proposed changes may not be related to organizational objectives or backed by top management, and so on.

Generating effective participation is an art in itself. Intraorganizational participation raises challenges as unit interests come into conflict. A host of group process considerations pertain, including team building, creative idea generation, task structuring, and consensus generation (see Schweiger and Sandberg, 1989; or organizational behavior texts such as Gray and Starke, 1988).

As difficult as participation within the organization is to achieve, it should be stressed that obtaining widespread input and support is much easier when everyone has the same chief executive officer and the same signature on their paychecks. Public participation and ultimate acceptance of the plan or at least the actions that are produced by it is often as critical to success as internal acceptance. Yet there are severe problems of credibility and communication when people with radically different perspectives come together.

3.2.2 Impact Assessment Principles

Chapter 2 has pointed out the intricate interdependencies between technology and society. Technology planning requires that these interdependencies be considered. Otherwise, those responsible for technological decisions may ignore critical results. Such myopia about impacts has led to devastating human tragedy as well as the financial ruin of major corporations. Concern arises when the effects, or impacts, of technology are not well managed. It is clear that the courts and the legislatures will no longer tolerate the undesirable results. Furthermore, the technology manager should view the broad implications of technology as his or her professional obligation and social responsibility.

Some of the issues seem too large and complex for any one person. Recent headlines express apprehension over such technology-intensive issues as

- Greenhouse effect (global warming, attendant sea level rise)
- Nuclear and hazardous waste disposal (siting, transportation, terrorist potential)
- Lagging industrial productivity
- Acid rain and other air pollution problems
- Waste disposal, especially when the wastes are toxic
- Loss of technology to other industrialized nations

Concerns likely to reemerge include economic growth and energy availability; industrial development and water quality; increasingly powerful computers and unemployment among the unskilled workers unable to adapt to the changes; expanding communication and the resulting unrest of those who feel deprived of the features of life-styles they are able to know more and more about; and control of nuclear, chemical, and biological arms. These issues evidence a common feature of technological advance—an intricate, spreading web of desirable and undesirable impacts.

Some criticize society's willingness to be driven by unconstrained technological growth. They call for careful assessment of all technological change. Langdon Winner (1977) objected to the "technological imperative" wherein implicit requirements of ever-advancing technology overwhelm socially chosen ends—energy-dependent societies base foreign policies on securing oil supplies rather than on meeting basic human needs. He complained of "reverse adaptation" wherein technological means supplant human ends—for example, tomatoes are bred to be strong enough to withstand mechanical picking and extended shipping, rather than for eating.

Winner and others urge societies to control technological development by assessing the effects of technological change with great care. Such assessment of technology considers general patterns of development. A society ought, for example, to ask questions such as these:

- Does China, with 85 percent of 1.1 billion people involved in agriculture, want to move as fast as possible to become an Information Age society like the United States, with some 65 percent of its people supported in service occupations? If so, how can China manage the transition? If not, how can China meet the expectations of people who now are aware of the benefits of living in a highly developed economy?
- What are the implications of the international competition treadmill for the highly developed nations? Must the United States boost productivity by computerization and new work roles to keep up with the Japanese? Must the European community adopt common standards and coordinated industrial enterprises to compete with the United States and Japan? Must Japan move to automated factories to maintain its competitive edge over more resource-rich nations? Should everyone strive to minimize human involvement in production? If so, what will people do?

Broad-scope technological changes implicit in these examples must be considered with great care. People ought to choose desired living patterns and guide technological development accordingly. The principle of *appropriate technology* means that would-be developers should make sure that a technology is suitable to the social context; they should consider meeting rural African energy needs by cheap solar cooking ovens rather than nuclear power plants. The emergent idea of *sustainable development* cautions society's technology managers to aim toward environmentally stable developments, weighing the merits of technologies over their full life cycles.

These types of issues form the "big picture" concerns that should shape the context of technological planning. Society's demands for protection from undesirable side effects result in design constraints that are every bit as binding as the laws of physical science. Successful technology managers will take a "proactive" approach. They will cooperate actively to find effective ways to meet social objectives at reasonable cost and at the same time work to find ways to improve their organization's position in the changing world.

Over the past two decades, efforts have been made to develop a methodology for considering "big picture" problems. Early work was oriented to the public sector, but it has become increasingly clear that assessing the impacts of technology is a critical part of technology planning in all types of organizations. The first name given to the study of techniques for these tasks was *technology assessment*, and today the United States Congress carries out such activity in the Office of Technology Assessment (OTA). More recently the name associated with broad-based analysis of the effects of technology is *impact assessment*, and Chapters 15 through 18 describe tools to carry out such a study. Here the purpose is not to show specifically how to carry out an impact assessment, but to demonstrate that it is a vital component of technology planning.

In spite of the name change, perhaps the best definition of impact assessment is the one that Joseph Coates (1976) developed for technology assessment. This early intellectual pioneer and continuing leader in the holistic assessment of technology said that technology assessment (or impact assessment) is

> the *systematic* study of the *effects* on society, that may occur when a technology is *introduced, extended, or modified,* with emphasis on the impacts that are unintended, indirect, or delayed. [italics added]

A "systematic" study involves an approach that is as orderly and repeatable as possible. Information sources and methods should be defined. The study focuses on the "effects" or "impacts" of the technology on society (although a comprehensive study will address the reverse effects of social forces on technological development as well). These effects are the result of the "introduction, extension, or modification" of a technology—the various ways in which a technology may change. "Unintended, indirect, or delayed" effects extend beyond the direct costs and benefits traditionally considered in technical and economic analyses.

Exhibit 3.1 captures the idea of indirect, or higher-order, impacts. Joseph Coates (1971) accompanied this tongue-in-cheek illustration by similar assessments of the impacts of refrigerators and automobiles—the sixth-order impacts of these also turn out to be "divorce." Serious determination of cause and effect turns out to be problematic and exceedingly difficult—even for so-called retrospective assessments of past technologies (Tarr, 1977).

■ Exhibit 3.1   The Effects of Technology

At times, technologies can have consequences that combine to create a serious impact undreamed of by their creators. The following table suggests how television may have helped to break down community life:

CONSEQUENCES OF TELEVISION

First Order:	People have a new source of entertainment and enlightenment in their homes.
Second Order:	People stay home more, rather than mingling with their fellows at local clubs and bars.
Third Order:	Residents of a community do not meet so often and therefore do not know each other as well.
Fourth Order:	Strangers to each other, community members find it difficult to unite to deal with common problems. Individuals find themselves increasingly isolated and alienated from their neighbors.
Fifth Order:	Isolated from their neighbors, family members depend more on each other for satisfaction of their psychological needs.
Sixth Order:	Frustration occurs when spouses are unable to meet the heavy psychological demands that each makes on the other; this leads to increased divorce.

[Excerpted from J. Coates (1971:228-229)]   ■

If determining past causes and effects is hard, what hope is there to determine future ones? Certainly not certainty—there is little hope to predict the exact effects of a change in technology, less the precise magnitude or timing of those effects, even less the manner in which those effects will interact among themselves and with other forces. It is more helpful to seek to reduce uncertainty. The technology planner can profit from identifying possible impact vectors. Knowing what is possible, and going further to assess what is relatively likely, can greatly facilitate planning. The alternative to assessing impacts is to cover one's eyes and jump into the future unguided. It is far better to "look before you leap," even if future vision is considerably less than 20/20.

The technology manager needs impact assessment to understand likely patterns of acceptance and resistance to a changing technology. Chapters 15 through 18 elaborate the important interrelationships and describe in more detail some techniques for conducting impact assessments. However, it would be useful here to examine the basic concepts and their relation to technology planning.

Some topics are part of any good impact assessment. These subjects also are often discussed in their own right. Because they are plagued with the use of acronyms, their shorthand names are also included.

Environmental impact assessment (EIA) began as a response to the need to prepare environmental impact statements (EISs) to meet the requirements of the United States National Environmental Policy Act of 1969 (NEPA). Since that time EIA has been applied throughout the world and is applied more broadly than just to the physical environment.

Social impact assessment (SIA) emphasizes the estimation of potential impacts on people and their cultures and institutions caused by a new technology or a particular project. For example, the introduction of modern technology in rain forests may shatter the way of life of indigenous people who live in traditional ways. However, social impacts and their problems are not restricted to developing areas. American communities have experienced severe adverse effects when new plants arrived. Exhibit 3.1 described possible effects of television on the family. Certainly, the innovation of cable TV has affected sports, and not everyone believes that all the money and attention has been good for college athletics. These are examples of the types of impacts that SIA attempts to discover and evaluate.

Risk assessment has probably been used most often for technologies related to food and drugs, but it also applies to nuclear power plants, chemical plants, highway safety, and many other public health issues. Although some may believe that risks like the carcinogenic (cancer-causing) effects of some industrial chemicals can be reduced to quantitative values of stochastic variables, Chapter 16 shows that these issues are far more complex and involve subjective judgments at least as much as the objective measurement of probabilities.

3.2.3   A Seven-Step Planning Approach

Maddox, Anthony, and Wheatley (1987) have provided a planning framework shown in Exhibit 3.2 that can be modified to serve the manager in technology planning. Although the steps appear to make the process unidirectional—from initial forecasts to operations decision—it should be stressed that a good planning process is usually iterative. New knowledge gained in step 4 or 5 may require that the analyst return to steps 1, 2, and 3 to adjust for the additional information.

■ **Exhibit 3.2 Maddox Planning Framework (Based on Maddox, Anthony, and Wheatley, 1987).**

1. *Forecast the Technology*: The starting point of technology planning. Both internally owned technology and that available in the marketplace will need to be projected over the planning period. Chapter 4 will describe technology forecasting as a management tool.

2. *Analyze and Forecast the Environment*: Identification of key factors in the organization's environment, potential future states of the environment, key uncertainties, and major threats (especially competition) and opportunities. The previous section's discussion of impact assessment is directly applicable to this step.

3. *Analyze and Forecast the Market/User*: Development of a requirements analysis that identifies the current needs of major customers, determines the likelihood that these needs will change, and specifies explicit demands that these needs make on the organization's products or services.

    Here the tools of market research and impact assessment will complement each other. However, analytical tools, no matter how sophisticated, will never be adequate. It is imperative that this step include direct contact with potential customers. Real quality is the fulfillment of customer requirements and desires (Crosby, 1979), and the best way to know those is to get close to the customer.

4. *Analyze the Organization*: Delineation of the major assets and problems; development of a catalog of available human and material resources; assessment of recent performance against stated objectives. Understanding the strengths and weaknesses of your own organization is critical, and objectivity cannot be overemphasized. This may be a great time to involve external consultants in order to avoid the errors that arise when members of an organization assess themselves.

5. *Develop the Mission*: Specification of critical assumptions; establishment of overall organizational objectives and specific target objectives for the planning period; specification of criteria by which to measure attainment of those objectives. This step will provide the central focus of the organization and should include as many participants as possible. Organizations have a much better chance for success when each member understands and feels a sense of ownership of the mission.

6. *Design Organizational Actions*: Creation of candidate actions; analysis and debate of these; development of a consensus strategy limited to a few key actions, possibly attendant on several key contingencies. This is another excellent time to apply the tools of impact assessment.

7. *Put the Plan into Operation*: Development of timely subobjectives, if appropriate; specification of action steps, schedule, and budget; development of tracking mechanisms; specification of control mechanisms in case performance falls below established standards. During this step, monitoring (see Chapter 8) can be very useful. Technological marketplaces are dynamic, and each firm must maintain a knowledge base of changes and customer reactions to them. ∎

3.2.4 Some Illustrations of Goal Setting and Mission Statements for Planning

Specifying goals leads to better performance than simply trying to "do your best." This result has been amply reaffirmed in many human performance studies (cf. Gray and Starke, 1988; Connolly, 1983). Goal setting works best when

1. The goals are specific and challenging.
2. The people involved accept the goals.
3. Feedback on attainment is provided.

The ability to set goals in this manner requires that clear criteria and measures be associated with objectives. The goals of any new technology also should be related to the long-run benefits of the firm that is developing it. This means first and foremost that the goals should be developed with customers in mind.

The suggested seven-step planning process is intended to provide a guide for goal setting. In particular, step 3 suggests a requirements analysis to help formulate organizational action plans (step 6). The development of the Designer's Associate—a system intended to offer intelligent computer assistance capabilities for the engineers trying to design aircrew stations (cockpits)—provides an example of such a market/user analysis. A vital step in this technology planning effort (by Search Techonology, Inc., Norcross, Georgia) for the Air Force was to interview over 20 aircrew station designers to find out how they presently performed their jobs and what they thought they would want from a computer "associate." This information was used in a requirements analysis that spelled out the functions the product should fulfill. The research effort then turned to a search for technologies that could perform these functions and would be reasonably likely to exist within the target time frame and be available at a tolerable cost.

Performance is usually better when very few goals are set. If nothing else, a smaller number of goals is easier to communicate to all organization levels so everyone is working in the same direction. Identification of a few "critical success factors" can therefore facilitate planning (Jenster, 1987). Compaq Computers has enjoyed meteoric success, partly credited to its management focus on two critical success factors: fast technological response in adapting new technology to customer needs and support of the company's dealer network as the vehicle to provide customer support and service (Buckley and Reed, 1988).

Another way to facilitate a goal-setting requirements analysis is by preparing a "mission statement." Exhibit 3.3 illustrates a mission statement in facilities planning. Similar statements of the relationships of mission, functions, activities, and tasks could be developed for other types of businesses or applications of technology. The essential ingredients are a set of objectives, clearly broken down to help planners decide on strategic and operational actions.

Many firms fail at strategic technology planning. This has been variously blamed on pressure for short-term payoffs leading to emphases on marketing and financial manipulations; technological illiteracy of upper management; organizational structures ill-suited to communicating about, no less managing, technology; and isolation of research and development (R&D) units. However, the effective development of mission statements can clearly contribute to success.

Ford (1988) offers a strategic technology planning framework under the rubric of a "technology audit." Exhibit 3.4 highlights Ford's approach, which is applicable to the hardware aspects of the seven-step planning process described here. The technology audit can be done on a one-time basis, but it is more useful on an ongoing basis (for example, yearly). Ford distinguishes three categories of technologies that are illustrated here using Black & Decker, Inc.:

- *Distinctive*: Manufacturing technology to produce fractional horsepower electric motors
- *Basic*: Assembly processes for small hand tools
- *External (peripheral)*: Plastic parts bought in from other companies

Like the Brauer suggestions, the Ford audit approach should be supplemented by additional features from the seven-step process. This approach has been used well by companies interested in long-range strategic technology planning. It lays out the options for acquisition through internal R&D, purchase of the technology from other developers, or purchase of its components. A technological development can also be exploited in a variety of ways. It can be introduced into a firm's own products or production processes, contracted out, used as the basis of a joint venture, or licensed out. Ford noted that a British company, Titanium Fabricators, considered licensing production technologies to firms in India, where technology lags a decade behind the British state of the art. A similar situation may arise in the newly open economies of central and eastern Europe. Technology that may be obsolete in the West could dramatically improve quality and productivity in industrial nations that are well behind the leading edge.

The issue of standing can be strategically important. Standing is equated with reputation—a firm doing its own R&D could lead to high standing. Exploiting the firm's own technological developments may not require any particular reputation, but getting others to license them does. Conversely, if the firm lacks capability in the supporting technologies needed to implement the target technology, its best bet is to contract out or enter into a joint venture. In some cases another organization's marketing strength or financial power may give it a better chance to successfully commercialize an innovation. Also, an impact assessment may identify problems

■ **Exhibit 3.3   The Mission Statement (Based on Brauer, 1986).** [1]

The mission statement posits a hierarchy:

1. Mission (overall goal or objective)
2. Functions (major classes of activities to accomplish the mission)
3. Activities (what is done to accomplish a function)
4. Tasks

First the planners specify the mission of interest, then identify the functions that must be performed to accomplish it. In one example, Brauer considers a community fire department's mission to provide life- and property-saving services to the community. Two of the seven corresponding functions are

1. Fire calls
2. Rescue calls

The planners then need to consider the requirements for each function. For fire calls, the requirements might be charted as follows:

FUNCTION = FIRE CALLS
REQUIREMENTS

Activities	Personnel	Equipment
1. Identify need.	1 dispatcher per shift	1 telephone switchboard 1 alarm network and indicator panel
⋮		
N. Put out fire.	N fire fighters per station	1 water delivery system (conventional) 1 water delivery system (high pressure) N chemical extinguishers.

Other requirements could be pinpointed, such as systems control, scheduling, and the like. Brauer focuses on facilities planning, so he emphasizes space requirements, adjacency, and so forth, in further detailing the analysis.   ■

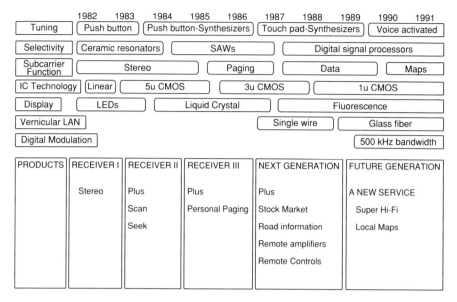

Figure 3.1   Technology roadmap matrix. (*Source*: Willyard and McClees, 1987. Reprinted by permission.)

with the technology in the developing firm's market area—problems that may not exist in other areas.

Motorola, Inc., has devised the "Technology Roadmap Process" that helps produce strategic product plans based on a combination of business and technology analyses (Willyard and McClees, 1987). One form of roadmap deals with a whole product line. It combines historical information with anticipatory analyses of changing technologies and the competitive environment. Motorola combines this information with appraisal of its resources to established priorities over time, as technologies develop and markets unfold. Features of this comprehensive process that hold special interest include:

- Product life cycle curves (composite plots of sales versus time for each product in a series help establish target dates to initiate development, introduce new products, and phase out earlier products)
- Experience curves (indicate cost or other data as a function of cumulative volume produced)
- Technology roadmap matrix (profiles the improvement sequence for the product and for each of its main component technologies—see Figure 3.1)
- Patent portfolio and licensing patterns (Chapter 8 discusses patent analysis)
- Minority reports (call attention to products or processes that may warrant investigation)

■ **Exhibit 3.4   The Technology Audit (Based on Ford, 1988).**

1. Identify the technologies (or technology clusters) important to the business. Consider both product and process (production) technologies.

2. Assess current and desirable strategies to acquire each of these technologies:

    A. Internal R&D—Appropriate for those technologies with a distinctive competitive advantage, associated with low urgency of acquisition (willing to take time) and an early position in the technology life cycle, implies highest resource commitment, appropriate to technologies in which the company has high standing

    B. Intermediate strategies—joint development ventures, contracted-out R&D, licensed-in technology

    C. Nonacquisition of technology—Buying products from others, having others produce parts using certain process technologies; suitable for technologies that are not basic or distinctive for your business; associated with high temporal urgency, no resource commitment involved in acquiring the technology at any stage in the technology life cycle; leading to no standing to the organization in the technology

3. Assess current and desirable strategies for exploitation of each of the company's technologies:

    A. Use of own production processes or products—Associated with distinctive technologies and highest resource commitment, no relative standing in the technology needed, low urgency to exploit, no gaps in required supporting technologies, early stage in the life cycle; yields narrowest exploitation

    B. Contracting out manufacture or marketing

    C. Joint venture

    D. Licensing out—Associated with noncritical technologies and lowest resource commitment, high relative standing in the technology, high urgency to exploit, low gaps in required supporting technologies, later stage in the life cycle; yields widest exploitation

    (Intermediate strategies generally fit within intermediate value parameters, except that contracting out and joint ventures tend to be suitable when the organization lacks critical supporting technologies.)

4. Assess current and desirable strategies for managing the technologies:

    A. Ensure internal technology transfer

    B. Integrate product and process technologies

    C. Develop and maintain the portfolio of technologies ■

3.3 TOOLS FOR TECHNOLOGY PLANNING

The remainder of this book describes techniques that can be useful in making good planning and implementation a reality. This chapter has shown the importance of understanding the outlook for technology and its impacts in relation to the operations of the firm. The dynamic world situation discussed in Chapter 2 requires that the technology manager know how to apply the seven-step technology planning process to his or her business. Because the process involves complex interacting relationships, the techniques outlined here are designed to handle multidimensional analysis and decision making.

The various planning techniques need to be integrated with one another and fit within the larger planning process. In the past, there was a belief that discovery and invention ("Build a better mousetrap and . . .") were the answers to the problems of businesses, other organizations, even nations. However, it is clear that the difficult, internationally competitive, and environmentally conscious world of the twenty-first century will guarantee that this is no longer true (if it ever was). Careful planning and implementation of technology will be critical to success. This may mean that understanding and properly evaluating technologies and their impacts are the real ways to get the world to beat a path to your door, rather than simply run over you.

EXERCISES

3.1. Consider an organization with which you are familiar (a fraternity, club, student organization, business, your school, or the like). How does this organization use planning? Does its planning exhibit the characteristics of the model suggested by Schmidtlein and Milton?

3.2. Select some business (or imagine one) and provide the answers to Drucker's three strategic planning questions. What other information will be needed to turn these answers into a plan?

3.3. What special features differentiate technology planning from other types of planning?

3.4. Why is impact assessment important for technology planning?

3.5. Select a new technology in which you have an interest. Show how the seven-step planning process might be applied.

3.6. Develop a mission statement (see Exhibit 3.3) for the technology selected in the previous exercise. How does the mission statement relate to the planning process?

3.7. Develop a technology roadmap matrix for the technology you selected. What parts of the planning process does it illustrate?

____4
FORECASTING

OVERVIEW

This chapter introduces forecasting as a tool for technology management and presents managerial requirements imposed on forecasting. A conceptual model presented here illustrates the probabilistic nature of forecasts. Targeted forecasts may address technology, social factors, or economic variables; all of which may interact as well.

4.1 INTRODUCTION

Some general observations about technology management are appropriate before exploring the relationship between forecasting and management. First, all industries manage technology, even if their management plan is to have no plan. Second, managing technology (or anything else) is inextricably linked with time. Industries seek to manage the technology they control, use, or produce to contribute to corporate goals *today*. They try to manage the development and implementation of technology to increase the realization of those goals *tomorrow*. To manage, they draw on the lessons of *yesterday* buttressed by management models developed from experience. In short, technology management draws on historical and future perspectives.

Forecasting is intended to bring information to the technology management process. It does so by trying to predict possible future states of technology and/or conditions that affect its contribution to corporate goals. For instance, technology forecasters might be requested to determine what computational power will be on a manager's desk in the year 2000, what the office will look like in the year 2010, or what the future holds for teleconferencing as a substitute for transportation. The forecasting process shares some attributes with the management process. For example, forecasting deals explicitly with the future but draws on the past and the present for guidance to predict the shape of the future. Forecasting, like managing, is an empirical science. Thus it involves elements of art, just as engineering, medicine, and management do. It is impossible not to forecast—even ignoring change involves the tacit forecast that change will be unimportant.

The trick is not so much to forecast as to do so in a way that will assist effective technology management. It is not enough to forecast or even to forecast accurately. The product of the forecast must contribute to the effective management of technology—that is, management for realization of corporate goals. Thus requirements are placed on forecasting by organizational goals and by the information needs of decision makers. These shape not only what is forecast, but also the format in which the forecast is presented. That format must be chosen to make the forecast one that the decision maker finds credible and usable. This implies that the forecaster needs a clear grasp of organizational goals and goals of the unit for which forecasts are prepared. It also is desirable that the forecaster know the individual decision makers and their information requirements and preferences. Effective forecasters, like effective communicators, must analyze their audiences.

Because forecasting is an empirical process, it embodies the strengths and weaknesses of other empirical sciences (such as medicine and engineering). Hacke (1972) noted that the foundations of an empirical science do not rest on reasoning deductively from self-evident axioms. Rather, the basis consists of rules, principles, models, or laws developed inductively from repeated observations. In other words, the foundations rest on generalizations inferred from experience. A speculation becomes a hypothesis, then a theory, and subsequently a law, as experience mounts to support it. Its robustness derives from the number and variety of observations that can be explained by reasoning deductively from it.

Technology forecasting has not attained the level of having "laws" because of the inherent complexity of the systems that are forecast and the difficulty of removing them from their technical, social, political, environmental, economic, and ethical contexts. The nature of these contexts greatly affects the growth, adoption, and diffusion of technology. In the jargon of some scientific disciplines, these are very "dirty" problems indeed. This "dirtiness" makes it virtually impossible to design a series of verifying experiments. Thus the forecaster must take historical examples, dirt and all, and attempt to compensate for contextual differences, often with no firm idea of what those differences may be. The forecaster, carefully considering the assumption of continuity, must examine the supporting context to see if changes may lead to a discontinuity in projected behavior.

The lack of significant data bases also has contributed to the failure to infer a general theory to support technology forecasting. Most methods depend on analyzing the behavior of relevant variables over a sustained period. Thus the lack of sufficient data is a major stumbling block. Economic forecasting has benefited from humankind's fascination with economic matters—a fascination that has ensured a rich and lengthy data base. However, measures of the performance or capacity of technologies have not been recorded so faithfully, and forecasters with more than a handful of data feel fortunate. Social forecasting is similarly crippled, and the social forecaster often cannot even be sure which measures are relevant.

4.1.1   The Future Is Uncertain

Despite these problems, forecasts will be made. However, they must be made with full realization of the inherent uncertainties—uncertainties that must be relayed to the decision maker. Uncertainty is a reality. All forecasters must contend with it. Fortunately, tools have been developed to help forecasters deal with uncertainty. While these tools cannot eliminate uncertainty and generally are based on restrictive assumptions about the nature of the uncertainty, they have been used successfully in many applications. Moreover, they often provide powerful ways to approximate the right answer. The basic framework for uncertainty can be illustrated by considering the general functional relationship presented in the following expression:

$$Y = f(X_1, X_2, \ldots, X_n) + e$$

This expression says that dependent variable Y is a function of the values of independent variables $X_1, X_2, \ldots, X_n$ on which it depends, plus an error, e, which results in uncertainty. This error arises from many sources, including incorrect specification of the variables (for example, not all factors that affect Y are included in the analysis), incorrect functional form (for example, linear in X even though Y may depend on $X^{1.37}$), and measurement error (such as inaccuracies in measuring the values of the Xs or Y).

The forecaster can use probability theory to enhance the predictive attributes of this functional relationship. First, the reasonable assertion is made that the expected value (mean) of e is 0—that is, while predicted values of Y will be sometimes high and sometimes low, the average of the predictions will be the same as the average of the true values of Y. It also is assumed that e is normally distributed about the mean value of 0. This implies that a graph of the probabilities of various values of e is bell-shaped with its peak over 0. Assuming this distribution enables the forecaster to predict ranges of the value of Y which will, with a given probability, contain the true value of Y. For example, suppose there is a 75 percent probability that the true value of Y will be between A and B. It is important to note that the forecaster still cannot be certain that the predicted value of Y is correct. Further, the forecaster must realize that, while a high probability can be achieved simply by widely spacing the values of A and B, the result probably will not be very helpful to the decision maker. These caveats aside, however, this approach can be extended to give a systematic measure of confidence to forecast values.

Lipinski and Loveridge (1982) proposed a model illustrating the probabilistic element of forecasting. In this model, the future fans out as a wedge-shaped terrain of possibilities, (see Figure 4.1). The probability of any one path through the terrain is small, but the sum of the path probabilities is unity. The terrain is uneven; peaks

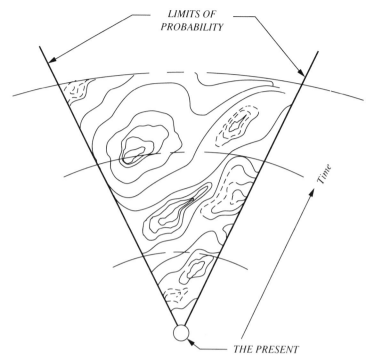

Figure 4.1 Contour map of the future. (*Source*: Lipinski and Loveridge, 1982. Reprinted by permission from *Futures*, vol. 14, no. 2, 1982.)

and valleys of opportunity and threat dot the landscape. The forecaster must explore this future and convey the lay of the land to the decision maker. A balance must be struck between the *probability* that a given piece of future terrain will be traversed and the *detail* in which the trip can be described—that is, the finer the detail of the forecast, the lower the probability of that detail occurring. At one extreme, a single path can be described in vivid detail; however, the probability of the future represented by such a path is small. At the other extreme, a future that is represented by expansive boundaries can be predicted with more certainty, but little useful detail can be given to prepare for the trip. A forecast that posits a single future is neither likely nor credible. A range of futures must be forecast, and the probability of the true future lying within the range must be measured. This range should be tight enough to give the decision maker useful guidance in managing the technology.

There are two ways that the corporate path through the future can be planned. In the first, the corporation makes travel arrangements so as to be prepared to grasp opportunities and to protect itself from challenges along the way. Planning for such a trip is called *extrapolative forecasting*. In the second approach, *normative forecasting*, the corporation stakes out some future high ground and plans its path from the present to lead to that objective. The corporate philosophy dictates the forecasting approach. The forecaster must understand this element of the corporate environment and forecast accordingly.

An examination of the model in Figure 4.1 also shows that as the forecaster ventures farther into the future, the area bounded by the limits of possibility grows steadily. It follows, therefore, that short-term forecasts necessarily are more detailed and more probable than longer-term forecasts.

4.1.2 Forecasting Targets

Effective management of a technology requires some idea of how its capacity may grow with time. The decision maker also must understand the emergence, growth, and diffusion of competitive technologies to anticipate how rapidly new technologies may substitute for old. This is equally important for corporations dependent upon the old technology and for those dependent on the new. Frequently, there is still significant profit potential in old technology even after substitution has begun. This is especially true if demand for the capacity provided by old and new technologies is growing (for instance, demand for computing power). Further, if the replacement rate is relatively low, corporate transition from one technology to another can be cushioned. For example, Marchetti (1987) forecasts that while coal demand peaked in about 1980 at 2.5×10^9 tons, demand still will be nearly 0.4×10^9 tons in 2050. A coal mining or equipment company could employ the forecast to plan strategies in the light of the projected decline in coal demand. The decline in demand, about 2.5 percent per year, is slow enough to allow a corporation to smoothly transition to a new service/product or to phase out operations gradually. Using a forecast in this way allows for extrapolative planning.

Corporations also can employ forecasts for normative planning. For example, a corporation might weigh its target market goals against the projected growth potential of several technologies to determine its development strategy.

Economic forecasting is, of course, highly relevant to the management of technology: decision makers must use the scarce resources available to their corporations to produce, use, and/or market technology. To do so they must have some idea of the state of the economy; the costs of technology, labor, and resources; and the price and demand structure that will evolve. The importance of economic forecasting has long been recognized, and economic forecasting techniques and data bases are well developed. Indeed, economic forecasts are characterized by a degree of sophistication that is inappropriate to social or technology forecasting.

It once was less apparent that effective technology management also requires adequate social forecasts. However, decision makers have come to recognize that even the most powerful technology is valueless if the social, political, and/or regulatory environments prevent it from being produced, used, or profitably sold. One only has to consider the problems of the nuclear power industry to realize the centrality of social context to technology management. A classic example of the interaction of social, economic, and technical factors appears in Exhibit 4.1.

■ **Exhibit 4.1   The Lock (Based on Drucker, 1985).**

A cheap lock exported to British India was an export firm's best seller. As personal income of Indians increased in the 1920s, lock sales declined. This was interpreted as unhappiness with the quality of the lock, so a new, superior model was designed to sell at the same price. The new lock was a disaster. It seems that the Indian peasant considered locks to be magic; no thief would dare open a padlocked door. Therefore, the key was unimportant, often misplaced, or lost. Thus, to gain entry to their own homes, peasants needed a lock that was easily opened. The new lock was not, nor was it strong enough to discourage a thief at homes of the more well-to-do. Therefore, the new lock could not be sold, and the firm went out of business four years later.   ■

This section has provided a brief rationale for the importance of integrated technology, economic, and social forecasting to aid effective management. Each type of forecast has its own requirements, variables, and techniques. Taken together, however, they sketch the contours of the future terrain that the forecaster must map for the decision maker. The forecaster must realize that failure to consider the future terrain may leave the decision maker in the position of the poker player with the second best hand: a lot may be wagered with no probability of return.

4.1.3   Criteria for Good Forecasts

Although the characteristics and techniques of technology, economic, and social forecasting differ, the requirements for a good forecast have much in common. A number of individuals have addressed these requirements (cf. Ascher, 1978; Culhane, Firesema, and Beecher, 1987; Martino, 1983; Porter and Rossini, 1987), and their observations are summarized in this section.

First and foremost, a good forecast must be credible to the decision maker and have utility for the decision-making process. It must be based on the best available information, and the methods used to process this information must be clearly described, methodologically sound, replicable, and logically consistent. The assumptions underlying the forecast also must be clearly defined and supported. When appropriate, the forecast should be quantified using legitimate units of measure. The significance, timing, and probability of forecast events must be noted, and the confidence levels that can be placed on the forecast must be stated.

Forecasts are made to aid effective technology management. If they are not credible to the decision maker or lack utility for the decision-making task, they fail to fulfill their purpose. Martino (1983) describes an interrogation model designed to evaluate forecast utility for decision making. Major elements of the model are presented in Exhibit 4.2. The interrogation model primarily is intended to evaluate the usefulness of an existing forecast. However, its joint application by forecaster and decision maker at the start of a forecasting process will help ensure both credibility and utility.

■ **Exhibit 4.2 Interrogation Model: Forecast Evaluation (Based on Martino, 1983).** [1]

This model consists of a sequence of questions that should be answered by forecaster and decision maker to ensure that a forecast has utility for the decision-making process.

I. *Interrogation for Need*
   1. What individual or organization will make the decision which the forecast is to inform?
   2. What information about the future is needed? (This broad question is asked initially to avoid omitting relevant factors.)
   3. How much is knowing the future worth? What is the cost of not knowing?
   4. Over what time horizon must the forecast apply?
II. *Interrogation for Underlying Cause*
   1. In what ways can the subject area change (magnitude, composition, character)?
   2. What can cause these changes (technological, economic, managerial, political, social, cultural, intellectual, religious/ethical, ecological factors)? Note that several may be contributory.
   3. Iterate questions 1 and 2 in Step II until a set of "fundamental" causes are found. (Produce a causation hierarchy or relevance tree.)
   4. What are the key elements (that is, the causes with strong influence or that appear frequently as "fundamental" causes)?
III. *Interrogation for Relevance.* Here the forecast is examined to see to what degree it deals with key elements from Step II. There can be three categories: key elements that are (1) relevant and included, (2) relevant but excluded, and (3) irrelevant but included. Category 2 elements are critical because they must be factored into the forecast. Category 3 elements can be ignored.
IV. *Interrogation for Reliability*
   1. *Methods* must be identified and clearly described. If they are not or cannot be categorized, the forecast is of unverifiable reliability.
      a. Are the methods replicable by others? If not, results, while not necessarily wrong, cannot be verified.
      b. Are the methods formally consistent? Do conclusions logically follow from the premises?
   2. *Assumptions* are used when facts are unknown and thus are crucial to the forecast.
      a. What are the assumptions? (Assumptions may include the following: the method used is appropriate to this case; a group will behave in a certain way; or a data sample is representative of the population.)
      b. Are assumptions adequately defined?
         (1) Is the assumption static or dynamic? Does it apply to a level or a rate?

[1]Reprinted by permission. © 1983 by Elsevier Science Publishing Co., Inc.

(2) Is the assumption realistic or humanistic? Is it independent of opinion or dependent on values, perceptions, and intentions?

c. How are assumptions supported? By a law, principle, or axiom? By
credible expert opinion? By evidence?

d. Are the assumptions directly supported by the available evidence? The
following criteria should help.

   (1) Static assumptions must be from the state or time specified by the
   assumption.

   (2) Dynamic assumptions must be gathered dynamically with timing
   and sequencing consistent with the assumption.

   (3) Realistic assumptions must be as free of human bias, prejudice, and
   evaluation as possible.

   (4) Humanistic assumptions must be obtained from the behavior of the
   group or individual in question.

   (5) A law or principle must be applied under circumstances for which
   the law or principle is assumed to apply.

   (6) Credibility of experts must deal with their knowledge regarding the
   case under consideration.

3. *Data* may be used to support assumptions or as raw material from which a
pattern is to be determined.

a. How accurate are the data?

   (1) Are there uncertainties of observation?

   (2) Are there uncertainties of measurement?

   (3) Is there a conscious distortion?

   (4) Is there an unconscious distortion?

b. Does the evidence tend to confirm or deny the assumptions?   ■

Forecasts are guides for *future* action. Thus their accuracy is unknown when they
are produced and used for management decisions. The time horizon of the forecast
has been found to be the greatest determinant of accuracy—the shorter the horizon,
the more accurate the forecast (Ascher, 1978). A recent forecast is preferred to an
earlier forecast, even if the recent forecast is cruder in its methodology and has less
content. Recalling the contour map in Figure 4.1, this is not surprising. Of course,
forecasts cannot always be limited to short time horizons. Since accuracy is harder
to achieve for longer horizons, close attention must be given to the interrogation
model in Exhibit 4.2. Attention to data selection, to selection of forecasting tools,
and to underlying assumptions are vitally important.

Occasionally a forecast is effective despite its inaccuracies. For example, Vanston
(1985) cites a 1978 survey by the Technical Association of the Pulp and Paper Industry. In this survey, manufacturers were asked to estimate the paper volume
that they would produce in 1985. When estimates were totaled, it was found that
predicted production would require that "every tree then growing in the United
States and Canada...be harvested every two years." Even though the estimate
was patently inaccurate, decision makers realized that the forestry program would
have to be expanded and other sources of cellulose would have to be found. Ex-

amples not withstanding, however, it is foolish to ignore the precepts of accurate forecasting simply because inaccurate forecasts are occasionally effective.

For many topics it will be appropriate to select a quantifiable measure to forecast. Typical examples are system performance measures (such as efficiency, seat-miles per hour, operations per second), market share, demand, population, gross national product, and number of workers. Sometimes representative measures can be single parameters (such as maximum miles per hour of fighter aircraft); in other instances, a composite measure (such as ton-miles per hour of cargo aircraft) is required to capture important system attributes. Measures either can be direct (such as horsepower and torque as measures of marine engine performance) or indirect (surrogate measures of a characteristic not directly measurable, for example, R&D investment as a measure of future corporate competitiveness).

Appropriate quantified measures are preferred to qualitative measures. Results generally are easier to interpret and less ambiguous than those based on qualitative measures. Quantified measures also allow powerful analytic tools to be applied. The forecaster must realize, however, that sophisticated manipulation of doubtful data based on unsupported assumptions will not produce a viable forecast. In fact, it has been found that methodological sophistication adds little to forecast accuracy (Ascher, 1978). Moreover, if the method employed is not easily explained or does not appear logical to the decision maker, the forecast will have little credibility.

Too frequently, a quantified measure is not legitimate or may not be accepted as legitimate by the decision maker. In such instances, qualitative measures must be used. There is a tendency to equate nonquantified measures solely with social forecasts. For example, the statement that "the political climate is too complex to allow a numerical measure" is often voiced by political decision makers. Examples of qualitative forecasts mixed with subjective probabilities of occurrence and projected cost information made at the beginning of the 1990s are shown in Table 4.1. These forecasts, and many others, appeared in *Future Stuff* by Abrams and Bernstein (1989).

TABLE 4.1 What's Ahead for the 1990s?

Trend	How Likely?	When?	How Much?
X-ray-less mammogram (breast exams using light rays)	100%	1995	$35/exam
Intracorneal rings (correct nearsightedness)	75%	1994	$2,000
Poison-ivy vaccine	50%	1994	Unknown
Walking TV (follows the viewer from room to room)	50%	1995	$5,000
Smart houses (computer controls all electronics)	90%	1997	$7,000–$10,000

Source: Abrams and Bernstein (1989).

Other popular forecasts include Adler and Leonard's (1989) predictions that robots will be cooking hamburgers in fast-food restaurants, mopping floors in shopping malls, and delivering meals in hospitals; however, the robots will not be able to make beds. Another robot feat may entail acting as a security guard in a warehouse with the ability to fire a weapon if an intruder is encountered.

In the same forecast, *Newsweek* predicts a reinvention of television. Examples include the combination of the TV, telephone, and computer. The transformation of television will occur in stages, with high definition television (HDTV) as the first step. The next step will be a powerful mixture of television and computer technology. This mixture of technologies will allow the user to play a travelogue for London, but select only museums of history for display, as an example.

4.1.4 Common Errors in Forecasting

Causes for failure are implicit in the interrogation model described in Exhibit 4.2. Chapter 19 discusses causes of errors; examples are all too abundant. No list, however can be complete: we will inadvertently discover new ways to fail. Nonetheless, there are fundamental sources of error that should be noted at this introductory stage and carefully avoided. These errors can be attributed to *contextual oversights*, *biases of the forecaster*, and *faulty core assumptions*.

Contextual errors arise because the forecaster does not consider changes in the social, technical, and/or economic contexts in which the technology is embedded. Changes in these contexts affect the assumption of continuity between the past and future that lies at the heart of empirical forecasting. Thus contextual changes can produce a seeming discontinuity in forecast behavior. For example, the more stringent regulatory practices produced by the environmental movement of the 1960s and 1970s have had a major effect on the growth of nuclear power generation technology. Conversely, more recent federal deregulation policies have produced major changes in the airline industry. Not infrequently, the development of a competitive or supporting technology can have a major effect. For example, Martino (1983) notes that the digital computer would not have been possible "without the transistor or something that shared its properties of low cost, high reliability, and low power consumption" (p. 230). Contextual errors arise from viewing the forecast subject too narrowly. They sometimes can be avoided by careful attention to the ways in which a technology can change and the potential causes of those changes (see Step II in Exhibit 4.2).

Forecaster bias can be conscious, but more frequently it is unconscious—a product of culture, personal history or attributes, group affiliation, viewpoint, and/or disciplinary training. Intentional bias is often the result of having a personal, political, ideological, or corporate "turf" to protect. While conscious bias is the most easily recognized, it is not always easy to correct. Unconscious bias is subtler simply because it is unrecognized by the forecaster and may be very difficult for the decision maker to evaluate. Occasionally, bias even may result from overcompensation for personal biases forecasters recognize in themselves (Martino, 1983). There are innumerable sources for unconscious bias, a few of which include: underemphasizing or overemphasizing recent data or trends at odds with historical behavior; the natural desire not to be the bearer of bad news; belief in or reaction to the "technological imperative"; and world view. The best safeguard is a careful search for intrusion of forecaster bias.

Assumptions fill gaps where no data or theory exist. Thus they are especially critical in forecasting. The forecaster will do well to internalize the common-sense observation that "It's not what you don't know that hurts, it's what you know is true that isn't." Problems deriving from core assumptions (that is, assumptions that derive from the basic outlook within the forecast context) are particularly troublesome. Core assumptions are similar to unconscious biases in the sense that they often are not recognized by the forecaster. However, they are so central to the forecast that they strongly influence the result. The assumption about the relation between lock quality and sales from Exhibit 4.1 can be interpreted as a faulty core assumption.

Ascher (1978) has noted that core assumptions are major determinants of forecast accuracy. If the core assumptions of a forecast are correct, the choice of method is either obvious or secondary; if they are not, the result cannot be corrected by method selection. If the forecaster begins with the notion that a given trend is taking place, whether the substitution of one product for another or the growth in performance of a technology, the data and methodology can usually be made to bear this out. As with many human endeavors, if we are not careful, the end is determined by where we begin.

4.2 TECHNOLOGY FORECASTING

Given the pervasiveness of the term "technology," it might be supposed that every person has an intuitive understanding of its meaning. That understanding can be surprisingly vague, however, and vagueness breeds misunderstanding. In this text the term is broader than might be supposed; therefore, it is appropriate to provide a definition:

Technology is the systematized knowledge applied to alter, control, or order elements of our physical or social environment. This definition includes not only the hardware systems usually equated with technology, but systems of analysis, regulation, and management as well. For convenience, we might refer to these as hardware and software technologies, respectively. Of course many technologies (such as communications) integrate elements of both.

It has become fashionable to talk of the accelerating rate of technological change. Indeed, the manager has been deluged by new technologies that have altered the ways in which even the most fundamental activities must be performed to be competitive. Such new technologies include: computer-aided design, drafting, and manufacturing; robotics; zero-defect quality control; and communication networks. Decisions about production, use, and/or sale of these technologies, and of others affected by them, require some conception of how they will change in the future.

The term "technology forecasting" designates forecasting activities that focus on changes in technology. Technology forecasts usually center on changes in functional capacity and/or on the timing and significance of an innovation. These forecasts should be distinguished from forecasts in which technology plays a role but is not the central issue. Ascher (1978) cites population projections as an example of a forecast in which technology plays a role; the focus is population growth, but technologies such as birth control are important.

To forecast technology, the forecaster must understand how a technology develops and matures. Unfortunately, the growth of technologies is strongly affected by changes in the social/political context in which they are embedded and by the growth of supporting and competing technologies. Not only is this context dynamic, it affects different technologies in different ways. Thus *there is no single growth pattern that describes the development and diffusion of all technologies*. There are general concepts of how technologies develop, however, and these can be useful guides.

4.2.1   Models of Technology Growth and Diffusion

The attributes of technology most often forecast are

1. Growth in functional capability
2. Rate of replacement of an old technology by a newer one
3. Market penetration
4. Diffusion
5. Likelihood and timing of technological breakthroughs

Regardless of the attribute to be forecast, it is important to understand both the technology and the process of conception, emergence, and diffusion that characterizes its growth.

The forecaster must understand the characteristics and attributes of the technology to anticipate how the technology might be used and to choose legitimate measures of functional capacity. These measures may differ for technologies that appear very similar. For example, maximum speed is a legitimate measure of fighter aircraft performance because of the nature of their mission. However, speed alone is not a legitimate measure for transport aircraft because it captures only part of the aircraft's functional capacity—to deliver a payload rapidly. Often the forecaster must understand not only the technology in question, but also earlier approaches that fulfill the same need. Such understanding is required to develop trends that are defined by successive technological approaches. A firm understanding of basic principles also is required to identify competitive technologies and the technologies necessary to support, or that may be supported by, the subject technology.

Understanding the process of introduction and growth of a technology also is important to forecasting. Bright (1978) proposed a model of growth that has been somewhat expanded by Martino (1983). The process, shown in Exhibit 4.3, extends from scientific discovery or the recognition of opportunity and need through widespread adoption and proliferation. Some of these stages may overlap while others may be difficult to pinpoint (for example, the first stage, sometimes called the Hahn-Strassmann point—Prehoda, 1972). Therefore, when the model is used to compare the growth of several technologies, care must be taken to employ consistent definitions of each stage to all technologies.

■ **Exhibit 4.3   Stages of Technology Growth (Based on Martino, 1983).**

Stage 1—Scientific findings; determination of opportunity or need
Stage 2—Demonstration of laboratory feasibility
Stage 3—Operating full-scale prototype or field trial
Stage 4—Commercial introduction or operational use
Stage 5—Widespread adoption
Stage 6—Proliferation and diffusion to other uses
Stage 7—Effect on societal behavior and/or significant involvement in the
        economy   ■

Bright (1978) suggests that it probably is not feasible to forecast technologies that have not progressed to at least the latter stages of full-scale development (Stage 3 of Exhibit 4.3). However, the decision of when to forecast depends on how critically the forecast is needed, what is to be forecast, and the method used to make the forecast. For example, in cases in which one technology is substituted for another, many forecast only after data indicate a 10 percent replacement (well into Stage 4). This decision is made because the Fisher-Pry model (see Exhibit 4.4) often employed is felt to overestimate takeover rate when only very early data points are used (Sharif and Kabir, 1976). Lenz (1985), however, notes that the most crucial decisions must often be made early in the substitution process when the replacement rate is very low. He suggests that the Fisher-Pry model should be applied if even as few as four early data points indicate a trend. This tentative forecast should then be revised as later data become available.

■ **Exhibit 4.4   The Fisher-Pry and Gompertz Models**

The Fisher-Pry and Gompertz models are mentioned in this chapter but are fully described in Chapter 10. The Fisher-Pry model is styled as a growth curve because it predicts characteristics loosely analogous to those of the growth of a biological system. It is also referred to as a substitution model because of its application in forecasting the rate at which one technology will replace another. When plotting a response as a function of time, the Fisher-Pry curve takes on an S-shape with a slow beginning, a rapid ascent, and a tapering off at the finish. A scale factor allows variation in the shape of the curve.

The underpinnings of the Gompertz model are quite different from those of the Fisher-Pry model. The Gompertz model is often referred to as a mortality model and is most appropriate in cases in which equipment is replaced because it is worn out rather than because it is technologically, obsolete. The Gompertz model also produces an S-shaped curve, but it looks different than the curve formed by the Fisher-Pry model. The difference is that the Gompertz curve rises more sharply but begins to taper off earlier. Like the Fisher-Pry curve, the Gompertz curve can be adjusted by the forecaster. Figure 4.2 sketches the two curves for cable television subscribers over time (this example will be developed in detail in Chapter 10). ■

Bright (1978) noted that some phases in the growth of new technologies may have been compressed, but that the full process usually takes 10 to 25 years or more—hardly a blinding pace. Although this may appear inconsistent with the concept of accelerating technological change, there is a 15-year lag between "rapid" and more leisurely growth times. Thus if a number of technologies shifted to more rapid growth patterns, the perception of pace could be greatly changed. Observa-

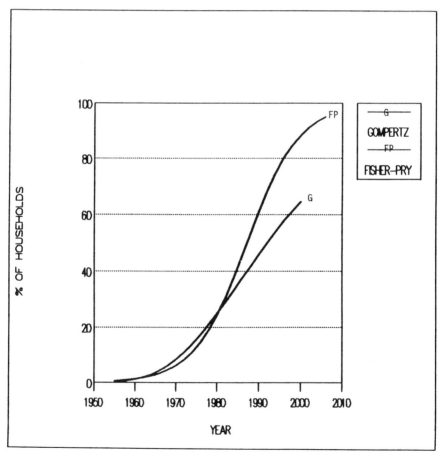

Figure 4.2 Fisher-Pry versus Gompertz growth models for cable television subscribers in the United States.

tions of rates of change in information technologies during the 1980s suggest a quickening of this pace, with typical maturation probably less than 10 years.

Technological breakthroughs come in response to latent need (Stage 1 of Exhibit 4.3) that can be met by new, clearly superior methods. Supporting technologies are often needed to accomplish such qualitative advances. Typically, the growth of a successful technology proceeds rapidly after start-up difficulties are overcome (Stages 3 and/or 4) and before potential is saturated (Stages 6 and/or 7). Finally, the process often can be shortened by making it a high priority and by assigning major resources to it. Thus it is probable that the deluge of technological change has been fueled by the speed and ease of communication across international boundaries; the larger number of players and the increased competitiveness of world markets; and the number of fundamental changes to personal and corporate activities (that is, the number of technologies entering Stage 7 that affect us directly).

The growth in capacity of many technologies exhibits an S-shaped (also variously specified as a sigmoid or logistic curve) pattern (see Figures 4.2 and 4.3). Growth in the early stages (Stages 2 and 3 of Exhibit 4.3) is apt to be slow and halting as fundamental problems in design and development are overcome. At commercial introduction (Stage 3), the new technology may not exhibit a clear performance superiority over existing technologies. Early problems of production, supply, and maintenance may slow adoption, and firms committed to older technologies may await performance improvements or cost breaks spurred by the new technology. If these start-up difficulties are overcome, the growth of the new technology (latter part of Stage 4 and Stages 5 and 6) can be very rapid (essentially exponential) as gains are consolidated and inexpensive improvements are made. Eventually, however, further improvement becomes progressively more difficult and expensive and the rate of improvement slows as performance nears theoretical or practical limits. Thus an S-shaped growth pattern is typically followed.

If demand for the service provided by the old technology remains, opportunity for a new approach exists. Then a new technological approach may well emerge and itself follow an S-shaped growth pattern. Typically, the new technology will begin to develop when limitations of the old approach are apparent but before it has reached its limit. The *envelope curve* depicts the cumulative development pattern for a family of technologies (see Figure 4.3). The overall progression is sometimes exponential for a considerable time period, as illustrated by improvements in computing power per dollar from the 1950s into the 1990s. Note that Figure 4.3 shows a linear pattern for the *logarithm* of capacity or performance versus time; that is equivalent to exponential growth.

When a forecaster selects an exponential growth pattern, he or she makes the core assumption that cumulative breakthroughs will spur the rapid development that is forecast. Although a number of technological families exhibit exponential growth, not all do. Martino (1983, pp. 79–80) indicates that technologies occasionally exhibit parabolic or higher-powered growth curves; however, if the cumulative pattern is not exponential, it is most frequently linear. Again, the most prominent pattern for a single technology's growth is an S-shaped curve.

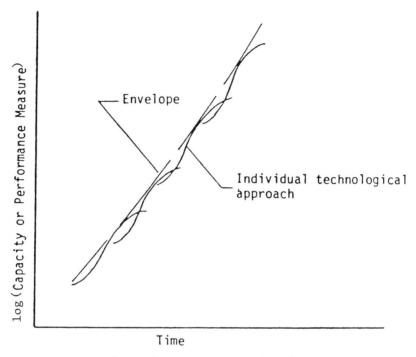

Figure 4.3 Capacity growth of a technology.

As Ascher (1978) notes: "The unique challenge of technological forecasting is to anticipate the consequences or capabilities of inventions not yet fully conceived or imagined" (p. 167). Therefore, forecasters of exponential growth need not be able to identify the specific technological approach that will extend capacity, but they must support the assumption that an exponential pattern is appropriate. This is most often done by identifying a clear opportunity or need for new technologies to provide that growth.

4.2.2 Technology Forecasting in Context

Dynamics of the economic and social/political contexts fundamentally affect development of a technology. Therefore, technology forecasts require background forecasts in economic, social/political, and other areas as well. The forecaster, lacking resources or expertise, may be forced to rely on existing contextual forecasts. This can cause problems. Wheeler and Shelley (1987), for example, investigated forecasts of demand for innovative, high-technology products (personal computers, artificial intelligence, fiber optics, satellites, genetic engineering, photovoltaics, etc.) and found them uniformly optimistic by 50 percent or more. Among other things, they attribute these results to a lack of forecaster expertise in consumer behavior. They also cite overenthusiasm for high technology and poor judgment.

Existing background forecasts may embody core assumptions that are not explicitly cited or are no longer valid—Ascher (1978) refers to the latter problem as "assumption drag." This difficulty is especially acute in social/political arenas that can exhibit considerable volatility. In such cases, the forecaster should assess the magnitude of errors that inaccuracies in the contextual analyses might induce in the technology forecast (see Steps II and III and Step IV, part 2 of Exhibit 4.2). For critical background forecasts, the forecaster should recall that recent projections are apt to be more accurate regardless of their methodological sophistication (or lack thereof). Thus rough, relatively inexpensive forecasts can be contracted or produced with the anticipation that they will be worth the investment despite their lack of sophistication. It is possible to produce several such forecasts at costs comparable to a single sophisticated forecast. Finally, the interdependence of forecasts suggests the need for several disciplinary specialties within the forecasting team.

Ascher (1978) notes that "the selection of a broad method and of more specific techniques is much more than a technical choice or a matter of convenience" (p. 196). For example, choosing an exponential envelope involves the core assumption that sequential technological approaches will emerge to sustain continual rapid growth. Forecasting by analogy requires the forecaster to identify an earlier technology whose historical growth the new technology will be assumed to follow. Using a lead-lag indicator method (discussed in Chapter 9) involves the assumption of some logical connection between the leading and lagging technologies. The use of formal models implies that the assumed structural relationships of the model will be valid for the technology studied over the forecast time horizon.

Ascher (1978) suggested that one should consider the following factors in preparing a technology forecast:

1. Dependence on basic scientific breakthroughs
2. Physical limits to the rate of development
3. Maturity of the science and applications of the technology
4. Sensitivity of the pace of innovation to high-level policy decisions
5. Relevance of R&D funding
6. Extent of substitutability by other products or by parallel innovations
7. Relevance of diffusion
8. Opportunities to borrow advances from related technologies

These guides and attention to the interrogation model presented in Exhibit 4.2 point the direction toward valid and useful forecasts.

4.2.3 Technology Forecasting Methods

Numerous schemes have been proposed to classify technology forecasting methods. All contain ambiguity; none are entirely satisfactory. For example, as discussed earlier, forecasting methods can be classified as either extrapolative or normative — that is, whether they extend present trends into the future or look backward from a desired future to determine the developments necessary to achieve goals. Although this is a useful distinction, many methods can be considered either normative or extrapolative depending on how they are applied. Further, the classifications can be confused with the perspective of the planning or decision-making activity the forecast is to inform.

Porter and Rossini (1987) suggest that the hundreds of forecasting techniques fit into five families:

1. Monitoring
2. Expert opinion
3. Trend extrapolation
4. Modeling
5. Scenarios

Although this categorization system is used in selecting methods for inclusion in this book (see Chapter 6), it does have limitations. The first family, monitoring, is not a forecasting method per se; rather it is a method for systematic accumulation and analysis of the data from which forecasts are made (see Chapter 8). Further, it is unclear where to place forecasts made by analogy or forecasts that employ lead-lag indicators. Cetron and Monahan (1968) suggest categorizing methods as extrapolation, trend correlation, analysis, and analogy, a scheme that has limitations similar to Porter and Rossini's categorization.

It may also be helpful to categorize methods as to whether they are direct, correlative, or structural. Table 4.2 indicates how some of the more common techniques might be categorized in such a framework. *Direct methods* are those that directly forecast parameters that measure functional capacity or some other relevant characteristic of the technology. These methods do not explicitly consider correlations with technological, economic, social, and political contexts, nor do they consider structural relationships within those contexts. It is clear that direct methods imply major assumptions about the nature and permanency of context and structure. It is important to note, however, that expert opinion methods do allow subjective consideration of contextual change through the implicit, mental model that each expert has internalized about the nature and likelihood of change.

Correlative methods relate development of the subject technology to the growth or change of one or more elements in its context or in contexts thought to be analogous. Lead-lag correlation techniques, for example, seek to identify a technology for which growth precedes that of the technology to be forecast. Martino (1983, pp. 100–103) presents such an analysis for combat aircraft (lead) and transport aircraft (lagged) speeds. Likewise, forecasting by analogy asserts that development of the subject technology will follow the pattern established by an earlier technology even though no scientific/technological tie of the kind assumed for lead-lag correlations exists. Scenarios often are used to forecast major portions of the context along

TABLE 4.2   Categorizing Technology Forecasting Methods

Category	Definition	Applicable Forecasting Methods
Direct	Direct forecast of parameter(s) that measure an aspect of the technology	Expert opinion Delphi Surveys Nominal group Naive time series analysis Trend extrapolation Growth curves Substitution Life cycle
Correlative	Correlative parameter(s) that measure the technology with parameters of other technologies or background forecast parameters	Scenarios Lead-lag indicators Cross impact Technology progress functions Analogy
Structural	Explicit consideration of cause-and-effect relationships that effect growth	Causal models Regression analysis Simulation models Deterministic Stochastic Gaming Relevance trees Mission flow diagrams Morphology

with the technology, although specific statements of the structural relationships are implicit. Cross-impact methods begin with a matrix that arrays some set of factors against another to examine their interactions. For instance, one might explore how gains in one energy technology (say, nuclear fusion) would affect prospects in another (say, nuclear fission). Cross-impact is explicit about the impacts of selected elements of technology and context but involves no explicit statement of the cause-and-effect structure that produces these impacts. All correlative methods proceed on some formal or informal assumption of the relationship between the forecast technology and one or more elements of its context. They also involve the implicit assumption that the relationship does not change (that is, correlative methods are structurally static).

Structural methods forecast by formally considering the interaction between technology and context. To varying degrees, they must be explicit about the structural relationships between the technology and other elements of its context. Some methods (such as relevance trees and mission flow diagrams) merely develop or portray the paths that connect the various elements to each other and to the technology. These methods are most often used in normative forecasts. Simulation models, however, also must quantify the relationships among elements. Regression analyses seek to structure those relationships. All, regardless of sophistication or complexity, simplify reality to make problems tractable. A structural method, therefore, is valid only if it retains the relationships critical to accurately predict the growth of the technology to be forecast. If a simulation model is chosen, not only the structure, but also the mathematical formulations it embodies must be valid. It is important to realize that a method may satisfy this condition for a technology at a given time but not at other times or for other technologies. Thus when a structural method is chosen, the forecaster makes the core assumption that the structure embodied by the model is appropriate and that it will remain so over the horizon of the forecast.

The most common models are those that assume change can be explained by factors internal to the system that produces the technology or by economic factors. However, there are models that consider social, political, and other factors, as well as the effects of policy interventions (see the discussion of KSIM and System Dynamics in Chapter 12).

Morphology is a technique used to probe the structure of a problem to help generate ideas for innovation and/or discovery (Shurig, 1984). Morphology is a bit difficult to define concisely, but picture it as a cross-impact matrix that depicts nonnumeric relationships. Unlike cross-impact, however, morphology is intimately concerned with structure and thus more truly structural than correlative in nature. Morphology has been used to investigate a range of diverse problems from possible jet engine types (Zwicky, 1962; 1969) to Mendeleyev's periodic table of elements and Kondratieff's long wave business cycle (cf. Volland, 1987).

4.2.4   Forecasts Impact Decision Making

Technology forecasts are made to assist decision making. In a poll of 1,114 corporate executives, Cetron and Ralph (1971) found planning, marketing, manufacturing, and programming and budgeting staffs were the most frequent users of technology forecasts. They also found that technology forecasts were either frequently or always used as planning aids by 73 percent of the respondents and were used as aids in resource allocation by 61 percent. Also, nearly 25 percent used technology forecasts to justify previous decisions and to help acquire a government contract. Several authors (cf. Martino, 1983; Vanston, 1985) also have considered the range of uses to which corporations put technology forecasts. The most common are

- To maximize gain and/or minimize loss due to actions/events either internal or external to the corporation; to plan production, inventory, or sales
- To guide resource allocation

- To identify and evaluate market opportunities or threats
- To guide staff, facilities, or capital planning
- To develop administrative plans, strategies, or policies—including the assessment of their risks
- To assist R&D management
- To evaluate new products or processes

Accurate technology forecasts can give a corporation, government agency, or nongovernmental organization great advantage. For example, Pratt & Whitney, a major aircraft piston engine producer at the close of World War II, recognized the growing importance of turbojet engine propulsion systems in military and commercial aircraft applications. Consequently, it moved piston engine operations to Canada and devoted U.S. manufacturing facilities to turbojet engine production. This decision played a major role in its prominence in today's market. The Whirlpool Corporation was able to increase market share when, after monitoring chemical and textile industry developments, it introduced permanent press cycles on washers and dryers nearly a year before competitors (see Chapter 8 for a thorough discussion of the Whirlpool example). Such benefits can often be achieved by assigning resources and priority to projects to compress the time span of the growth curve characteristic of a technology. Corporations also can reap benefits from increased operational efficiency by taking advantage of projected technological developments. On the governmental side, technology forecasting began with efforts by the U.S. Department of Defense to maintain a technological edge in the cold war of the 1950s.

The penalties of inaccurate forecasts can be severe. For example, Wheeler and Shelley (1987) note that when artificial intelligence sales, forecast as high as $443 million for 1984, turned out to be less than $200 million, many companies were forced to lay off employees and eliminate pay raises and stock dividends. Likewise, the failure of the projected growth of the nuclear power generation capacity to materialize has brought some public utilities to the brink of disaster.

4.3 SOCIAL FORECASTING

When the forecaster considers the underlying causes of change relevant to a technology forecast (Steps II and III, Exhibit 4.2), he or she often will identify causes that are neither technological nor economic. These generally are classified as social or social/political. (Chapters 15, 16, and 18 expand the list of factors interacting with technology.) This may seem somewhat arbitrary since both technology and the economy are elements of the social context; however, the division is convenient because forecasting social/political (sociopolitical) factors involves different concepts and problems than projecting technological or economic ones.

Social forecasts often deal with deference values, such as respect and power, rather than with welfare values, such as wealth or well-being. There are five issues that make these factors more volatile and therefore more difficult to forecast (Ascher, 1978):

1. The factors often can be easily altered through human volition since material resources frequently are only marginally important.
2. There is seldom a consensus on a preferred direction of sociopolitical change.
3. Social attitudes are far less cumulative than material growth patterns.
4. Single, discrete events are often the central focus.
5. A single factor is apt to be meaningless without reference to the entire sociopolitical context.

These issues make it difficult to assess the validity of sociopolitical forecasts. For example, since social attitudes are not especially cumulative, the relaxed interpretation of the U.S. Environmental Protection Agency's mission in the 1980s might lead to stricter interpretations of environmental regulations rather than to less strict ones. Likewise, a forecast that environmental regulations will be more rigidly enforced is meaningless unless it also includes forecasts of the nature and restrictiveness of those regulations.

Sociopolitical forecasts are likely to rely more heavily on qualitative approaches. Exceptions include social indicator and demographic projections, regression analyses, and certain simulations. Indeed, even simulation models that produce quantitative output (such as KSIM) rely on quantifying qualitative input about the interaction of important variables.

In 1978 Ascher suggested that two sociopolitical forecasting techniques had special promise: scenarios and social indicator projections. To these we would add expert opinion methods (such as Delphi) and note that simulation models can be useful in some contexts. Although no method is without problems or limitations, these seem most promising.

A *scenario* is a descriptive sketch intended to produce a more or less holistic view of a future social state. The great strength of these "future narratives" is their richness—a richness that can set the context within which the meaning of individual sociopolitical factors can be assessed. Scenarios (see Chapter 13) may be largely verbal or be composed of an array of quantified forecasts, which in total convey a holistic perspective. The former are more common. However, it is not unusual for the narrative to be "hung on hooks" provided by one or more quantified forecasts. Whatever form they take, scenarios blend insight with storytelling skill to provide a relatively complete picture that illustrates possible outcomes. It is common to provide a set of three scenarios that span what the forecaster believes to be the range of possible futures: (1) a surprise-free projection, (2) the worst case projections, and (3) the best case projections.

Social indicators are aggregate measures of various phenomena that, taken collectively, represent the state of society or some subset of society (such as a community). They can be used to measure economic, demographic, educational, welfare, employment, health, and criminal justice factors, as well as other factors important to sociopolitical description and forecasting. Social indicators are analogous to measures of the functional capacity of a technology. In addition, they possess three important characteristics: (1) they are numerical and thus can be used with trend analysis and other quantitative methods; (2) they lend themselves to geographic disaggregation and analysis (Ferriss, 1988); and (3) they are widely collected.

Since social indicators are numerical measures they can be used with a wide range of direct, correlative, and structural methods. A comprehensive overview of techniques applied to demographic projections is given by Branch et al. (1984). For example, regression techniques have been used to forecast a wide range of trends—from murder and car theft through the 1990s (Cohen and Land, 1987) to cancer mortality from 1980 through 2020 (Greenwald and Sondik, 1986). When using forecasts produced by regression analysis, the forecaster must assess the continued relevance of the independent variables employed, and the continued irrelevance of variables omitted, over the time horizon of the forecast. Further, since indicators are frequently projected singly, the forecaster must assess their dependence on context.

Social indicators are collected by a wide variety of organizations (Ferriss, 1988). Data can sometimes be disaggregated by successively smaller geographic areas: nation, region, state, district, county, or even census tract and block. Social indicator data may be gathered by groups that consciously or unconsciously employ biased techniques (Cook, 1985) or, by organizations for which the task is not a high priority. Thus inaccurate or careless techniques may have been used to gather them (see Sechrest, 1985). The forecaster should also ensure that social indicators used as surrogate measures of other phenomena are used validly. In this regard, surrogate social indicators are no different than surrogate technological or economic measures. Forecasts based just on social indicator trends do not provide as rich or as integrated a picture of a changing societal context as scenarios can.

4.4 CONCLUSION

Given the uncertainty inherent in the future, expectations placed upon forecasts should be realistic. Do not expect forecasts to be perfectly accurate or highly precise. On the other hand, do expect them to delimit the extent of uncertainty, establish correct vectors of change, and provide valuable information to the would-be manager of technology. Forecasts should indicate ranges of future possibilities, not point values.

Technology forecasting relates to economic forecasting and to social forecasting. They share many common characteristics of what constitutes a good forecast. Change in one domain affects the others. Forecasting requires application of technology, economic, and social forecasting models in conjunction with appropriate data and suitable techniques. The key models, data issues, and techniques introduced in this chapter will be detailed in later chapters.

EXERCISES

4.1 How has the introduction of the personal computer affected societal behavior?

4.2 Enumerate the criteria for good forecasts using Section 4.1.3 as a starting point.

4.3 Explain the difference between extrapolative and normative forecasting; give specific examples.

4.4 Which forecasting method is the best?

4.5 List possible causes of forecaster bias.

4.6 Give examples of each attribute of technology most often forecast (see Section 4.2.1).

4.7 Examine literature from the 1970s or before on forecasts for technology by the year 2000. How close does the current state match the path from the date of the forecast to the present? Can you make quantitative or qualitative statements about the breadth of the contour map (see Figure 4.1) from the date of the forecast to the year 2000?

4.8 Choose five variables on which an extrapolative forecast could be made (such as energy use, GNP per capita, etc.). How could normative forecasts be made for the same variables?

4.9 Search the literature for a technology forecast in an area that interests you. Apply the interrogation model from Exhibit 4.2 to evaluate the forecast.

4.10 List ways in which contextual oversights, biases, or faulty assumptions could have affected the forecast evaluated in Exercise 4.9. Think of ways to avoid such sources of failure.

4.11 List four examples of technology that have appeared in the twentieth century (such as TVs, automobiles, computers, robots, etc.). Find suitable data and plot the changes in the number produced over time; compare the plot to the S-shaped curves discussed in this chapter.

4.12 Apply the stages of technology growth (see Exhibit 4.3) to each of the plots in Exercise 4.11. Are all of the stages evident? What additional information is needed to specify the stages?

4.13 Write three 10-year scenarios for a technology about which you have some knowledge: (1) a surprise-free scenario, (2) a best case scenario, and (3) a worst case scenario. How might a technological manager use these scenarios?

___5
MANAGING THE
FORECASTING PROJECT

OVERVIEW

Managing a forecasting project differs in some ways from managing other projects. This chapter emphasizes the special management requirements of the forecasting project. First information requirements implied by the questions usually posed for technology forecasts are discussed. Next the crucial importance of scheduling is emphasized, and the Project Evaluation and Review Technique (PERT), the Gantt chart, and the Project Accountability Chart (PAC) are detailed. Finally, the characteristics of multiskill projects such as forecasting are discussed and different models of team structure and communication are considered.

5.1   INTRODUCTION

The technology manager must organize and manage the search for information upon which to base sound decisions. Much of that information will deal with future conditions. For example, the manager may wish to know how technological advances will affect the profitability of existing plants, equipment, or products; what new technologies offer opportunities or challenges to the firm; what technologies can be brought to market sooner by increasing R&D resources; or what the competitive or regulatory environments of the future hold. The information necessary to make management decisions involves stretching present knowledge into the future—forecasting—and, since the future does not yet exist, the decisions also involve uncertainty.

The manager must understand that forecasting is not the same as decision making. Forecasts alone will not produce decisions. They provide the raw material—

information—from which to fashion wise decisions. Thus the manager should not begin a forecasting project with the expectation that it will produce the answer to the problem he or she must decide. To reach a decision, the manager will need to synthesize forecasts with his or her knowledge of the needs and characteristics of the organization.

The manager must determine if the decision will be best served by extrapolative or normative forecasting perspectives (see Chapter 4). The former asks what the future may bring if trends remain as they are; the latter asks what actions, advances, or breakthroughs may be needed to shape the future. Some decisions will require both perspectives. Early in the project the manager also must decide what information is needed, the degree of specificity required, and the uncertainty that can be tolerated. Further, he or she must assess the potential benefits of having the information so as to decide the costs justified in obtaining it. It is important as well to decide when information will be needed. Even the best information is of little value after the decision has been made. Finally, the manager must determine the human and financial resources that will be available. Perspective, information needs, potential benefits, timing, and resources shape the project and the choice of methods.

Managing a forecasting project demands most of the same management qualities required of other projects: sound goals, objectives, and constraints; careful scheduling and cost accounting; and good communication and people skills. However, there are differences because of the uncertainty involved in forecasting and the characteristics of the people needed for the task. For example, some people deal better with the uncertainties of extending existing knowledge than do others. Further, the forecast may require individuals with substantive knowledge in a variety of disciplines (such as science, engineering, economics, and social sciences). Differences in disciplinary approaches, vocabulary, and perspectives can complicate communications and cooperation. For these reasons and others, forecasting projects may require different organizational and communication structures than other projects.

5.2 INFORMATION REQUIREMENTS

One of the first tasks faced by the manager is to determine the information needed to make the forecast. To do this effectively, the forecast must be "bounded"—that is significant thought must be devoted to the factors that will be considered. "Bounds" are strongly influenced by the information requirements of the decision the forecast is to inform and by the timing of that decision. They also are affected by the nature of the factors that will be forecast. For example, forecasting sales potential for a new technology implies that the dynamics of the marketplace peculiar to it must be considered. The study must be bounded early to allow work to begin. However, initial bounds should be set broadly and remain flexible as long as possible. This provides the opportunity to incorporate factors whose importance is recognized after work has begun. Maintaining breadth and flexibility will require willpower; the pressures exerted by time and resource constraints are relentless.

Vanston (1985) suggests that there are five types of information required by the technology manager. Because of the present climate of social and environmental accountability, the authors have added a sixth:

1. Projections of rates at which new technologies will replace older ones
2. Assistance in managing technical research and development (R&D)
3. Evaluation of the present value of technology being developed
4. Identification and evaluation of new products or processes that may present opportunities and/or threats
5. Analysis of new technologies that may change strategies and/or operations
6. Probable responses of regulatory agencies and society to a new product, process, or operation

Each of these types is discussed in the following paragraphs. Examples cited for illustration are taken from Vanston (1985), unless otherwise noted.

To forecast the rate at which a new technology will substitute for an old, basic characteristics of both must be understood. Further, substitution must have proceeded long enough to establish a trend. (The specific information sought—the rate of substitution—will be discussed extensively in Chapter 10.) This substitution rate is as important to old technology producers as to producers and potential producers of new technology. All need it to allocate resources for production and staffing. New technology producers will be especially concerned about financing and expanding plant facilities and developing strategies to speed substitution. Potential producers also need to time their market entry well. Old technology producers may wish to develop strategies to retain profitability as long as possible or to plan an orderly production halt and the introduction of new products. Substitution forecasting is an established tool of technology management and there are many examples of its use. The Firestone Company, for instance, employed it to forecast the substitution of radial for bias ply tires. United Technology Corporation has used substitution forecasts to plan introduction of new aircraft engines. Recent examples of the substitution of one technology for another include fiber optic cable for copper wire and office facsimile machines for overnight delivery services.

The manager should realize that multiple substitutions can occur. For example, Martino (1983, pp. 240–246) discusses substitution in metal beverage containers. Older three-piece steel cans were being replaced, in this instance, by two-piece aluminum and two-piece steel cans. The manager could hypothesize three scenarios: (1) two-piece aluminum cans are being substituted for older three-piece cans while two-piece steel cans are being substituted for aluminum cans; (2) the older can is being replaced by two-piece cans and either aluminum or steel will dominate depending on cost; or (3) aluminum cans are being substituted for the older cans and two-piece steel cans are merely a steel industry tactic to slow their loss of market share. The manager must be sure that he or she does not implicitly assume the nature of the substitution and then forecast to confirm that assumption. Such assumptions can critically shape the forecast.

Forecasts of technology growth usually are required to make informed decisions about R&D projects (the technology progress function introduced in Chapter 10 can help). Whether decisions are motivated by a need to efficiently employ company resources or by competition, the manager will need to know expected growth or development rates of more than one technology. Targets not only include the R&D technology but also competitive technologies and breakthroughs that provide better ways to achieve R&D goals. Forecasts for other R&D projects and for existing or proposed products farther along the development cycle are important as well to evaluate resource allocations and target desirable project milestones.

Such forecasts are common in industry and government. They are used to target product introduction and to determine the resources necessary to achieve those targets. The FMC Corporation, for instance, uses trend extrapolation (see Chapter 10) to specify equipment development goals. The U.S. Air Force uses it to specify goals for weapon development and to prioritize funding (see the AIMTECH case mentioned in several chapters). In some instances, forecast information may suggest discontinuing work on a project entirely and reallocating resources to more promising technologies.

Vanston (1985) notes that it is common for firms to assign a monetary value to a technology during its development. This provides a measurement of potential worth that can be used to decide resource allocation. He poses three questions that encompass the information needs:

1. How soon can the technology be commercialized and at what cost?
2. How big is the potential market and what rate of takeover can be anticipated?
3. How will profitability be affected by competing technologies and nontechnical factors?

The first two questions usually can be addressed by analogy to the past timing, cost, and market penetration of similar technologies. The final question involves forecasting rates of growth and development of competing technologies. However, it also requires forecasts of likely social, political, environmental, regulatory, and economic parameters over the life of the technology. These are perhaps the most complex and uncertain of all forecasting targets. Examples from industry include AT&T, which carried out a program to evaluate the present value of existing and developing technologies, and GE, which completed a similar corporatewide study.

Evaluation of new products and processes that can be supportive of or competitive to the firm's technologies is vital. The information required here is much the same as that discussed in previous paragraphs. However, uncertainty is magnified because the most important technological developments may occur in areas remote from existing product lines of the firm and its competitors. The Whirlpool Corporation's experience in monitoring synthetic fabric development is a good example. As a result of the monitoring program (see Chapter 8), Whirlpool was able to forecast the introduction of permanent press fabrics and market washers and dryers with appropriate cycles a full year before competitors (Porter and Rossini, 1988). Many others have used a variety of forecasting techniques to evaluate new products and techniques, including General Housewares (Roper and Mason, 1983).

Often an organization wishes to evaluate improvements that a new technology offers to its internal operations and/or strategies, rather than evaluating the products or services it markets. Although vendors readily supply information about their products, the manager still must forecast their likely effects on the organization. He or she also must assess the probable growth in functional capacity of the technology and the likelihood that it will be replaced by a competitor. Many of the same kinds of information already described are required; however, the information must be specialized to the organization and its probable growth over the technology's life. Union Carbide has used alternate scenarios (see Chapter 13) as a tool to improve personnel utilization. In another arena, Rose-Hulman Institute of Technology has used Delphi forecasting and the nominal group process (both presented in Chapter 11) to develop curricular and institutional changes to leverage advances in computer software and hardware (Commission for Computer Integration, 1989).

Today organizations are under increasing pressure to account for social and environmental consequences of their activities. Sound responses to these concerns are every technology manager's ethical and moral responsibility. On a different level, the viability of any technology may strongly depend on the response of society and its regulatory agencies. But what are the likely positions of society and regulators during the product life? What new concerns about quality of life and possible environmental and health effects are likely to emerge? What will be the impact of new technologies on the ecosystem? The answers to such questions pose some of the most vexing and important information any manager is likely to need. Predictions of such social and political factors are among the most uncertain forecasting tasks, and the cost of being wrong can be high (ask anyone from the nuclear power industry). Yet there are examples of efforts to provide such information. For instance, Gulf Oil has used the nominal group process, modeling (see Chapter 12), and other approaches to identify and evaluate sociopolitical factors that may affect shale oil development. Monsanto has used various forecasting techniques to predict possible regulatory and public responses to biodegradable beverage containers. Also, the state of Virginia has used the nominal group process and scenario construction to develop a decommissioning policy for commercial nuclear power plants (Lough and White, 1988).

5.3 PROJECT SCHEDULING

Scheduling is important to any project to ensure that it is completed on time and within budget. However, scheduling is even more critically important in forecasting for several reasons. First, forecast results must be available before the decision deadline to be of value. Second, dynamics between team members (see Section 5.4) may be different and less time-efficient than in most conventional projects. Third, information demands are difficult to predict, and iteration of the forecast usually will be required. Fourth, input frequently will be required from individuals and sources outside the company, which makes it difficult to exercise precise control. Finally, the project manager usually will wish to forecast by several methods to increase confidence in the result. All of these combine to make tight scheduling control of forecasting projects vital.

Three scheduling tools are discussed in this section: the Program Evaluation and Review Technique (PERT), the Gantt chart, and the Project Accountability Chart (PAC). Most readers with experience in project work will be familiar with the first two. The third combines the concerns of PERT and Gantt methods with information that depicts responsibility for each task involved in the project.

5.3.1   PERT

PERT depicts the flow of the project and indicates the interdependence of tasks. A PERT flowchart for a simple forecasting project is shown in Figure 5.1. To construct such a chart, the manager must first list the tasks (or activities) upon which completion of the forecast depends. The manager must be careful not to be overly detailed or unproductively general. Second, the manager must construct the chart to show which activities must be completed prior to others that depend on their output. These two steps normally will require iteration.

On the PERT flowchart, dependence is shown by arrows and tasks are represented by circles. The number in the upper half of the task circles is the task number. Tasks that need to be only partially completed before information from them is used to begin other activities should be represented by more than one circle. Iterations are not shown as loops, instead, sequences for the first iteration are repeated to show the second.

Once the flowchart is finished, the manager must estimate the time that will be required to complete each task—that is, the task duration. These estimates are usually made in units of both person hours and working days. Since it usually is not possible to be precise about these estimates a weighting system such as the following is used:

$$\text{Estimate} = \frac{\text{Optimistic time} + 4 \times \text{Probable time} + \text{Pessimistic time}}{6}$$

The estimates are recorded in the lower half of the task circles as shown in Figure 5.1.

Next the manager determines the longest path from project start to project finish. This is called the Critical Path (CP). The CP determines the shortest time in which the project can be completed as planned. If the flowchart is quite complex, computer assistance may be required to determine the CP (the TOOLKIT contains software to analyze PERT charts). If the CP exceeds the time available for the project, either some task must be shortened or eliminated or additional resources must be assigned to complete it more quickly than estimated.

Even if the PERT process is taken no farther, the manager will have gained a more thorough understanding of requirements. He or she will better understand

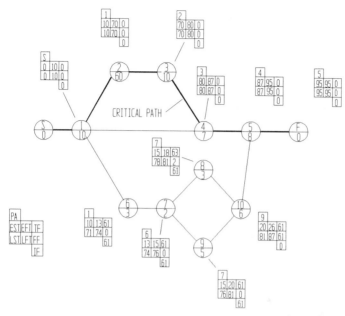

TASK #	DESCRIPTION	DURATION days	(Person Hr.)
S	Project start	0	(0)
1	Conduct & Analyze Poll of Internal Experts	10	(160)
2	Conduct Poll of External Experts	60	(32)
3	Analyze & Summarize Results of Task #2	10	(80)
4	Reconcile Results of Tasks #1 & #3	7	(48)
5	Integrate Results of Tasks #4 & # 10	8	( )
6	Rough Market Penetration Analysis	3	( )
7	Adjust Results of Task #6 For Regulatory Factors	2	( )
8	Obtain Additional Market Penetration Data	5	( )
9	Review & Revise Results of Task #6	3	( )
10	Iterate Tasks #6 & #7	6	( )
F	Project Finish	0	(0)

Critical Path Length = 95 working days

Figure 5.1   Example project PERT chart.

resource requirements and will know which tasks can be worked on simultaneously. More useful information, however, can be developed about the delay that can be tolerated in completion of certain tasks without delay to project completion.

To determine the length of delay that can be tolerated, the manager must first determine the Earliest Start Time (EST) and the Earliest Finish Time (EFT) for each project task. The EST of a task is defined by activities that precede it and that must be completed before it can begin. The EFT is merely the EST plus the estimated duration of the task. These times are computed by moving forward through the flowchart. For example, in Figure 5.1, Task 1 can begin immediately (time = 0) and takes 10 days to complete—its EST is 0 days and its EFT is 10 days. Since Task 2 cannot begin before Task 1 is complete, it has an EST of 10 days and an EFT of $10 + 60$ days. Once these computations are complete, the manager works backward through the flowchart to determine the latest that each task can be completed (Latest Finish Time or LFT) without lengthening the project and the latest it can begin (Latest Start Time or LST). The LST is the LFT minus the estimated duration of the task. Note that for tasks on the CP, the EST is equal to the LST and the EFT is equal to the LFT.

The completion of tasks not on the CP often may be delayed without lengthening the project. This delay is called *float*. The Total Float (TF) that can be allowed is simply the difference between the EST and the LST recorded for the task. It is composed of two different types of float. Free Float (FF) is the time that completion of a task can be delayed without delaying the EST of a subsequent activity that depends on it. Interfering Float (IF) is the total delay in completing an activity that can be tolerated without delaying project completion. Mathematically, $IF = TF - FF$. Clearly, FF is the more desirable, since IF in one task uses some of the TF available in subsequent tasks. The various start, finish, and float times usually are displayed as shown in Figure 5.1.

The completed PERT chart gives a visual representation of the forecasting project and indicates clearly which tasks are critical to timely completion. It also indicates which tasks can be delayed and how the project might be shortened or how lost time may be made up. For example, Task 4 in Figure 5.1 might be completed in less than seven days if more people are assigned to it. While this information is valuable, the PERT chart gives no clear visual clue about elapsed time. That is provided by the Gantt chart.

5.3.2   Gantt Chart

The Gantt chart is a bar chart representation of information generated for the PERT chart. In Figure 5.2, which presents a Gantt chart of the example forecasting schedule shown in Figure 5.1, tasks are represented on the vertical axis and time is represented on the horizontal axis. Floats can be represented using cross-hatching, and partially completed tasks can be indicated by filling in an appropriate portion of the task bar. The Gantt chart gives a more direct visual representation of progress and timing than does the PERT flowchart. Unfortunately, this is obtained at the cost of a clear representation of the sequencing of tasks. Neither representation indicates with whom the responsibility for completing the task lies.

5.3.3   Project Accountability Chart

The Project Accountability Chart (PAC) was suggested by Martin and Trumbly (1987). It combines a visual representation of task responsibility with aspects of scheduling provided by the PERT and Gantt charts. The horizontal axis of the PAC, like that of the Gantt chart, is time. The vertical axis, however, represents respon-sibility, either organizational or individual, for executing various components of the project. Graphically it is best to cluster those with the largest number of re-sponsibilities near the center of the vertical axis. Tasks with shared responsibility

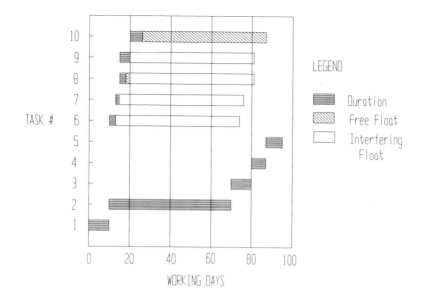

Figure 5.2 Example project Gantt chart.

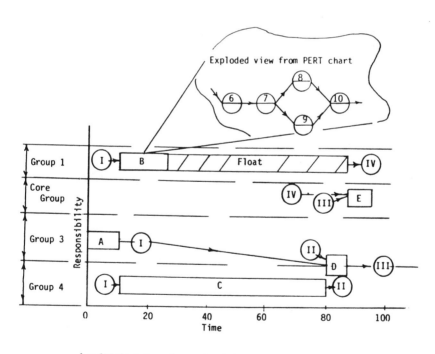

A = Obtain internal expert opinion
B = Market penetration analysis
C = Obtain external expert opinion
D = Reconcile internal & external expert opinions
E = Integrate opinion & market penetration results

Figure 5.3 Example project PAC.

can appear twice or, if units sharing responsibility are adjacent on the chart, as overlapping organizational boundaries. A PAC for the example project is shown in Figure 5.3.

Project sequencing is displayed on the PAC using the PERT flowchart with two significant changes. First, several tasks may be combined into a single identifiable activity for which one entity has responsibility. For instance, in Figure 5.1, Tasks 2 and 3 might be combined under the title "obtain external expert opinion" if both are the responsibility of a single organization or individual, and represented by one symbol or node on the PAC. Second, the symbol for the combined activities would extend from the EST of one task to the LFT of another (in the example here, the EST of Task 2 to the LFT of Task 3).

The PAC displays responsibility, timing, and the interdependence of groups of project tasks; however, this is obtained at the expense of detail about the tasks themselves. This deficiency can be remedied by exploding individual nodes to show PERT and Gantt diagrams for the tasks they include. Regardless of the approach chosen, however, the intent is to schedule important aspects of the project and to give a quick visual reference for its progress.

5.4   MANAGEMENT AND COMMUNICATIONS

Forecasting projects usually involve factors that fall into categories traditionally thought of as within the scope of different disciplines. For instance, market and economic factors generally are studied by economists and technological factors are studied by engineers and scientists. Social and political considerations introduce other disciplinary specialties. Such projects have been characterized as "multiskill" (Porter and Rossini, 1986). Management arrangements and communications between team members are of critical concern in multiskill projects, especially if team members have no history of collaboration. Successful schemes of management and communications for multiskill projects are somewhat different than those for other projects.

5.4.1   Managing the Forecasting Project

Activities in which a number of individuals with different disciplinary backgrounds cooperate to produce an integrated information product have been characterized in several ways. Differences in these characterizations primarily rest on the degree to which individual work has been integrated to produce a "seamless" result. A more productive approach seems to be to emphasize the intellectual skills involved and their organization rather than the disciplines—hence the concept of the multiskill project suggested by Porter and Rossini (1984). They maintain that such work can best be characterized in terms of five factors:

1. Substantive knowledge
2. Techniques

3. Range of intellectual skills
4. Administrative structure
5. Number of personnel and the permanence of their relationships

The first two factors address the intellectual skills required for the project, and the third factor addresses the range of those skills. The fourth factor recognizes the importance of administrative arrangements between organizations involved in the project or, in the case of a single organization, between subunits. The final factor could mean a single individual, an ad hoc group, project team, or permanent organization. While this characterization is based on studies of multiskill research in the academic arena, it provides a good insight for industrial forecasting project managers as well.

Porter and Rossini (1984) note that substantive knowledge and techniques can be exercised at either the expert level (that is, generation of new knowledge) or the journeyman level (that is, application of common knowledge in a straightforward way). The range of knowledge required can be a single "grand category" (such as physical sciences) or can cross grand categories. Moreover, this range could be provided by several people or, more rarely, by a single individual. The typical forecasting project will require both substantive knowledge and techniques, which most often will be applied at the expert level. This does not mean that some techniques cannot be applied in a straightforward way, but their selection, direction, and interpretation will require substantive knowledge applied at the expert level. Nearly without exception, forecasting projects will cross grand categories of knowledge. The required range of intellectual skills occasionally may be provided by a single individual, especially in small organizations. However, the complex interplay between technological, economic, social, and political factors on one hand and organizational needs and characteristics on the other usually will require several people, often from different units of the organization. Even in single-person forecasts, internal or external experts usually play the roles of surrogate team members.

These characteristics suggest that forecast team members often will be individuals with very different disciplinary perspectives, methods, and world views. Although it is true that disciplinary differences are exaggerated by the structure of academic organizations, even in industry there are marked differences in perspective, method, and viewpoint between individuals from different units (for instance, individuals from production and R&D). Thus the forecasting project must be managed to allow these individuals to cooperate and communicate effectively.

When a relatively large number of individuals are involved, the manager usually will find it expeditious to establish a core (or coordinating) group to direct activities. Ideally members of this group are individuals expert in the various areas required for the forecast. If multiple units from an organization are involved, however, political realities usually dictate that the group be composed of individuals designated by the various units. Regardless of how the core group is constituted, the manager must work hard to foster communication and cooperation among its members. He or she also must ensure that they understand the special requirements of the forecasting project.

Porter and Rossini (1984) suggest that there are two distinctly different kinds of administrative organizations: open and closed. The structure of the former is devised to tolerate, even foster, a range of substantive knowledge and techniques. The structure of the latter discourages such diversity. Whether or not the organization is open or closed is usually determined by the nature of the problems it must solve. For example, open structures are characteristic of organizations (such as forecasting groups) that deal with problems acknowledged to straddle grand categories of knowledge. Closed structures often are found in groups that consider a more narrowly focused problem to great depth (for instance, heat transfer in the turbine blades of a jet engine).

It is clear that the structure of a forecasting project group must be open. It is perhaps less obvious that even within an open structure a pecking order based on factors such as the quantification or "practicality" of individual skills may exist. Thus a group structure that appears open may actually function more nearly in a closed mode, denying itself important contributions from members whose skills are implicitly discounted. When this happens, forecasts can be unbalanced or seriously flawed. In organizations dealing with technology, the internal pecking order often discourages input from so-called soft disciplines. The manager must ensure that this situation is avoided, for unpleasant surprises await those who use forecasts in which social or political factors are dominant but underplayed.

The manager must account for internal and external openness in developing a method for organizing the forecasting project. Rossini and Porter (1981) identified four methods used in structuring technology assessment projects, which share many of the same concerns as forecasting: (1) common group learning, (2) modeling, (3) negotiation among experts, and (4) integration by leader. These methods are represented schematically in Figure 5.4.

In *common group learning* the information product is generated by the group, which learns and acts as a whole. Thus the forecast becomes the common intellectual product of the group (Kash, 1977). In this approach, the forecast is first bounded and then tasks are divided among group members, usually according to substantive skills and personal interest. Preliminary analyses are generated by these individuals and then critiqued and modified by the full group. Finally, each task is redone by a new individual, often not expert in the area. This iterative process continues until the group and the manager are satisfied with the result. The final forecast is well integrated, but the process is time-consuming and tends to achieve integration at the expense of technical sophistication and depth of individual analyses (Rossini and Porter, 1981). In large projects only the core group generally employs the common group learning approach.

Models are simplified representations of reality. Rossini and Porter (1981) note that they "can provide a common ground where disciplinary contributions can meet" (p. 17)—that is, construction of a new model or operation of an existing one can provide a platform upon which to integrate the dynamics that will shape the technology. Models sometimes can be highly quantified representations that require significant computer resources. Although such models provide focus and a platform, they also can narrow perspective as they tend to undervalue factors that are difficult

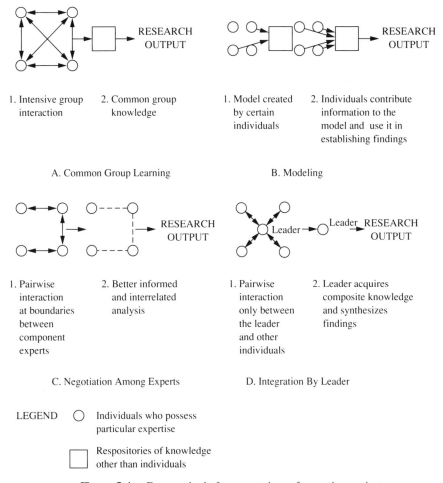

1. Intensive group   2. Common group
   interaction   knowledge

A. Common Group Learning

1. Model created   2. Individuals contribute
   by certain   information to the
   individuals   model and use it in
                               establishing findings

B. Modeling

1. Pairwise   2. Better informed
   interaction   and interrelated
   at boundaries   analysis
   between
   component
   experts

C. Negotiation Among Experts

1. Pairwise   2. Leader acquires
   interaction   composite knowledge
   only between   and synthesizes
   the leader   findings
   and other
   individuals

D. Integration By Leader

LEGEND   ◯   Individuals who possess
               particular expertise

          ▢   Respositories of knowledge
               other than individuals

Figure 5.4   Four methods for structuring a forecasting project.

to quantify. Few forecasting topics encountered by the technology manager will be amenable to highly quantified models that can be employed off the shelf, nor will time and resources generally permit their construction. When such models are used, however, the manager must guard against the spurious credibility sometimes granted their output because of the sophistication of their computational techniques. The most useful models often take the form of graphical representations—trees or flowcharts that capture interactions between factors germane to the forecast. The process of constructing such models can be very useful in understanding and directing the forecast, and they require few resources to create. However, iteration is absolutely necessary if a satisfactory model is to be created.

Many forecasting projects will demand more or less co-equal participation by several units of an organization (such as R&D, production, and marketing). In such instances, politics usually will dictate a project structure that involves *negotiation among experts*. In this approach, tasks are divided among members of the core group on the basis of expertise or unit responsibility. Predictably, each of the initial results reflects the substantive knowledge peculiar to the unit each analyst represents. These results are integrated by negotiation, and tasks are redone to reflect the results of the negotiation. This approach tends to preserve depth and expertise but usually at the expense of full integration. However, it can build broad-based support for the decision ultimately based on the forecast.

The *integration by leader* management structure often is employed by strong managers who feel it necessary to maintain tight project control. In this structure, all tasks are assigned by the manager and he or she becomes the sole integrator of various components of the forecast. Typically, there is little interaction between team members working on disparate aspects of the study. This method requires that the manager assimilate and understand each contribution before integrating them; thus it makes major demands on a single individual. Like common group learning, integration is achieved at the expense of depth of analysis, because the manager is unlikely to grasp the details of the variety of areas of substantive knowledge required for the forecast (Rossini and Porter, 1981). However, the tight control exercised by the manager increases the probability that the project will be completed on time and within budget. The integration by leader arrangement works best for small, tightly bounded forecasting studies.

These four methods are archetypical. Few real-world projects employ management structures that are solely one or the other. For example, the core group may function in a common learning or negotiation among experts mode while supporting groups use the integration by leader or modeling structures. However, these methods provide useful insight into the strengths and weaknesses of typical management structures. This insight is important because of the differences between forecasting and traditional industrial project tasks and the complexities of the multi-skill projects. In many instances, the management structure will be inherited by the technology manager and may benefit from modification. In others, ad hoc project teams formed to make a single forecast may have no established structure and the manager will have to create one. Either way, an understanding of the function and importance of the management scheme will be critical to a successful forecast. The manager should organize the project structure to leverage personnel strengths and accommodate established ways of doing business within his or her organization.

5.4.2 Communications Patterns

The communications patterns established within a forecasting project team are strongly influenced by the management structure chosen. If the pattern is incompatible with the management scheme, it can produce a de facto structure that functions quite differently than the manager intends. (An example of de facto structure is the "old boy network" that is independent of organizational charts and impervious to reorganization.)

Rossini and Porter (1981) list three types of communications patterns that may exist: (1) all channel, (2) hub-and-spokes, and (3) any channel. In the first pattern, everyone communicates with everyone else. This arrangement is most compatible with the common group learning management structure. In the hub-and-spokes pattern, individuals communicate with the project manager but not with each other. This pattern is encountered in management structures comprised of a strong manager, centralized responsibility, and team personnel who are not located in close proximity or in structures that allow a single manager to control the input from a group of outside experts. In any channel communications patterns, all channels of communication are open but are used only as needed. This pattern is often encountered with negotiation among experts and modeling management structures.

The any channel communication is likely the most appropriate for the typical forecasting project. It provides for the sharing of important knowledge, information, and perspectives necessary for multiskill tasks without the lost effort inherent in the all channel pattern or the isolation produced by the hub-and-spokes arrangement. However, effective use of the pattern does assume that individuals know when they need to communicate with each other. Real-world communication systems are nearly always combinations of these three patterns. For example, the core group may employ all channel communications while subgroup members employ one or more of the other patterns but do not communicate directly with core group members other than their leader.

It is important for the technology manager to note that effective communications patterns do not just happen; nor do they always evolve in a fashion compatible with the management structure specified by the organizational chart. They must be fostered by the manager by contriving meetings and by assigning tasks and responsibilities in a way that forces communication to occur in the pattern he or she desires until it becomes a natural part of daily activity.

5.5 CONCLUSION

Managing forecasting projects requires many of the management qualities necessary for managing other projects; however, because forecasting differs in essential ways from other projects, there are management differences as well. The manager must carefully bound the forecasting task and determine the information required by the decision the forecast is intended to support. Careful scheduling is required to ensure completion of the project on time and within budget. Dynamics among team members tend to be more complex and iteration and forecasting by several methods are generally required to produce a satisfactory product. Finally, because the forecast is a multiskill product, special attention must be given to management structures and communications patterns that acknowledge the need for substantive knowledge and contribution from a variety of fields.

EXERCISES

5.1   Consider the tasks shown in Figure 5.1. Analyze the example assuming the task times are as follows:

Task	Days	Task	Days
S	0	6	5
1	10	7	4
2	20	8	6
3	5	9	8
4	7	10	9
5	8	F	0

5.2   Consider the project shown in Figure 5.1. Analyze the project assuming Task 2 is eliminated.

5.3   Prepare Gantt charts for the projects in Exercises 5.1 and 5.2.

5.4   Prepare a PERT chart and a Gantt chart for some portion of your classwork or for a project in which you are involved.

5.5   Use the TOOLKIT to solve for the PERT charts in Exercises 5.1, 5.2, and 5.4.

5.6   Research and write a report on the history of PERT and the technology it helped to spawn.

5.7   Choose a small project, perhaps associated with your class. Carry out the project using two of the four structures shown in Figure 5.4. Contrast the project operation and the communications patterns that developed. Speculate about the results that might have been obtained in a larger, more complex project.

PART II
METHODS

___6
GENERAL METHODOLOGICAL ISSUES

OVERVIEW

This chapter gives an overview of methods and techniques for forecasting sociotechnical change. The intellectual bases of inquiry are stressed, and alternative approaches to developing knowledge are presented and compared. The potential for using data and theories of sociotechnical change is also considered. Next the role of technique in forecasting is explored and the range of alternative techniques is discussed. With this groundwork, the five most commonly used general techniques, their strengths, weaknesses, and uses are discussed.

6.1   INQUIRING SYSTEMS

Forecasting is a means of generating knowledge—in other words, a form of inquiry. Most professionals, however, consider forecasting much less reliable than scientific inquiry. What are the intellectual bases of inquiry? What are the different systems? How can they be compared and contrasted? This section presents a general approach to inquiring systems first articulated by Churchman (1971). A fundamental tenet of this approach is that different systems of inquiry produce different types of knowledge. These, in turn, refer to experience in different ways and are "guaranteed" by different mechanisms. Churchman introduced five systems that span most of scientific inquiry and some other areas of inquiry as well. Mitroff and Turoff (1973) summarized the systems and their problem-solving capabilities (see Table 6.1). This discussion emphasizes their role in forecasting.

TABLE 6.1   Five Underlying Philosophical Approaches to Knowledge Generation

Approach	Description	Best Suited for Problems
A priori	Formal models from which one deduces insights about the world with little need for raw data	Well defined conceptually
Empirical	Beginning with data gathering, one inductively builds empirical models to explain what is happening	Well defined with available data
Synthetic	Combines the a priori and the empirical so that theories are based on data and data gathering is structured by preexisting theory or model	More complex and ill-structured
Dialectic	Opposing interpretations of a set of data are confronted in an active debate, seeking a creative resolution	Ill-structured and when conflict is present
Global	A holistic broadening of inquiry by questioning approaches and assumptions	Nonstructured and that require reflective reasoning

Source: Based on Mitroff and Turoff, 1973. Reprinted by permission. ©1973 IEEE.

The first type is the *a priori inquiring system*. In this system the inquirer builds logical structures or models that relate to the "real" world. These are intended to represent the major features of some part of the world that is of interest. The model may be expressed in many ways: sophisticated computerized simulations, simple boxes and arrows diagrams, theories, equations, or scale models. It is not necessary that the forecaster model the entire "real" system or even every feature of some part of it: models are simplifications. It is important, however, that the model incorporate the features and dynamics that are central to the inquiry. Thus the modeler must be able to identify those aspects of the world that are important to the inquiry. Knowledge is generated by the process of model building and by observing the behavior of the model under various conditions.

In this approach the assumptions on which the model rests are critical. However, they may come from any source as long as they are credible to the modeler. For example, opinions generated by experts could be used to develop assumptions for a system dynamics model (see Chapter 12). An obvious weakness of the a priori approach is that it does not necessarily require data. Thus models may be weakly founded and open to criticism on empirical grounds. Still, in areas for which data are lacking, modeling can provide a means of approximation that often is quite useful. Examples of a priori inquiring systems are operations research, models used in wind tunnel testing, and some models built by theoretical physicists. Specific models useful in forecasting will be discussed later.

The second type of inquiring system is the *empirical system*. In its pure form, this system consists entirely of data that relate to some aspect of the world. Its strength stems from the fact that these data are concrete and closely linked to very specific features of the world. Moreover, the empirical system is very close to the system being studied. The source of its major weakness is that data alone do not provide principles or rules of structure and selection. When data selection is arbitrary, its usefulness for inquiry may not be great. Issues about the categories of data that should be measured, relationships among data, and the conditions of measurement cannot be settled either by the system or by the data alone. Yet the incredible richness of data makes them enormously powerful in inquiry. Examples of empirical inquiring systems are field biology and cultural anthropology. In forecasting, empirical inquiring systems are used to muster past data in order to project future trends (see Chapters 9 and 10) and to generate expert opinion about the future.

Individually, the a priori and empirical inquiring systems provide useful bases for inquiry. Yet each lacks something that the other could provide. The *synthetic inquiring system* combines the a priori and empirical systems to overcome some of their limitations. This third inquiring system provides an interplay between the first two in which frameworks, concepts, relationships, and variables, are determined a priori yet measured empirically. Frameworks guide measurement of the data, and the data are then analyzed to modify the framework within which empirical work takes place. This iteration between theory and concept, on the one hand, and measurement and observation, on the other hand, allows inquiry to move forward systematically. Fields in the natural sciences for which empirical and theoretical approaches are highly developed, such as physics and chemistry, are good examples of synthetic inquiring systems. In forecasting, synthetic inquiry is useful to develop assumptions for model building and to integrate forecasts that are made by different techniques or that require complex parameter selection. Synthetic inquiry also is useful in monitoring approaches because it helps to structure and filter the data so as to avoid information overload.

The *dialectic inquiring system* is a fourth type. In this system an extreme view of the world is posed and then countered by a diametrically opposed extreme. The goal is to synthesize these views in a way that resolves their intrinsic conflict and moves understanding to a new plane. This system stresses the role of conflict in developing knowledge. Its use is exemplified by our legal system. In forecasting, the dialectic inquiring system plays a role in some forms of expert opinion forecasting (see Chapter 11), such as Delphi, where the extreme forecasts of group members and their supporting rationale are fed back to the entire group. Panels also provide opportunities for diametrically opposed positions to be presented and defended. The resulting forecast may represent a synthesis of conflicting views.

The fifth inquiring system is the *global system*. In this approach information from a wide variety of diverse sources is swept into the system. The global inquiring system establishes a wide perspective on the subject being studied. It offers the potential of a broad synthesis without entering the thicket of detail. The coherence of the pattern developed in the inquiry and the overall robustness of the knowledge base help to build user confidence and guarantee forecast viability. However, the global system is weak in inquiry that requires detail since it can provide little. In forecasting, the global inquiring system is used by monitoring systems (see Chapter 8) and often is employed in scenario construction (see Chapter 13).

The five inquiring systems presented here are neither exclusive nor exhaustive. Moreover, they are archetypical; thus any real inquiry will combine or approximate these systems (and possibly other systems as well). The discussion in this section is designed to develop a consciousness of the different ways of approaching knowledge generation, the strengths and weaknesses of each, and how they can be used to complement each other. In typical academic and professional training, only one inquiring system is studied. This limitation leads to the mistaken belief that there is a single best approach to developing knowledge. This belief must be countered so that the forecaster and technology manager may use a variety of inquiring systems, adapted and applied as the situation requires. The resulting range and flexibility provide the philosophical perspective and generic tools to design the most appropriate forecasting system.

6.2 SCIENTIFIC INQUIRY AND FORECASTING

Scientific inquiry, at its strongest, rests on the twin pillars of theory and empirical data—that is, on the synthetic inquiring system. As we have seen, theory and data are used interactively in this system. The result has been progressively developing scientific knowledge about the world. When observations are made that do not fit a theory, it may be necessary to construct a radical new theoretical framework. These "paradigm changes" (Kuhn, 1970) explain the radical discontinuities that scientific thought has undergone over the years. Of course, this is a very idealized view of scientific development; the true system of inquiry can vary dramatically from field to field and from time to time.

The synthetic inquiring process is effective because the scientific community is confident that it can both develop conceptual frameworks and gather data that relate to them. If confidence in the ability to accomplish either of these tasks is lacking, or if conceptual frameworks and empirical data do not relate to each other, the potential for traditional scientific inquiry is weakened. As discussed in Chapter 2, conceptual frameworks for the study of sociotechnical change are inadequate, and available data, usually social indicators, often do not relate to those frameworks. Moreover, it is not clear that the synthetic inquiring system approach is necessarily appropriate for forecasting sociotechnical change. Perhaps a different approach is required that explicitly deals with the unique character of the sociotechnical future.

The approach used in forecasting sociotechnical change substitutes technique for theory and data. As practiced, this approach incorporates human judgment to a very high degree. Judgment is important in scientific work as well. However, its role is largely ignored in the reconstructions that comprise the technical literature and educational curricula that equip students with frameworks for inquiry. In other words, formal course work and the literature do not explain how science actually gets done. Only in the apprentice relationship of graduate school is the student permitted to see the pivotal role that human judgment plays. Therefore, the forecaster or technology manager with a scientific background may have to broaden his or her perceptions of inquiring systems to function effectively.

The techniques that have dominated technology forecasting do not require a very robust data environment or extremely sophisticated models. Over the years, formal and quantitative techniques have been steadily supplanted by more informal and qualitative techniques. Indeed the most recent trend is to keep techniques in the background and rely on interaction with the client to communicate forecasts developed using them. The professional forecaster no longer simply presents a forecast. Thus the burden of the work and the key to making forecasts credible has increasingly been assumed by the communications process rather than by the more abstract qualities of technical analysis (J. F. Coates, personal communication, 1989).

6.3 STRATEGIES AND TECHNIQUES FOR FORECASTING

Forecasting strategy can be divided into two parts—the strategy used to conduct technical aspects of the forecast and the strategy employed to communicate the forecast to the user and relate it to his or her assumptions and needs. These strategies must complement each other to produce an outcome that is both valid (insofar as a forecast can be said to be valid) and useful.

6.3.1 Strategy for Technique Selection and Application

As a general rule, the more complementary ways in which a forecast can be made, the greater the confidence that can be placed in its results. The five most frequently employed families of forecasting techniques are: monitoring, trend analysis, expert opinion, modeling, and scenario construction. Before a forecasting strategy can be implemented, these techniques must be evaluated. Exhibits 6.1 through 6.5 briefly describe each type of technique, its strengths, weaknesses, and uses. In this section, conditions appropriate to the use of each type are discussed and ways to integrate them are considered. Specific examples of each type are discussed in subsequent chapters (Chapters 8, 9–10, 11, 12 and 13, respectively).

Strictly speaking, *monitoring* is not a forecasting technique. However, it is by far the most basic and widely used of the five general types because it provides the background material upon which forecasts are based. Thus it is fundamental to almost all forecasts. In its purest form, monitoring is an empirical inquiring system. It can be used to attack all sources of information with vigor and to produce a rich and varied body of information. The primary sources for information are technical and trade literature. In addition, interviews with experts may be conducted and other nonliterature sources can be tapped. Yet the very richness and variety of the data produced by monitoring can cause information overload for the forecaster. Thus filtering and structuring the information flow are prerequisite conditions for success. Although the techniques of monitoring can be mastered by an intelligent child, they can produce widely accepted and extensively used results. For example,

■ **Exhibit 6.1 Monitoring**

Description: Monitoring is the process of scanning the environment for information about the subject of a forecast. It is not really a forecasting technique, but rather a method for gathering and organizing information. The sources of information are identified and then information is gathered, filtered, and structured for use in forecasting.

Assumptions: The environment contains information useful for a forecast, and that information can somehow be obtained.

Strengths: Monitoring can provide large amounts of useful information from a wide range of sources.

Weaknesses: Information overload can result without selectivity, filtering, and structure.

Uses: To maintain current awareness of an area and the information with which to forecast as needed. To provide information useful for structuring a forecast and for the forecast itself.   ■

the work of John Naisbitt and his firm is essentially monitoring. It has produced a substantial consulting business and at least two best sellers, *Megatrends* and *Megatrends 2000*.

Expert opinion techniques are based on the assumption that an expert can forecast developments in his or her field better than an outsider. However, individual experts often have produced amazingly poor forecasts. For instance, Lord Rutherford, the leading nuclear physicist of the mid-1930s, forecast no serious future for nuclear energy in his lifetime. Yet, within 10 years he had been proven dramatically wrong by the nuclear bomb and subsequently by the commercialization of nuclear power production. Instances such as this have led forecasters to believe that forecasts made by groups of experts are safer than those produced by individual experts. The main condition for using expert opinion is to be able to identify and enlist a group of experts in the appropriate field. If this cannot be done, it is best not to use expert opinion techniques. The more independent indicators there are that an individual is an expert, the safer the selection of that expert. (Methods to identify and select experts are discussed in Chapter 11.) The forecaster must be aware, however, that consultation with experts does not guarantee a good forecast. Experts can be as guilty of perpetuating conventional wisdom as laypeople.

Trend analysis depends on reliable time series data that relate to well-defined parameters. When these do not exist, as is often the case, trend analysis is ruled out. However, when adequate data do exist, there are statistical trend analysis techniques that will allow reasonable projections to be made. These techniques range from simple bivariate regression to more complicated and sophisticated techniques such as Box-Jenkins. In technology forecasting, data often cover limited time periods and/or are expressed in terms of somewhat arbitrarily defined parameters. Thus sophisticated techniques may prove to be overkill. "Eyeball fitting" or straightforward regression are generally the most useful techniques.

■ **Exhibit 6.2   Expert Opinion**

Description: The opinions of experts in a particular area are obtained and analyzed.

Assumptions: Some individuals know significantly more about parts of the world than others, thus their forecasts will be substantially better. If multiple experts are used, group knowledge will be superior to that of an individual expert.

Strengths: Expert forecasts can tap high-quality models internalized by experts who cannot or will not make them explicit.

Weaknesses: It is difficult to identify experts. Their forecasts are often wrong. Questions posed to them are often ambiguous or unclear, and design of the process often is weak. If interaction among experts is allowed, the forecast may be affected by extraneous social and psychological factors.

Uses: To forecast when identifiable experts in an area exist and where data are lacking and modeling is difficult or impossible.   ■

An example of a trend analysis and some broader observations are useful. Roessner et al. (1985) described studies that involved forecasts of information technology. It soon became obvious that well-defined parameters and reasonably good time series data existed for hardware but not for software. Therefore, trend analysis was used to project computer hardware development and to show that it was on track. More qualitative forecasts were necessary for software projections, and these revealed a more uncertain picture. The investigators concluded that software development would be the critical factor in the development and utilization of computer systems in the near to intermediate term. This has been borne out during the intervening period.

■ **Exhibit 6.3   Trend Analysis**

Description: Trend analysis uses mathematical and statistical techniques to extend time series data into the future. Techniques for trend analysis vary in sophistication from simple curve fitting to Box-Jenkins techniques.

Assumptions: Past conditions and trends will continue in the future more or less unchanged.

Strengths: It offers substantial, data-based forecasts of quantifiable parameters and is especially accurate over short time frames.

Weaknesses: It often requires a significant amount of good data to be effective, works only for quantifiable parameters, and is vulnerable to cataclysms and discontinuities. Forecasts can be very misleading for long time frames. Trend analysis techniques do not explicitly address causal mechanisms.

Uses: To project quantifiable parameters and to analyze adoption and substitution of technologies.   ■

Models typically encountered in forecasting are either computer based (such as simulations) or judgment based. In either case, the quality of the assumptions that underlie the model are critical to its success as a forecasting tool (Ascher, 1978). Therefore, important factors expressing the assumptions must be incorporated in the model. Quantitative parameters typically are used in computer-based modeling. Thus qualitative subtleties that may have substantial effects elude the modeler. Judgment-based models rely on the forecaster's ability to build good assumptions and to make sound judgments about how they affect the forecast. If there is no available theoretical framework within which to develop a model or if one exists but it is difficult to make sound assumptions, it is best not to use the technique.

■ **Exhibit 6.4   Modeling**

Description: A model is a simplified representation of the structure and dynamics of some part of the "real" world. The dynamics of a model can be used to forecast the behavior of the system being modeled. Models range from flow diagrams, simple equations, and scale models to sophisticated computer simulations.

Assumptions: The basic structure and processes of parts of the world can be captured by simplified representations.

Strengths: Models can exhibit future behavior of complex systems simply by isolating important system aspects from unessential detail. Some models offer frameworks for incorporating human judgment. The model building process can provide excellent insight into complex system behavior for the modeler.

Weaknesses: Sophisticated techniques may obscure faulty assumptions and provide a spurious credibility for poor forecasts. Models usually favor quantifiable over nonquantifiable parameters, thereby neglecting potentially important factors. Models that are not heavily data based may be misleading.

Uses: To reduce complex systems to manageable representations.   ■

Scenario construction is a technique that can be used whether good time series data, experts, and useful models exist or not. Scenarios are sets of snapshots of some aspects of the future and/or sets of credible paths leading from the present to the future. They can integrate forecasts produced by other techniques and engage quantitative data as well as qualitative information and values. Scenarios may employ literary artifice and imaginative descriptions or even multimedia techniques to deliver forecasts effectively to diverse audiences. Further, scenarios can be used when no other technique is available. Moreover, for forecast users who are not technically trained, scenarios are a technique of choice.

Now that the most commonly used types of forecasting techniques have been described, a rationale for selecting among them can be outlined. In making their selections, the forecaster and technology manager should use as many different approaches to forecasting as possible, keeping in mind their resource limitations. Monitoring can provide the information base for the forecast. It is usually the

■ **Exhibit 6.5   Scenarios**

Description: Scenarios are sets of snapshots of some aspect of the future and/or future histories leading from the present to the future. The scenario set encompasses the plausible range of possibilities for some aspect of the future.

Assumptions: The full richness of future possibilities can be reasonably incorporated in a set of imaginative descriptions. Usable forecasts can be constructed from a very narrow data base.

Strengths: They can present rich, complex portraits of possible futures and incorporate a wide range of quantitative and qualitative information produced by other forecasting techniques. They are an effective way of communicating *forecasts* to a wide variety of users.

Weaknesses: They may be more fantasy than forecast, unless a firm basis in reality is maintained by the forecaster.

Uses: To integrate quantitative and qualitative information when both are critical, to integrate forecasts from various sources and techniques into a coherent picture, and to provide a forecast when data are too weak to use other techniques. They are most useful in forecasting and in communicating complex, highly uncertain situations to nontechnical audiences.   ■

---

starting point from which judgments about constructing forecasts are made. If experts and good time series data are available, expert opinion and trend analysis techniques can be used effectively. If models are available that incorporate the main features of a forecast topic, and if the forecaster is confident in the quality of the assumptions that drive them, then modeling is viable. Scenarios can be used to integrate, communicate, and/or present whatever information is available to the users of the forecast in a nontechnical, literary manner, or this technique can be used if no other techniques are available.

6.3.2   Strategy for Communicating Forecasts to Users

The strategy for communicating forecasts to users depends on the character and needs of the user. If the user simply wants a forecast, then a report or multimedia presentation will suffice. If, however, the user wishes to evaluate the assumptions about the future incorporated in the forecast or the implications of those assumptions or of the forecast for his or her organization, then substantial and continued interaction may be necessary.

In the latter case, the first task is to unpack the assumptions. These may or may not be explicit. In a difficult situation, the explicit assumptions may constitute the ideology of the organization, which may significantly differ from the operating assumptions that guide it. In such a case, the forecaster should redirect the interaction so that both explicit assumptions and behavior are considered. In all cases, the assumptions and the conclusions/outcomes resulting from them should provide the bases of the interaction. The project should be extended to provide other sets of assumptions and conclusions/outcomes to the client and to evaluate the entire group. Formal forecasts and impact assessments, which may or may not be shared with the client, constitute the basis for the forecaster to determine the outcomes of alternative sets of assumptions.

The purpose of interaction is to determine the needs and the knowledge state of the user. The overriding goal, as always, is to satisfy the user. Dialectic and global inquiring systems may be used in these interactions. Creativity techniques (see Chapter 7) may be appropriate to reveal the underlying world view and assumptions of the users, which typically are entangled with their needs. This interaction drives technique-based forecasting, which is used as a tool within the context of interaction rather than as a separate project.

6.4 CONCLUSION

Strategy for selecting methodology rests on the inquiring systems chosen to deal with the demands of the forecast as defined by the user. Useful inquiry systems range from the scientific to the dialectic and global. Choosing either of the latter will imply a context in which user interaction creates the forecasting environment. In such cases, technical analyses are used to support the broader pattern of interaction and not as independent forecasts.

There are five types of common forecasting techniques. Monitoring is the basic source of relevant information and so is almost always used. Expert opinion, trend analysis, and modeling depend on the existence of experts, good time series data, and appropriate modeling frameworks and assumptions, respectively. Scenarios provide a "fall back" position when conditions for applying other techniques are lacking. However, even when other techniques can be used, scenarios provide a way to integrate their forecasts with other useful information and are excellent vehicles for communicating forecast results to users of all backgrounds.

The bottom line is that forecasts are demand-driven. Thus they are primarily targeted at meeting and, when necessary, identifying user needs.

EXERCISES

6.1 Consider the following fields: law, physics, civil engineering, marketing, and your own, if not one of the above. What inquiring system(s) guide the generation of new knowledge in each field? The members of these various fields differ in their willingness to address the future. Describe these differences and how the future is addressed in each field. What are the information bases for any forecasting done in each field?

6.2 Obtain a technology forecast (such as a company report, a consulting firm report, or a public article in the *Futurist* or a similar journal).

   (A) Describe what methods underlie the forecast. Would the use of other methods augment that forecast? How?

(B) How effectively does the document communicate the forecast? What additional mechanisms might enhance communication? To which audiences?

6.3 Suppose your firm is engaged in producing telecommunications equipment. What methods would you use in forecasting the likely market penetration of a new local area network that your firm plans to introduce in the next year for use within organizations? What methods would you use in forecasting the potential for technological innovations in intraorganizational communications over the next 15 years?

___7
THE STIMULATION
OF CREATIVITY

OVERVIEW

Forecasting requires the capacity to envision what the future might hold. This chapter describes methods to enhance the forecaster's and technology manager's creativity and to increase their ability to visualize alternative futures. First, methods of stimulating individual creativity are described. These include lateral thinking, metaphors and analogies, checklists, attribute analysis, morphological analysis, and the use of random words. Finally, group techniques including brainstorming, Crawford slip writing, and Synectics are considered.

7.1 FIVE ELEMENTS OF CREATIVITY

J. P. Guilford's research into creative behavior established the basis for much of our current understanding. This research began shortly after World War II as a project funded by the U.S. Navy. Guilford (1959) identified five key elements of creativity: fluency, flexibility, originality, awareness, and drive. An understanding of these elements removes some of the mystery surrounding creativity and paves the way for us to encourage its growth.

Fluency usually is thought of as the ability to express thoughts in flowing, effortless style. In creativity, however, *fluency* is the ability to provide ideas in volume. A simple test might be to see how many uses of a ball bearing can be devised in a limited time. Clearly, fluency is important in forecasting to help ensure that all useful alternatives have been identified.

Flexibility is the ability to bend familiar concepts into new shapes or to jump from old concepts to new ones. "Nimbleness" is an apt synonym. For example, a creative person will consider uses beyond the ball bearing's intended purpose, coming up with less conventional ones, such as sling shot ammunition. Flexibility can be measured by the number of categories included in a stream of ideas. Many individuals will exhaust one category before moving to another; others will list only a few related ideas before moving on. Practice can increase flexibility.

Originality relates to the unusualness of ideas, while an individual with *awareness* has the imagination to perceive connections and possibilities beyond the obvious. Throughout the ages, some people have been able to look at one thing and "see" another: to look at a bird and see an aircraft or at a fish and see an undersea boat. Recently, imaginative engineers have begun to look at biological systems and see models for computer electronics and other engineering systems.

Individuals with *drive* have "stick-to-itiveness" or motivation. It is a common misconception to equate creativity with instantaneous, blinding flashes of inspiration. Like genius in Edison's famous quote, creativity often is one part inspiration, nine parts perspiration. However, drive should not be confused with the blind application of brute force. Confronted with a brick wall, the creative person will not attempt to batter it down but will employ fluency, flexibility, originality, and awareness to find another way.

The techniques described in the following sections are designed to enhance the five key elements of creativity. Raising the level at which individuals or groups apply these elements of creative behavior will increase their forecasting and management skills.

7.2 INDIVIDUAL CREATIVITY

There are many techniques for enhancing individual creativity, five of which will be detailed here (several are discussed further in Chapter 11): (1) lateral thinking, (2) metaphors and analogies, (3) checklist procedures, (4) morphological analysis, and (5) the use of random words as triggers for new ideas.

7.2.1 Lateral Thinking

Our senses provide our minds with a continuous barrage of information without which we could not make decisions. Since we cannot process all the information, we create patterns from which codes are established. The mind needs to process only enough input to recognize the appropriate code in order to react. Reflex action is one response to the coding and recognition process. For instance, the reflexive response of most men to a warning about pickpockets is to check their billfolds. The pickpocket's response is to note where they check.

Despite the obvious advantages, there is a downside to the pattern/coding/response process. Although our brain readily forms patterns, these can become difficult to restructure, especially if they are repeatedly used. Our minds try to sort information into existing patterns even if it does not fit. Further, the patterns we establish depend on the sequence in which we happen to receive information, and this is unlikely to be optimal. Finally, even though patterns may differ only slightly, one will be selected and the other ignored. This can produce errors and/or missed opportunities.

Established patterns tend to be clustered into groups that grow larger and larger, eventually becoming dominant patterns themselves. Lateral thinking provides a way to restructure and escape from old patterns and to provoke new ones (de Bono, 1970). Thus it provides a way to increase creativity. Lateral thinking encourages full use of our natural pattern-making capacity without hindering creativity.

Vertical thinking is selective. It seeks the most promising path. Lateral thinking is generative—that is, it generates new paths simply for the sake of finding the range of alternatives. Since it is not a building process, lateral thinking moves by leaps and bounds rather than sequentially. Far from excluding irrelevant information, it welcomes distraction as a stimulus to restructure old patterns and reveal new approaches.

Vertical thinking applies judgment to find the best path or idea. Thus some approaches are "good" and others "bad." Lateral thinking, however, does not involve judgment and dictates that all pathways remain open. Instead of following the most likely paths, lateral thinking may follow the least likely paths, seeking new perspectives, perceptions, and patterns. Categories, classifications, and labels are never fixed because new perspectives may reveal different frames of reference. Vertical thinking guarantees at least a minimal solution; on the other hand, lateral thinking improves the chances for an optimal solution but makes no guarantee.

Vertical and lateral thinking are fundamentally different but complementary processes. Lateral thinking enhances vertical thinking by providing more approaches to a problem. Vertical thinking justifies lateral thinking by developing the ideas it generates. To be more creative, we must not only understand the principles of lateral thinking, but also examine techniques that use these principles. A discussion of several such techniques follows.

Suspended Judgment. The need to be right is sometimes the greatest obstacle to creativity. It inhibits idea generation because we are afraid of being wrong, and it restricts the chances for improved solutions by ruling out ideas that can not be immediately justified. Thus the application of judgment, whether approbation or condemnation, too early in the search for ideas can cripple the creative process. If judgment can be suspended while the search is conducted, the chances for a creative solution are increased.

Judgment may be applied internally by the individual or externally by the group. By suspending both internal and external judgment, we can help ensure that

- Ideas will survive longer to breed more ideas
- Individuals will offer ideas they would have rejected
- Ideas will be accepted for their value as stimulation
- Ideas that are bad in the current frame of reference will survive long enough to be evaluated in any new frames that emerge

There are several guidelines to be followed if the potential benefits of suspended judgment are to be reaped. First, never rush to evaluate an idea: exploration is more important. Second, when an idea is obviously wrong, shift the focus from why it is wrong to how it can be useful. Third, delay discarding any idea as long as possible. Let it provide the maximum possible stimulus to the generation process. Finally, follow behind an idea rather than forcing it in the direction that judgment dictates.

Fractionation. The more unified a fixed pattern, the more difficult it is to visualize in different ways. Fractionation can help escape this inhibiting unity by dismantling the problems into parts or fractions. The object is to look at a problem less complex than the original and possibly solve it in parts. The fractions are restructured into larger fractions, and the larger structured problem is solved when possible.

Suppose you wish to develop a new technology that will reduce the number of lost pieces of baggage on domestic airline flights. The problem might be fractionated along functional lines as follows:

Tagging luggage	Unloading the aircraft hold
Conveying luggage	Placing luggage on baggage cart
Scanning tags	Delivering luggage to claim area
Routing to baggage cart	Unloading baggage cart
Delivering to aircraft	Placing on carousel
Loading into aircraft hold	Pick up by passenger
Transporting to destination	

To restructure you might combine "delivering to aircraft" and "loading into aircraft hold" and substitute "containerizing." You also could combine "unloading the aircraft hold" and "placing luggage on baggage cart" and substitute "removing container." Very large aircraft use this procedure already. Another possibility would be to combine "routing to baggage cart" and "delivering to aircraft." Similarly, "placing luggage on baggage cart" and "delivering luggage to claim area" could be replaced by "moving to baggage claim area." Transportation options for sending luggage to the aircraft and moving it to the baggage claim area, such as using conveyors and using automated guided vehicles, could be investigated. If two or four of these functions are replaced, then the others also can be evaluated by further fractionation or by combination.

In this example, fractionation is based on the *activities* that must be accomplished. Fractionation also can be based on the *attributes* of an object, idea, or technology. Such a method is referred to as *attribute analysis*.

Reversal. In this method, the problem is turned around, inside out, upside down, or back to front to see what new patterns emerge. The goal, as with all lateral thinking, is to find different perspectives by forcing us to adopt a new vantage point.

For example, suppose a vendor is paid by a $10,000 check from a buyer who has only $9,000 in the bank. The accepted pattern underlying this problem is that the buyer deposits money in the bank to cover all checks and the seller withdraws funds from the bank when cashing the checks. This pattern suggests that in this case the check will be returned due to insufficient funds. This will be the beginning of a lengthy and perhaps expensive process to force the buyer to make good—a process that could result in no payment at all. Reversing the accepted pattern, the seller might deposit $1,000 to the buyer's account and then cash the check and net $9,000. Such a reversal provides a new perspective and perhaps a new solution.

7.2.2   Metaphors and Analogies

Metaphors are words or phrases applied to concepts or objects that they do not literally denote (such as the system has bugs). *Analogies* express recognition of similarities between otherwise dissimilar things (such as the car rides as smooth as silk). Either invites comparisons that reject established patterns and force us to adopt a different perspective that can produce new solutions to old problems.

For example, von Oech (1986) describes consulting with a computer firm whose sales had not increased in a time of growth for the computer industry. Since the firm provided a full range of products, von Oech used an analogy between the firm and a full-service restaurant. The product line (menu) was very large, but there were so many restrictions that customers had to seek different sales personnel for different products ("I'm the wine steward, I ain't no waiter."). There were numerous restrictions on orders ("Ya get's either soup or salad, one or the other."). By viewing the problem from this perspective, management was able to quickly develop solutions.

Visual metaphors and analogies are usually much more effective than abstract ones, and it is important that they incorporate action. For example, when working with a team to develop a new roofing material, Gordon (1961) suggested that they visualize water running off a duck's back.

The best metaphors and analogies involve activities that are performed so frequently that we do not have to think about what to do next. For example, the analogy between fishing and recruiting management personnel used by de Bono (1970) follows a chain of activities that would be second nature to a fisherman. First, you need to find a "good stretch of water for fishing." By analogy, you need to find the best places to search for managers (perhaps college campuses or other firms). Next you need "fishing tackle" (advertising media, word of mouth, contacts at other firms) and "bait" (salary, benefits, stock options). If no managers are found, maybe the "waters are overfished," you're using the "wrong kind of bait," or you're not "fishing from sunup to sundown." The idea, of course, is to force you to look at the problem from another perspective, one in which the familiarity of an analogous activity will help key a creative solution.

7.2.3   Checklists

Checklists are a familiar part of everyday life: grocery lists, things to be done, personal calendars. They also are important parts of many technological tasks: takeoff and emergency checklists for aircraft, checklists for environmental impacts, and so forth. Building checklists can spur creativity, forcing us to think of possibilities and providing a framework that suggests completeness and consistency and that highlights omissions. Alex Osborn (1953), one of the pioneers of creativity techniques, provided the following checklist for new ideas:

- Put to other uses? (New ways to use as is? Other uses if modified?)
- Adapt? (What else is like this? What could I copy?)
- Modify? (Change meaning, motion, sound, form, shape?)
- Magnify? (Stronger, longer, heavier, exaggerated? Add an ingredient?)
- Minify? (Shorter, lower, miniaturize? Subtract an ingredient?)
- Substitute? (Other materials, processes, power sources, approaches?)
- Rearrange? (Interchange components? Other pattern, layout, sequence?)
- Reverse? (Turn it backwards, upside down, inside out? Open it? Close it? Transpose positive and negative?)
- Combine? (Blend, alloy, ensemble? Combine units, purposes, processes?)

Checklists are simple but very powerful devices for freeing creativity; however, they must be carefully constructed to allow the user to exercise latitude and imagination. The power of checklists is easily examined. Suppose it is the 1940s and you are trying to assess the potential of a technology that uses a round electronic tube to show black and white pictures of activities that take place at a remote site—some call it television. Use Osborne's checklist to guide your retrospective tour of imagination. Pay special attention to the questions relating to magnification, minification, and substitution. Extend your considerations into the future of high-definition television.

7.2.4 Morphological Analysis

Morphological analysis combines features of fractionation and checklists and expands them in a powerful new direction. In morphological analysis, fractionation is applied to choose the parameters of importance to a concept, the alternate possibilities for each are defined, and a checklist is created by making an exhaustive list of all combinations of the possibilities. Each of these combinations is examined in turn. Although some will be meaningless, some may already exist, and some may be eliminated for other reasons (such as impracticality or expense), others may merit serious consideration.

When there are two parameters, the possibilities form a plane. Three parameters form a cube. Each is relatively easy to represent and visualize. If there are four parameters, visualization is trickier, but there are several approaches that can be taken. For instance, any parameter could be chosen and a cube built for each alternative possibility. While a computer could be used to generate all possible combinations, an obvious limitation is that the combinations increase rapidly with the number of parameters and alternatives generated. Suppose, for instance, that there are three parameters, each with five possibilities. The number of combinations is $5 \times 5 \times 5 = 125$. Adding a fourth parameter with five possibilities raises the number of combinations to 625.

Suppose you are exploring the possibility of new mass transit technologies. You might select parameters such as the power source, the transport medium, and the guidance mechanism. The list of alternative possibilities under each parameter might appear as follows:

Power Source	Transport Medium	Guidance Mechanism
Hydrogen	Roadway	Driver
Gasoline	Air	Towed
Diesel	Water	Guided path
Steam	Underground	Electronic map
Electricity	Conveyorized	Collision avoidance
Battery	Rail	None
	Magnetic levitation	

This brief list produces 216 combinations. One is an electricity-underground-guided path, which already exists. Another combination, diesel-underground-driver, would probably be rejected unless a method could be developed to eliminate the effects of engine emissions. Each of the combinations is examined in an analogous manner to complete the morphological analysis. The TOOLKIT contains a program that aids in performing morphological analysis. This program allows up to 5 parameters and as many as 16 possibilities for each parameter. Combinations generally are made at random.

Koberg and Bagnall (1974) combine attribute analysis and morphological analysis with a twist. Rather than examining all possible combinations of attribute alternatives, they suggest making random choices and examining the new forms that develop. As an example, they examine the possibilities for improving the common ballpoint pen. As parameters, they choose shape, material, the cap, and the means used to contain the writing material. One possible new form is a faceted, glass pen with no top and a permanent cartridge that could be used to display advertising on each face.

Morphological analysis is intended to provide a disciplined framework for creativity. Simply, it provides a kind of accounting system for an array of possibilities too extensive for the mind to track. As with other creativity-enhancing techniques, morphological analysis encourages us to abandon preconceived patterns. Through it, we are forced to develop possibilities that we might otherwise overlook or reject and to consider ways to implement possibilities that we might be inclined to eliminate.

7.2.5 Random Words

We all have had conversations in which a random word sparked a completely unrelated thought in our minds. Random words often bring about a fresh association

TABLE 7.1   Random Trigger Words

01–010	Knife	Insect	Robot	Pan	Crown	Banana	Accent	Bottle	Violin	Computer
11–020	Pants	Dress	Grill	Tree	Peach	Motor	Buffalo	Floor	Plastic	Leopard
21–030	Barn	Town	Bingo	Club	Class	String	Lot	Gold	Trailer	Butterfly
31–040	Pine	Nose	House	Spice	Button	Key	Auto	Oven	Jungle	Picture
41–050	Staff	Bat	Paper	Lock	Brain	Face	Mask	Nail	Sight	License
51–060	Boat	Board	Cellar	Purse	Lime	Copy	Border	Vein	Milk	Window
61–070	Rose	Muscle	Mirror	Stove	Bed	Park	Fire	Line	Bone	Alligator
71–080	Wasp	Pail	Tribe	Nap	Court	Child	Stomach	Glass	Ring	Attorney
81–090	Photo	Lion	Magnet	Bow	Iron	Suit	Emblem	Car	Train	Stadium
91–100	Arrow	Water	Gym	Race	Voter	Pitcher	Chair	Ice	Razor	Highway
01–110	Ankle	Tower	River	Torch	Elm	Hawk	Circle	Test	Tie	Mountain
11–120	Twins	Snow	Flag	Factory	Track	Joker	Ghost	Play	Network	Building
21–130	Vault	Monkey	Bank	Skates	Rock	Cook	Pearl	Cover	Sling	Battery
31–140	Exit	Hotel	Street	Road	Alley	Sheriff	Top	Meter	Bottom	Jaguar
41–150	Apron	Fox	Fork	Clamps	Blender	Basket	Book	Peak	Union	Station
51–160	Guitar	Grass	Scale	Brush	Shell	Coach	Radar	Branch	Melon	Soda
61–170	Pole	Roll	Star	Oil	Cement	Torpedo	Piano	Smoke	Paint	Escalator
71–180	Mail	Zoo	Needle	Yard	Watch	Belt	Point	Badge	Gorilla	Handle
81–190	Bus	Candle	Comet	Fan	Knee	Spider	Role	Oval	Anchor	Stereo
91–200	Axe	Fiddle	Desk	Door	Back	Fist	Tent	Apple	Moose	Machine
01–210	Mat	Message	Officer	Port	Jockey	Sea	Ship	Seal	Trap	Weight
11–220	Bucket	Chariot	Agent	Garlic	Plate	Gate	Home	Ink	Helmet	Kitten
21–230	Chorus	Laser	Lungs	Pizza	Moon	Worm	Cream	Sink	Cloud	Magazine
31–240	Glove	Winter	Dance	Drum	Friend	Rug	Shoe	Radio	Zebra	Elevator
41–250	Roof	Knot	Folder	Fund	Bride	Glue	Grade	Hammer	Horse	Teacher
51–260	Ruffle	Artery	Wall	Tray	Rodeo	Vase	Ruler	Salad	Cup	Envelope
61–270	Camp	Dice	Bell	Cord	Escape	Judge	School	Present	Song	Football
71–280	Sailor	Puppet	Whale	Wheel	Flash	Colt	Turkey	Coupon	Deer	Flower
81–290	Plant	Crane	Record	Temple	Boxer	Team	Saddle	Athlete	Stunt	Telescope
91–300	Booth	Candy	Party	Organ	Tub	Diamond	Mouse	Jazz	Ocean	Hospital
01–310	Camera	Saloon	Rake	Flute	Ticker	Gas	Halo	Waiter	Hay	Calendar
11–320	Horn	Sole	Script	Energy	Garden	Pantry	Light	Jacket	Lodge	Canteen
21–330	Goose	Marble	Level	Noose	Elbow	Stamp	Memory	Boot	Farm	Elephant
31–340	Tiger	Storm	Kite	Ladder	Fawn	Globe	Spear	Turtle	Rope	Sweater
41–350	Peanut	Shrimp	Oak	Sand	Money	Maze	Cactus	Orange	Swan	Periscope
51–360	Map	Cake	X-ray	Dock	Goat	Chip	Perfume	Chain	Pipe	Chairman
61–370	Bear	Tooth	Polish	Lantern	Skull	Lap	Shark	Sugar	Label	Knuckle
71–380	Snake	Heater	Stage	Eagle	Wolf	Lash	Fly	Potato	Camel	Rooster
81–390	Menu	Pool	Cobra	Towel	Sky	Stool	Table	Eye	Lemon	Armadillo
91–400	Sponge	Rocket	Soap	Scarf	Statue	Poodle	Tack	Police	Pencil	Telephone

of ideas and trigger new concepts or new perspectives of familiar ones. In a way, they provide verbal links that help us look at one thing and see another. Table 7.1 provides a list of "link-rich" words similar to a table devised by von Oech (1986). These words are familiar to all of us so that many connections and similar concepts can be generated by using them.

The procedure is merely to select a word at random from Table 7.1 and then try to force a connection between it and the problem being considered. The TOOLKIT also contains a program to generate the random words shown in Table 7.1. You could generate a random number between 1 and 400 to select the word from the table or simply put your finger on any number. For example, suppose the random number is 301, camera, and the problem is how to limit graffiti in public places. An obvious connection this suggests is to use television cameras at high-risk areas. Other less obvious connections might include interesting graffiti artists in photographic art, offering cameras as rewards for the capture and conviction of graffiti artists, or requiring merchants to photograph anyone buying spray paint. Or, we could play with the word camera to see if novel ideas arise—came ra . . . come rah—perhaps a rally with community leaders could be organized to support the end of graffiti.

The point is not whether the sample ideas are good, practical, economic, or legally feasible. The point is that they represent different paths to solving the problem that would never emerge from our preconceived notion of its pattern. Other ideas may be better, but the quest of creativity is to multiply the paths available for reaching a solution.

7.3 GROUP CREATIVITY

Since the technology manager and the forecaster often work as members of a group, they must be concerned with ways to increase group creativity. Many of the techniques that stimulate individual creativity can also contribute to group creativity. However, the concerns of individual creativity are intertwined with concerns about the dynamics of group interaction as well. Although many aspects of the group techniques presented in this section will seem familiar, they address additional concerns. Other techniques, such as the nominal group process (see Chapter 11), also stimulate a different perspective for addressing a problem.

7.3.1 Brainstorming

Although brainstorming is a very old concept, its formalization as a group creativity process is largely the work of Osborn (1953). The members of a brainstorming group are asked to respond to a central problem or theme. Emphasis in the process is on generating a large number of ideas (fluency), and criticism or evaluation is deferred (suspended judgment). Thus brainstorming is a group implementation of the concepts of lateral thinking, and as such the results of brainstorming eventually must be linked to vertical thinking.

The brainstorming session is self-consciously unstructured. Four general guidelines are observed:

1. Criticism is ruled out.
2. "Free-wheeling" and wild ideas are welcome.
3. A large number of ideas is sought.
4. Participants are encouraged to combine ideas into new or better ideas.

The setting for the process should be relaxed and isolated, and participants should be encouraged to verbalize their responses as quickly as they come to mind. The session should involve at least six but no more than twelve participants with perhaps one-third of the participants directly involved in the topic under consideration. It may be important not to have a superior and his or her subordinates in attendance. Subordinates often do not feel free to generate "far out" ideas for fear that their superiors will think them silly. By the same token, superiors sometimes wish to avoid appearing foolish in front of subordinates. A broad range of backgrounds and interests should be represented in the group to ensure a rich range of ideas and perspectives.

Sessions should not last too long. Many authors suggest that one to six hours is ideal, but 30 minutes to one hour seems to work better. If participants are not familiar with the technique, a warm-up session dealing with a familiar but unrelated problem, can be useful. In a productive brainstorming session, the ideas may flow so rapidly that it is difficult to keep track of them. Thus some means of recording ideas must be provided, whether shorthand, video or audio tapes, or a condensed, longhand version, to ensure that the ideas are not lost. "Play back" ideas to participants, perhaps using tear-off sheets or a blackboard.

The chairperson must keep ideas flowing smoothly and control traffic so that only one person talks at a time and everyone has an opportunity. He or she also must assure that no evaluation takes place during the session. The chairperson may occasionally need to control the pace, slowing things down for the notetaker or jumping in with ideas if the session slows prematurely. Finally, as with any meeting, the chairperson is responsible for organizing the session (reserving space, issuing invitations, preparing the problem definition, naming a notetaker, etc.).

Problem definition is an extremely important part of the brainstorming process that often is given too little attention. The problem should be stated clearly but not too narrowly: a narrow statement invites a narrow range of ideas and may inadvertently bound out the very richness that is sought. For example, a firm concerned that too many trips are being taken to branch offices might state the problem in positive terms: "How can all forms of communication with branch offices be enhanced?"

Brainstorming is a useful process, but it is not without problems. For example:

- Delayed evaluation may cause some participants to lose focus.
- Dominant individuals may influence other participants and try to monopolize the floor.

- Bandwagon and other "group think" phenomena can undermine creativity.
- It is difficult to prepare reference material in advance because the ideas that will be generated are unpredictable.
- Some participants cannot help becoming emotionally involved, thus stifling the participation of others.

In contrast, brainstorming offers the positive benefits of suspended judgment, lateral thinking, and the use of random key words in a group setting. Further, since the group members "own" the ideas generated during the process, their support may be greater for implementing the solutions derived from them.

7.3.2   Crawford Slip Writing

Crawford slip writing can be used to generate dozens of ideas in a matter of minutes. The process is named for Professor C. C. Crawford of the University of Southern California who first used it to help prepare training manuals.

Each participant is given three-by-five-inch index cards or slips of paper, and the problem is stated (statements made in the how-to form seem to work well). Participants are told to refrain from judgment and write as many answers to the problem as possible in the time allotted. Each idea is written on a separate card or slip. The ideas generated during the listing process are then sorted into various categories. They may be evaluated and regrouped using categories such as utility, cost, implementability, novelty, chance for success, anticipated return, and so on. In Crawford's training manual application, persons who were to use the manuals listed difficulties with the explanations of specific operations in how-to form (such as "I don't understand how to disconnect the widget before the framus is removed"). These problems were fed back to the manual writers and other knowledgeable people who generated ideas in response to the perceived problems.

Unlike brainstorming, this technique preserves the anonymity of the individual. The participants who may be less dynamic or shy and more reserved than others can more easily enter the idea generation process.

Crawford slip writing may also be used for forecasting. In such an application, knowledgeable persons could be asked to forecast growth or breakthroughs that might occur in a technology over the next two to five years. Their responses could be categorized, sorted, and fed back to the same group or a different group for comment.

7.3.3   Other Group Techniques

There are many other group techniques for creativity stimulation. *Synectics*, developed by Synectics, Inc. of Boston, for instance, is a technique with both contrasts and parallels to brainstorming. This technique is much more involved, however, and can require several days and perhaps a consultant to complete. The basic concept is that only the individual or group with the problem (client) can implement a solution, so the goal of the group should be to inspire the client. Thus Synectics is not so much geared to producing ideas as to providing effective interaction within the group so that ideas will be implemented.

Synectics sessions and brainstorming groups are about the same size; however, Synectics sessions are typically calmer and shorter (45 minutes) than brainstorming sessions. The goal is not fluency, rather it is the generation of a few ideas at a time. The thought process is more vertical than lateral and seeks to expand and improve one of a few original ideas. Thus judgment (evaluation) cannot be suspended. It can be softened, however, by allowing two positive comments to be made prior to a negative or critical response (reservation). The process continues until the client is satisfied with the solution.

The nominal group process (Delbeq and Van de Ven, 1971) is another effective technique. It combines the fluency of lateral thinking with the judgment and refinement characteristic of vertical thinking (see Chapter 11 for a detailed discussion of this technique). The Delphi polling process offers elements of group creativity as well (see chapter 11 also). Model-building techniques such as KSIM can provide a creative climate by offering a checklistlike framework within which to attack complex problems (see Chapter 12). The literature abounds with variations on the themes represented by the techniques described in this chapter.

7.4 CONCLUSION

The five characteristics that mark creativity are fluency, flexibility, originality, awareness, and drive. Whether they are used by an individual or a group, creative processes are designed to enhance these characteristics. All seek to eliminate preconceived views of a problem or situation and to encourage a new pattern of perception. In other words, they encourage creative solution by helping us to look at one thing and see another.

Most of the processes presented in this chapter emphasize fluency—the creation of many ideas—and reserve evaluation for a later time. Thus they emphasize lateral over vertical thinking. All the techniques are designed to foster an atmosphere in which the flexibility, originality, and awareness of the individual or group are emphasized. Drive, however, must be provided by the group leader and the individual participant. The processes described in this chapter are excellent vehicles to apply in the first stages of a project or forecast to ensure that the forecaster or technology manager does not narrow the range of potential ideas and/or solutions too early. This is important as the constraints of time and resources will narrow the process soon enough.

EXERCISES

7.1 Consider the problem of developing security for a computer network (that is, limiting access to authorized users).

    (A) Use Crawford slip writing to generate several possible ideas.

    (B) Select a good idea and use attribute analysis to improve it.

7.2 Use the checklist technique to determine future possibilities for the facsimile machine.

7.3 Use morphological synthesis (that is, morphological analysis coupled with attribute analysis) to forecast possible advances in VCR technology.

7.4 Use brainstorming to forecast what an office might look like 10 years from now.

7.5 Clip articles that describe creative products or services from newspapers and magazines for a week. Analyze the creative aspects of these using a checklist.

7.6 Select a common task from your workday. Use fractionation to develop improvements that would make the task "doable" for a blind person.

7.7 Take four minutes to list all the possible uses for discarded plastic milk jugs. Take four more minutes and add to the list. Select a random word from Table 7.1 and spend four minutes listing additional possible uses that are generated by association with the trigger word. Analyze your fluency in the various stages. Now carry out a brainstorming meeting with three friends and compare the results.

7.8 Consider a major world problem such as acid rain, global warming, or poverty.

    (A) Use morphological synthesis to develop possible solutions to the problem.

    (B) Use reversal to attack the same problem (for example, if you are from an area generating sulfur oxide view the problem from the perspective of someone who lives near lakes that have been damaged by acid rain).

7.9 Obtain a catalog from a gadget company such as Brookstone. Pick several gadgets and apply the checklist procedure in reverse to reveal the procedure that the inventor might have used.

7.10 Choose a common metaphor (such as bugs in a computer program). Examine it closely and describe the changes in perspective that it creates for you.

7.11 Forecasters make an analogy between the growth of a technology and that of a biological system. Examine this analogy and note the changes in perception it inspires.

7.12 Expand the checklist used to assess the potential uses for television (see Section 7.2.3). Now apply attribute analysis for the same purpose.

7.13 Describe how concepts such as lateral thinking and suspended judgment might be used to improve your educational experience.

7.14 Suppose you own 500 acres of land 12 miles from a large metropolitan area. There are several derelict oil-pumping rigs on the land. Use morphological analysis to forecast the potential use that might be made of the land.

7.15 Prepare a study of the factors that contributed to the creativity of famous inventors such as Edison, Tesla, Westinghouse, Diesel, Fulton, Carver, or others of your choice.

7.16 Write a paper on the development of an important invention such as Teflon™, nylon, the transistor, the personal computer, the typewriter, the turbojet engine, or some other notable product.

7.17 Read one of the references cited and prepare a report about what you have learned and how it would be useful for a technological manager.

_____8
MONITORING

OVERVIEW

Monitoring uses diverse sources of information to set the stage for forecasting, to feed information to the forecasting and planning processes, and to update and improve the projections upon which management decisions are made. This chapter presents the assumptions underlying monitoring, the various types of monitoring, and examples of the use of monitoring in industry. Special emphasis is placed on monitoring the context of technological changes as well as the changes themselves. Techniques to develop, interpret, and communicate monitoring results are presented, and suggestions are offered on its integration within an organization.

8.1   INTRODUCTION

> Monitoring is to watch, observe, check, and keep up with developments, usually in a well-defined area of interest for a very specific purpose (Coates et al., 1986, p. 31).

This description captures the essence of this simple technique. Monitoring means scanning the appropriate environment for pertinent information. That information may pertain to a particular technology—*technological monitoring*—in which case one may want historical information on the technology's development, current information on the state of the art today, and/or information pointing directly to future prospects. Alternatively, monitoring may consider the context in

114

which technology develops—*contextual monitoring*, or *issues management*.[1] This approach derives from the assumption that technological change is foreshadowed by changes in other technologies and/or in the socioeconomic environment. Thus it should be possible to monitor signals in these environments, analyze them, and forecast technological development. For instance, concerns about carbon dioxide emissions promoting the greenhouse effect may presage the return to favor of nuclear power.

The last part of the monitoring definition emphasizes "specific purpose." This emphasis is vital to avoid accumulating vast quantities of ill-focused information with no organizing principle. According to Coates et al. (1986, p. 31), possible objectives in monitoring include:

- Detecting scientific/technical or socioeconomic events important to your company
- Defining potential threats for the organization implied by those events
- Seeking opportunities for the organization implied by changes in the environment
- Alerting management to trends that are converging, diverging, speeding up, slowing down, or interacting.

These generic objectives—looking for events and trends that imply threat or opportunity—can be broken down further: identifying competing or supporting relationships, flagging critical needs, and ascertaining potential breakthroughs and their effect on the organization. Whatever the objective, careful definition of purpose and focus are essential to worthwhile monitoring.

Successful forecasting based on monitoring involves more than merely gathering data. The forecaster must sift the information for meaningful signals and envision their implications. This is best done through a systematic monitoring procedure that directs information search and interpretation. Finally, results of monitoring must be synthesized and communicated effectively to generate appropriate action. Indeed, substantive work may be dwarfed by the effort to orchestrate, package, and sell the information to achieve the necessary action (Coates et al., 1986).

Monitoring is one of the most useful techniques in forecasting. Applications depend on what the user needs. For instance, an Air Force research organization built a technological forecast for artificial intelligence largely on an expert/literature monitoring effort to systematize research funding plans (Reitman et al., 1985). Search Technology Inc. has used a monitoring system to help design a new tech-

---

[1] *Issues Management* is defined by Coates et al. (1986) as "the organized activity of identifying emerging trends, concerns, or issues likely to affect an organization in the next few years and developing a wider and more positive range of organizational responses toward that future" (p. ix). Issues management can help companies devise positive, anticipatory (rather than merely reactive) responses to new technologies, possible governmental constraints, or potential confrontations. Issues management emphasizes contextual monitoring (law-making, social controversies, etc.), but it can include technological monitoring, too.

nology called the Designer's Associate (DA)—a computer-based system to assist cockpit designers—which was prepared for the Air Force. This required an assessment of what technological alternatives might be available to accomplish given DA functions in 1995. Monitoring contributed in the following ways (Porter, 1988):

- It ascertained technological capabilities as of 1988 to help "configure" the DA.
- It alerted management to possible new technological capabilities for 1995.
- It assessed the risk in designing a system based on those potential capabilities for 1995.

Monitoring also can serve those on the lookout for new products or new processes. Vanston (1985) has sampled such American corporate applications. For example:

- ALCOA watches for new market areas and examines threats to present markets.
- Kraft analyzes changes in home eating patterns to determine effects on new food products.
- Owens-Corning Fiberglas seeks new products and processes.
- Johnson Controls predicts competitors' entrance into new technical areas.

Exhibit 8.1 provides a case illustration of the implications of monitoring.

■ **Exhibit 8.1   Whirlpool Beats the Competition (Based on Davis, 1973).**

Whirlpool Corporation tracked developments in the chemical and textile industries (not their own business domains) to identify permanent press fabrics prior to their commercialization. They acted quickly on this information to generate the first washer and dryer permanent press cycles, beating their competition to market by about a year. This resulted in a substantial increase in market share. The major events involving permanent press products and their impact on Whirlpool are as follows:

Winter 1963-1964	Rumors in textile industry of new "delayed cure" process for resin application.
April 1964	Confirmation of rumor
May 1964	First glimpse of process
May–June 1964	Education of Whirlpool personnel about permanent press garments; forecast of doubling of dryer sales in a year
August 1964	"Development conference" arranged for vice-president, laundry, at large fiber manufacturer; permanent press garments shown at Whirlpool for first time
September 1964	Textile industry introduces permanent press

January 1964   Permanent press cycles on Whirlpool washers and dryers; first by appliance industry

March 1965   Research project for a "new concept ironer" dropped; monies diverted to other uses   ∎

In a quite different context, through its National Center for Science and Technology for Development, the Chinese State Science and Technology Commission maintains a significant technology forecasting effort. Monitoring new technologies available from developed nations, such as the United States and Japan, serves its immediate industrial needs. Monitoring for emerging technologies helps to guide China's own technological development efforts by avoiding dead ends and seeking realistic niches.

Policymakers also can be served by the monitoring of emerging trends and possible breakthroughs in technology as well as by contextual monitoring. The Congressional Clearinghouse for the Future provides this service for the U.S. Congress.

Monitoring also provides a base for technology forecasting using other techniques. It provides data and background for choosing and using forecasting tools. Section 8.2 considers the principles for sound monitoring; Section 8.3 offers a systematic monitoring process; Section 8.4 addresses implementation issues.

8.2 PRINCIPLES

This section describes the basic assumptions that support monitoring and the monitoring methods that are commonly applied.

8.2.1 Assumptions

Monitoring is an opportunistic technique. If patterned technological development is expected to continue, monitoring tracks this historical development as a basis for projection. If patterns are not as well-defined, monitoring relies on current information on the state-of-the-art and expert assessments of future advances. Such advances may be incremental or breakthroughs—monitoring can tackle either. In some cases, monitoring contributes more to current awareness than to true forecasts, but it is an extremely robust technique as long as the information gathered is cautiously interpreted.

Both technological and contextual monitoring must be tailored to fit the developmental stage of the technology, as well as to fit user needs. Table 8.1 presents a version of the "suspect" linear model of change (recall the innovation process models discussed in Chapter 2)–not to argue for linearity, but to suggest what information may be most salient to monitoring at various stages of a technology's development.

Some have tried to develop a monitoring framework based on technological characteristics. In particular, there have been attempts to distinguish functional and structural measures. For instance, Knight (1985) differentiates main structural units of the computer (memory, computation, control, input-output) and key functional

TABLE 8.1   Monitoring Information as a Function of Technological Development

Stage of Development	Key Issues	Information
Scientific discovery	Scientific uncertainty	Enduring interest within the concerned segment of the scientific community Sporadic interest in public media
Applied research & development	Technological uncertainty	Information may not appear in the open literature, depending on the developers Trade magazines may be a prime source of what information is useful "Inside" experts may be critical to the process of verifying information Concern about possible impacts of the technology may arise in scientific or public media
Initial applications	Economic uncertainty	Widely available from diverse sources Market assessment of key concern Impact assessment may become a serious issue Policy development and assessment issues
Widespread adoption	Social uncertainty	Technology evaluation Policy assessment Renewed emphasis on technological development, seeking to project successor technologies Market assessment to determine potential, niches, and competitors

attributes (computing power, cost, and reliability). He then cumulates qualitative and quantitative measures on each of these over extended time periods. His own assessment is that the approach does a good job of depicting the evolution of the computing industry, but is not detailed enough to isolate specific technological changes. This approach is appealing, but there are concerns about its application. For instance, over the long term, structural frameworks may change (for example, with the advent of personal computers or neural computing) and functional parameters may evolve (for instance, increases in computing power may result in qualitative changes in criteria, or new applications such as artificial intelligence may rely on nonmathematical processing power). It makes the most sense to consider function and structure, or other organizing principles, in light of the scope, duration, and purposes of the monitoring effort.

8.2.2   Types of Monitoring

Table 8.2 poses a series of overarching choices that must be made about any monitoring process. The basic focus, whether one is monitoring technology or contextual factors, has been discussed in Section 8.1.

The first choice is the time frame. In some cases, a technology or a particular context is being studied in an exploratory or preimplementation mode (for instance,

TABLE 8.2   The Overarching Choices of Monitoring

Time Frame
   Preimplementation vs. Imminent Decision vs. Postdecision

Monitoring Process
   One-time study vs. Ongoing monitoring

Focus
   Technological monitoring vs. Contextual monitoring

Breadth
   Macro vs. Micro

Purpose
   Choosing vs. Forecasting

Developmental Stage
   Invention vs. Innovation
   Established vs. Emerging technology

to help decide if the company wants to pursue that technology, enter a certain market, etc.). In this instance, monitoring is likely to have limited resources and uncertain management commitment; it is also likely to require strong selling to convey an effective punch line. In contrast, imminent decision monitoring suggests an attentive and anxious audience whose specific concerns will be directly addressed in the monitoring report. The postdecision time frame implies evaluative monitoring (that is, determining what happened by focusing on actual events directly involving the organization). This may be required by outside authorities (as a way to ensure that negative effects are mitigated) or may be undertaken to help the organization manage a specific technology. In any event, this postdecision monitoring clearly differs from preimplementation monitoring.

The second choice is whether to undertake a one-time study or to set up an ongoing monitoring system. The former must ensure that it does not narrow the monitoring focus prematurely and miss significant influences. The latter has the advantage of compiling contextual information over a long period that can be used to form more precise inquiries (see Exhibit 8.2). The monitoring system must have a clear focus and precise objectives to avoid becoming irrelevant to the organization.

Technological monitoring approaches will differ depending on the level of focus. Macro technology monitoring could encompass a whole spectrum (such as the information technologies, as per Exhibit 8.2). Alternately, micromonitoring might focus on one technology (the computer) or a particular type of that technology (personal computers). Another choice would be to focus on systems (office automation), individual devices (personal computers), or components (microprocessor chips). Whatever level is appropriate for the study, it is incumbent upon the

■ **Exhibit 8.2   A "Two-Stomach" Information Technology Monitoring System (Based on Neste, 1988).**

Search Technology implemented an ongoing monitoring system to keep abreast of changing information technologies in early 1989 (Neste, 1988). Essential criteria for this system included:

- Minimal effort required for dispersed participants to contribute information to the system
- Convenient access to information for users with quite different needs
- Low cost
- Ongoing system that facilitates generation of up-to-date reports as needed

One technical staff member takes responsibility to maintain the system and evaluate its performance (observe usage and suggest improvements). One clerical employee helps input information by copying documents, filing, and performing other related tasks. Six or so technical staff members each take lead responsibility for several technologies. They process the information as needed and generate suitable syntheses and forecasts. Another 20 or so technical people contribute to the system as they find materials pertaining to the targeted 20 (changing over time) technologies. Users include project managers designing a particular technological system; management personnel proposing to provide monitoring information per se to clients or drawing on the system for general background information; and technical personnel writing articles or reports who want information on one or more of these technologies.

Initiation required that management approve the system, that the target technologies be identified, that the six staff members agree to take responsibility for particular technologies, and that all staff be alerted to the new system. Staff prepared pilot forecasts on certain key technologies and generated initial technological maps to help identify what other technologies should be monitored. Management and project users were interviewed to determine their needs. Staff were briefed on how to contribute information at a meeting and through a memo, and a system manual was prepared. A special bulletin board alerted staff to key items.

The physical setup was based on a dual filing system—the two stomachs. One folder for each technology contained all raw (undigested) materials, including copied articles, book references (indicating where the book could be found), clippings, and other such material. A second folder for each technology contained monitoring reports prepared by Search Technology staff on an as-needed basis (every six months or so) plus important supporting sources (well digested). When a report was prepared, that input folder was emptied, and only vital materials were maintained in the report folder. System materials were kept in a designated set of easily accessible filing cabinets.   ■

analyst to consider how changes at other levels might impact the focal technology. For example, the person monitoring office automation cannot ignore improvements in computer storage that might substitute for networking arrangements.

Sometimes monitoring serves more than just forecasting purposes; it may also help in choosing a future course. As mentioned, the Chinese monitor technology developments in other advanced nations. The state of Japanese or Western technology provides an excellent menu from which to choose their own development programs. Obviously, monitoring existing technologies greatly reduces the uncertainty, but it must also consider technology transfer issues, including cultural and political factors.

Invention and innovation are concepts that denote the issues attendant to different stages of development (see Table 8.1). However, these concepts also represent a choice of monitoring emphasis. For certain uses, interest may be invention-oriented—that is, focused on how to reduce technological uncertainty (for instance, it may be necessary to determine technological capabilities by a given date). For innovation-oriented uses, the key is really economics—Will a technology succeed in some market? Invention focuses on critical technical "milestones," their prerequisites, and the likelihood of success by a given date. Innovation, on the other hand, focuses on whether a technology will meet needs of some users at an attractive price and/or in a way that is better than competing technologies. It is often easier to predict the emergence of technical capabilities (particularly when improvements come as a stream of incremental gains rather than discrete breakthroughs) than to predict the date of the actual innovation (that is, success in some market). These two types of technological monitoring emphasize different information sources (e.g., engineering versus market knowledge).

The monitoring of an established technology can draw upon historical pattern information concerning generally agreed-upon parameters. Martino (1987) offers a process to monitor established technologies that uses precursors (see Exhibit 8.3). In contrast, the monitoring of an emerging technology must draw on more diffuse information (also likely to imply more diffuse sources) with a less well-defined target. The emerging technology is likely to be more sensitive to contextual influences, and the time frame for it is likely to be longer, with attendant uncertainty in any forecasts.

Motorola devised a systematic monitoring/forecasting approach that used a technology roadmap process to sharply differentiate between emerging and established technologies (see Chapter 3). A very different approach is to track emerging technologies in general. For example, the National Institute for Standards and Technology (NIST, formerly the National Bureau of Standards) identified 12 emerging technologies for the 1990s (Technology Administration, 1990):

Emerging Materials

1. Advanced materials
2. Superconductors

■ **Exhibit 8.3A Precursors of Technological Change**

Martino (1987) illustrates this approach by compiling entries from the Engineering Index (and its on-line Compendex data base) for selected technologies from 1970 to 1987. The resulting qualitative histories can be used to anticipate coming changes. Martino recommends watching for

- Incomplete inventions that need other elements before they can be deployed economically
- Development of performance-improving supporting technologies needed by a basic technology
- Development of cost-reducing supporting technologies needed by a basic technology
- Development of complementary technologies that mesh with the basic technology to make it useful (such as sensors for automated controllers)
- Use of a technological advance in a prestige application before transition to general use (such as transition of technology from aviation to automotive uses—early uses in racing cars)
- An incentive for use, such as reduced cost or elimination of externalities (for instance, fuel economy or reduction of automotive emissions)
- Different origins for leading indicators (for instance, motor car manufacturer advances are nearer-term indicators of coming automotive change than are advances coming from parts suppliers)

Martino provides the accompanying chronology of plastic auto body shells to illustrate one such precursor trail. (His report also provides another trail of improvements in steel auto body shells to "counterattack" the threat posed by plastics.)

Sometimes, a leading indicator relationship can be found. Martino illustrates this relationship using the development of new aluminum alloys, followed by their first application in an aircraft, as an example. In that case, a mathematical relationship (by using regression, for example) may be devised to anticipate when to expect the application of new alloys (average lag is 4.2 years). Martino suggests treating such leading indicator relationships by devising probability distributions. In the alloy–aircraft case, he uses the method of maximum entropy (Tribus, 1969) with only the mean specified (to yield a geometric distribution) or with the standard deviation of the lag specified (to yield a Gaussian distribution truncated on the lower end—lag of zero). In this case, the Gaussian distribution is most informative, predicting virtually no chance of application of a new alloy in less than three years and almost certain availability within six years. Such information would be quite helpful to an aircraft manufacturer planning materials for a new plane. ■

■ **Exhibit 8.3B   Events in the History of Plastic Auto Body Shells**

1978

Ford built demonstration Graphite Fiber Reinforced Plastic car; test results reported in 1983.

ICI America introduces resins and urethane thickeners for producing of structural grade sheet molding compound (SMC). Actual field use still needed to demonstrate usefulness for auto applications.

1980

Ford Motor Co. reports design, development, manufacture of lightweight, one-piece SMC hood for Econoline Van.

Laminates of steel sheet with plastic core have same formability as steel, lower cost, lighter weight.

1981

About 100 kg polymers now used in medium-sized car.

Reports of several experiments with plastic for body panels, interior components.

1982

FIAT builds demonstration car with steel skeleton, plastic panels for body shell. Skeleton has only 800 welds as compared with 3,000 typical for all-steel car. More resistant to corrosion, better formability for low drag shape, saved 20% weight over all-steel car.

1983

1984 model Fiero bolt-on body allows ±.005 tolerances in finish and assembly of bodies; precision is equal to or better than steel.

1984

Production of Pontiac Fiero, steel "space frame" with all-plastic body skin. Flexible plastic "Bexloy" used in rear. Proves plastic can match surface finish of steel. Special painting arrangements; separate lines for space frame, for plastic panels, for flexible front and rear fascias. Painting panels before mounting reduces paint repair work.

Plastics use in autos, triples that of a decade earlier, now 10% of car total weight.

1985

Finite-element methods used to design steel chassis which carries all torsional loads, no stress on plastic body.

Successful in-plant coloring of ABS body plastic, reducing inventory costs by eliminating need to store colored plastic.

1986

2,000-ton plastic molding machine produced for auto plants.

Ford Sierra plastic grille panel is body-color painted and in same plane as painted metal panels, demonstrating that plastic can be used in conjunction with steel.

Source: Martino, 1987. Reprinted by permission of Elsevier Science Publishing Co., Inc.   ■

Emerging Electronics and Information Systems

3. Advanced semiconductor devices
4. Digital imaging technology
5. High-density data storage
6. High-performance computing
7. Optoelectronics

Emerging Manufacturing Systems

8. Artificial intelligence
9. Flexible computer-integrated manufacturing
10. Sensor technology

Emerging Life-Science Applications

11. Biotechnology
12. Medical devices and diagnostics

In contextual monitoring, macro and micro distinctions are also made. At the macro level, various entities (typically governmental bodies) compile indicators—ideally, time series data—on various socioeconomic factors of general interest (such as educational achievement test scores, unemployment rates). Private-sector firms may also compile such data, monitor literature, and/or prevail upon experts to identify contextual trends of note. For instance, the Roper Organization produces "The Public Pulse"—a report on American preferences. One issue (Roper's, 1987) offered 31 major trends shaping the future of American business. Two examples from that report are:

> *The Culture of Convenience*—the fast-rising number of two-income households lies behind the takeoff of the "convenience industry," projected to continue. . . . this also explains the tremendous potential of prepared take-out foods and of appliances like microwave ovens. . . .

> *Permanent Damage to Nuclear Energy Industry*—growing environmental concerns, combined with the Chernobyl disaster, have retarded further developments in nuclear energy for the foreseeable future. . . . Public worries and opposition to new nuclear plants are high and rising. . . . The nuclear power industry probably cannot recover in the short term. . . .

It appears that the first sample projection is holding true in the 1990s, but the second one could be overturned by other environmental concerns (such as acid rain and the greenhouse effect).

Micro-level contextual monitoring focuses on those elements of the environment perceived to impinge more directly on an organization's interests (or a technology of interest). For instance, computer firms monitor announcements, patents, and rumors concerning their competitors' activities.

Contextual monitoring can address a vast array of possible changes. At times it may be useful to check coverage of all domains likely to influence the organization as regards its monitoring objectives. The following list will generally provide a comprehensive catalogue of 10 areas for investigation:

- Technological (scientific)—the 12 NIST technologies to watch
- Health—AIDS rates, life span, nutritional levels
- Institutional (legal, organizational)—governmental regulation and agency power, expansion of multinational corporations, industrial concentration in a sector, public participation
- Social (behavioral)—advent of the Information Age, changing educational levels, urbanization, birth rates, expanding elderly population, minority interests
- Political—governmental stability, national and state party power, organizational politics
- Economic—recession versus prosperity, union strength, job displacements, service sector growth
- Cultural (values)—changing work ethic, mass media, cultural homogenization, rising aspirations, consumerism, women's movement, acceptance of fractured families
- International (development)—world economy, changing patterns of competition, financial instabilities, OPEC
- Ecological—greenhouse effect, aquifer pollution, resource shortages, nuclear waste disposal
- Security—changing patterns of hostility, evolving alliances, military funding, SDI

8.3 MONITORING STEPS

Given the diversity of possible forms of monitoring, any suggested set of steps must be carefully adapted to specific needs. However, the following steps are generally useful:

1. Determine the monitoring objectives and focus
2. Describe the technology and map the pertinent context
3. Adapt an appropriate monitoring strategy
4. Interpret and communicate results

8.3.1 Monitoring Objectives and Focus

Sections 8.1 and 8.2 have discussed a range of monitoring information types, possible uses, and prospective users. The first task of the would-be monitor is to determine which elements of these ranges fit the situation. Specific objectives should be spelled out (monitoring has a tendency to generate overwhelming amounts of information unless objectives are kept strictly in view), and a focus must be established. To undertake technological or contextual monitoring, you need to resolve the choices posed in Table 8.2.

Objectives and focus should be discussed by those engaged in the monitoring effort and with potential users. Agreement should be reached on issues such as scope, temporal extent, personnel and financial resources, timetables, whether the effort will be ongoing, and interest in trends or breakthroughs.

8.3.2   Technology Description and Contextual Mapping

As discussed in Chapter 4, you must describe a technology before you forecast it. Monitoring is, however, likely to generate an iterative situation. It begins with a presumptive definition/description of the technology. Monitoring enriches understanding of the technology, possibly leading to its redefinition in terms of level (for instance, addressing microprocessors instead of microcomputers), the pertinent technological system, and the critical parameters or milestones (critical qualitative or quantitative improvements) to monitor.

Mapping, whether graphic or text, can help identify relationships among technologies (see Figure 8.1 for selected information technologies perceived pertinent to computer systems). Likewise, it can help identify vital contextual socioeconomic influences (see the discussion of technology delivery systems in Section 2.3). Without such mapping, critical signals may be missed because monitoring may be wrong or too narrow.

Technologies and contextual influences should also be linked to organizational interests. Brown (1980) suggests composing matrices that indicate the likelihood of significant change on one axis and its likely impact on the company on the other (see Figure 8.2). Such a matrix allows quick identification of which changes would constitute milestones for the organization.

8.3.3   Monitoring Strategy

A number of strategic monitoring issues have been mentioned (such as making monitoring choices, meeting different objectives, making do with different resource levels, and performing under accelerated or extended time schedules). In this section, monitoring approaches are differentiated by how familiar you are with the subject. (This discussion assumes technological monitoring, but the implications generally fit contextual monitoring, too.) As a guideline, it is useful to contrast three levels of familiarity:

Level 1 Cold
Level 2 Warm
Level 3 Hot

Figure 8.1 Map of technologies relating to computer systems. (*Source*: Prepared by S. Cunningham, 1990.)

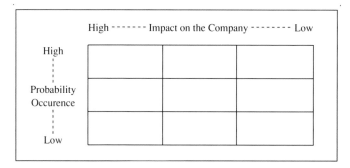

Figure 8.2 Impact and probability of occurrence matrix. (*Source:* Brown, 1980, p. 31.)

At Level 1, you start "cold"—that is, you are unfamiliar with the subject. At this level, the immediate questions include:

- What is the technology? How is it defined and described? What is the state-of-the-art?
- How do other technologies relate to it? What other contextual factors effect it?
- Who are the key players (individuals, organizations, suppliers, regulators, users)?
- What are plausible future development pathways?

To accomplish Level 1 monitoring:

- Use a "shotgun" approach to gather information—that is, grab anything that is convenient and may be pertinent
- Emphasize recent literature and review state-of-the-art articles or books
- Locate one or two accessible professionals with sufficient expertise in the technology to point out information sources and to help ensure that the monitoring does not drift wide of the mark

Level 2 suggests that you have either completed a Level 1 introductory effort or that you have some familiarity with the subject. Objectives become more focused; pertinent questions include:

- What are the driving forces propelling this technology?
- Can important interdependencies with other technologies or with socioeconomic factors be mapped?
- What are the key uncertainties along the development path?

To accomplish Level 2, the sources of information shift:

- Literature searches become more focused; on-line searches are likely to be more fruitful; and historical searches now may make sense as a way to identify leading indicators of progress and significant influences.
- Locating other forecasts for the focal technology can help answer the pivotal questions and help ground forecasts.
- Networking can be used to identify experts with various perspectives on the technology.

It now makes sense to begin to synthesize the information obtained by formulating an image of what is happening to the technology.

At Level 3, you are very familiar with the subject. Level 3 objectives are even more focused; pertinent questions include:

- Can you specify the key factors to be watched?
- What is the most likely development pattern for the immediate future? For the longer term?
- Can you offer specific projections?
- What recommendations can you make to help your organization manage development?

Actions undertaken to accomplish this level of monitoring include:

- Extending the information search to be as comprehensive as feasible with regard to the key factors
- Developing a conceptual model of what drives and what impedes development of this technology
- Seeking direct confirmation of this model and a review of your projections from experts
- Generating a credible forecast by integrating the monitoring results with other forecasting techniques
- Possibly establishing a structure for an ongoing monitoring system

These three levels suggest different requirements. Subject expertise is not essential at Level 1, but it is necessary at Level 3. Comprehensive information access (such as access to on-line systems) is not critical at Level 1, but it is at Level 3 to ensure thoroughness. Commitments of time and effort also increase at the higher levels. It is sensible to commit only to a Level 1 effort in investigating a technology to ascertain its relevance. The results of such an introductory examination may be sufficient to conclude that the subject is not critical to the organization. On the other hand, if the subject does seem sufficiently important, you may need to revamp the monitoring program to involve experts (from inside or outside the organization), boost resource commitments, and set forth a full-blown monitoring system.

8.3.4 Interpretation and Communication

Data do not speak for themselves; therefore, it is very important that key points be flagged in such a way as to get management's attention. Possible mechanisms to foster communication with potential users include maps, milestones, and multiple channels.

Graphical presentation of technology maps (imagine Figure 8.1 with all of the linkages shown) quickly conveys complex relationships, especially to visually oriented individuals (engineers are great candidates). Technology delivery system sketches fall in this same category, particularly if a few key contextual influences can be highlighted. Any model of what drives the development of a given technology should be interpreted to point out the few, critical factors that should be watched.

Monitoring may generate extensive historical profiles of events and/or trends. These should be offered with further interpretation. The notion of milestones for which to watch can be quite effective. Table 8.3 illustrates innovation milestones in the history of cement manufacturing in the United States. Milestones may reflect events or trends advancing dramatically so as to improve performance significantly.

Multiple communication channels improve the chances of successfully transmitting information. Both parallel channels and serial repetition (same or similar messages repeated at different times) may make sense. For example, the monitoring system discussed in Exhibit 8.2 has multiple channels. Some channels work well for some types of information, but not for others. Paper folders fulfill the configuring objectives quite well, but they do not accomplish the alerting or forecasting functions. Wall maps of relationships (see Figure 8.1) seem excellent for configuring and helpful for alerting. A bulletin board provides a nice alerting mechanism, but little more. Reports appear best for presenting forecasts, but they are too infrequent to alert organizational staff in a timely manner. Finally, meetings can help round out rough spots by providing rich, two-way communication opportunities, but they are costly in personnel time.

In addition to the inherent strengths and weaknesses of the communication channels, consider the individual quirks of the providers and users. Some people favor graphical presentations; some want it written; others thrive on verbal exchange. When feasible, it may be best to use multimedia schemes to reach as many as possible.

8.4 IMPLEMENTATION

Implementation involves three issues: Who should perform the monitoring? How can they put together a monitoring program? How can they integrate monitoring results with other forecasting techniques?

8.4.1 Who Should Conduct Monitoring?

Informal monitoring can be done by anyone who alertly keeps track of information obtained from routine reading of journals, casual browsing, and other such activities. Some sources provide a sort of ready-made monitoring. For instance, *IEEE Spectrum* publishes a "newslog" each month of items pertaining to electronics developments. *Predicasts Forecasts* compiles forecasts from a wide range of government reports and open literature; these are compiled quarterly (and maintained on-line as well) using Standard Industrial Classification (SIC) codes. An example

TABLE 8.3 Significant Technological Discontinuities in Cement Manufacturing

					Locus of Innovation		
Industry	Year	Event	Importance	Type of Discontinuity	New Firms	Existing Firms	Probability
Cement	1872	First production of Portland cement in the United States	Discovery of proper raw materials and importation of knowledge opens new industry	Niche opening	10 of 10	1 of 10	
	1896	Patent for process burning powdered coal as fuel	Permits economical use of efficient rotary kilns	Competence-destroying	4 of 5	1 of 5	.333
	1909	Edison patents long kiln (150 ft)	Higher output with less cost	Competence-enhancing	1 of 6	5 of 6	.001[a]
	1966	Dundee Cement installs huge kiln far larger than any previous	Use of process control permits operation of very efficient kilns	Competence-enhancing	1 of 6	7 of 8	.000[a]

Source: Tushman and Anderson, 1986. Reprinted from "Technological Discontinuities and Organizational Environments" by Michael Tushman and Philip Anderson published in *Administrative Science Quarterly* (vol. 31, no. 3) by permission of *Administrative Science Quarterly*.

[a] $p < .01$

Note: Fisher's exact test compares the pool of firms that are among the first to enter the niche with the pool of firms that introduce or are among the first to adopt a major technological innovation. The null hypothesis is that the proportion of new firms is the same in each sample; probability is the probability of obtaining the observed proportions if the null hypothesis is correct.

issue contains some 95 items pertaining to industrial robots, reporting projections for robot sales in units or dollars, related employment, prices, or numbers in use (specifying years that the projections address and providing source information on the projections). Any person with a professional level of expertise related to the topic should be able to select pertinent information from these sources.

Formal monitoring varies so greatly in scale that formulating rules would be foolhardy. Instead, consider the range of possible performers. Outside information services may be able to provide the information desired. Options include targeted newsletters, clipping services, trade associations, or contracted monitoring studies done by research firms. Possible disadvantages include less knowledge of the organization's needs, weaker communication bonds to organizational management, and less control over the accessibility of information to others. Advantages include possible strong subject expertise and established access channels.

Within the organization, a single staff person could be assigned to monitor a subject; alternately, that person might be supported by some outside information services. Although the single-person approach is efficient, it may be less effective than a more broadly based approach.

A study by Allen (1977) on where and how scientists and engineers obtain information has pointed to the importance of "gatekeepers." These professionals (perhaps, but not necessarily, managers) serve as critical bridges to the outside world. Most of the flow of ideas in technical development takes place within project groups. According to Allen, relatively few individuals in R&D read much technical literature or, especially in the case of engineers, actively participate in technical meetings. The gatekeeper funnels in a wide sweep of information for use by the project team in their relatively narrow application focus. Gatekeepers tend to emerge, rather than to be created by their managers. They should be well networked within their organizational unit (whether this is a single project team or a large company) and be avid information hounds. Gatekeepers make ideal monitors.

The next logical step is to assign an ad hoc team, from some central staff unit, to conduct monitoring. Alternatively, a team might be set up on a permanent basis to run a monitoring system (see Exhibit 8.2), or monitoring could involve more widely dispersed personnel (see Exhibit 8.4). The dispersed effort can directly involve those with subject expertise and decision-making responsibilities, but it requires incentives to participate and checkpoints to ensure performance and communicate results. Coates et al. (1986) elaborate on possible arrangements.

■ Exhibit 8.4 Dispersed Monitoring in a Food Processing Company

Technology forecasting is a new venture for this company. In the past, innovation has not been a high priority for this marketing-oriented firm. Now, however, management is making new technology a major corporate objective to save on costs through process improvement and to expand markets through new products and packaging.

This company is pursuing a diffused monitoring strategy. The director of engineering has initiated a comprehensive search strategy that categorizes possible application areas for new technology. Within each category, the director has specified technologies to monitor, For instance:

- Pumps, agitators, and mixers
- Ultrafiltration
- Microprocessors
- Wastewater treatment
- Wire-guided vehicles

One engineer is assigned responsibility for each area. That person is charged with monitoring the technology, identifying potential innovations for the company to pursue, and developing specific proposals for action. Such proposals can come forth at any time. (In addition, regular semiannual progress reports are due.) Superior proposals become development projects with specific objectives and budgets.

The great advantage of this approach is that the engineers responsible for development and implementation do the forecasting. No separate "sales pitch" is needed to convince engineers to act on the forecast. ■

8.4.2 How Should Monitoring Be Conducted?

Traditionally, monitoring relied heavily on printed materials. In decades past, technology monitoring relied on composing journals with entries interpreted with respect to prescribed parameters and concerns (see Bright, 1972). That appears too restrictive for ongoing systems that try to track rapidly advancing families of technologies.

Advances in electronic information accessibility represented by the some 5,000 on-line databases (as of 1990), the gateway services (such as BRS and DIALOG), and the wide availability of networked personal computers offer attractive options. Instead of maintaining a monitoring file with journal reports keyed on certain identified parameters, on-line searches on old or new parameters can be generated. As of early 1990, however, searches are still not satisfactory for many purposes. Hard copy systems supplemented by on-line searches to capture additional new items on specific topics appear to be most cost-effective (see Exhibit 8.2). However, as data bases become increasingly "full text" and include more graphic information, electronically based monitoring should emerge as the method of choice. Watch for intelligent information retrieval mechanisms that go beyond Boolean and probabilistic searches (using systems such as SIRE, for instance) to help the user select valuable information (Wiederholt et al., 1989). Indeed, artificial intelligence–supported electronic monitoring could, one day, help the user pose the monitoring questions and come back with suitable answers, not just unprocessed sources.

Today, it may be worthwhile to develop monitoring files on the computer for ease of cross-referencing and/or browsing (that is, using hypermedia systems). Input and maintenance of such a system is more time-consuming than use of hard copy files, but the payoff can be worthwhile gains in effectiveness in some large monitoring efforts with many people involved and many cross-linked interests (such as those suggested by Figure 8.1).

8.4.3   Integrating Monitoring Results with Other Forecasting Techniques

Monitoring provides the basis for most technology or socioeconomic forecasting. Sometimes it provides all that is needed or affordable. Most of the time, however, monitoring provides one component of a forecast. If monitoring is restricted to mean gathering of information from published sources, then it must usually be enriched by expert opinion. For instance, one forecast of intelligent information retrieval (Wiederholt et al., 1989) began with a casual discussion with in-house experts, continued with consolidation of published literature, and then formally requested review and elaboration by outside experts. The process continued with revision and further monitoring over several months before the final report. In such a process, monitoring is tightly integrated with expert opinion methods. Monitoring may initiate forecasting studies that utilize other forecasting techniques as well. For instance, monitoring of developments in personal computers has uncovered salient time series data, suggesting trend extrapolation.

Monitoring also provides a vehicle to keep forecasts alive. Rather than allow a study to be published and become out of date, assign someone to track developments and update the study periodically (this can be done at relatively low cost). Should such monitoring uncover milestone changes, then reexamine the forecast. In the previous paragraph, monitoring was cast as the precursor to a forecast; here it is the follow-on activity.

Because changes in different technologies are interdependent and technological change depends on contextual social influences, it is attractive to consider forecasting such sociotechnical systems rather than focusing on a single technology. Again, monitoring constitutes the main activity of such forecasting efforts. If monitoring turns up a critical change then initiate a more focused study. Such integrated, monitoring-based forecasting systems serve the organization best.

8.5   RECOMMENDED SOURCES

Here are the general categories of published information sources with examples of each:

- Journals: *The Futurist*, *Technological Forecasting and Social Change*, *Futures*, *Science*
- Trade journals: *PC Week*, *Iron Age*
- Newsmagazines: *Time*, *Business Week*

- Newspapers: *New York Times*, *Washington Post*, *The Wall Street Journal* (especially indexes of these, on-line or hard copy)
- Government activity reports:

    Legislative reports—Congressional Research Service studies, Office of Technology Assessment studies, CIS Index (Congressional Committee hearings and reports), Congressional Record, Congressional Quarterly

    Executive reports—weekly compilation of presidential documents, monthly catalog of government reports, agency indexes (such as NASA, DoE)

- Indices: Engineering Index (known as Compendex on-line), Science Citation Index (also Social Science Citation Index), Trade and Industry Index, Computer Index (all these are available on-line), Library of Congress Catalog of Books by Subject
- On-Line Database Gateways: services such as BRS and DIALOG provide access to multiple electronic data bases; college library collections increasingly are cataloged on-line and may provide reduced price (or free) access to outside data bases
- Information Services: International Data Corporation (Framingham, Massachusetts), Bacon's Clipping Bureau (Chicago)

EXERCISES

8.1 Exhibit 8.1 chronicles Whirlpool's successful monitoring of advances in chemicals and textiles to affect home appliances. Consider the following industries:

(A) A major American steel company

(B) A Japanese home entertainment company

(C) A multinational oil company subsidiary in the pesticide business

What domains would you monitor for these industries today and why?

8.2 Monitoring can be configured very differently as a function of its purposes and the situation in which it is occurring. Imagine yourself formulating a monitoring effort in three of the following situations:

- Information technology developments—for the government of a small developing country in Latin America
- Information technology developments—for a major American bank
- New materials developments—for a "Big Three" American automaker
- Genetic engineering developments—for a Congressional Committee
- New drug possibilities—for a multinational pharmaceutical company
- Socioeconomic contextual changes—for a multinational pharmaceutical company

(A) Identify the key type of uncertainty for each of these three and your major information acquisition concerns (refer to Table 8.1).

(B) Following Table 8.2, make each of the pertinent choices and briefly explain your reasoning for these three situations.

8.3 For one of the situations in Exercise 8.2 (or another of your choosing), lay out the "big picture" aspects of a monitoring program in the form of a proposal to top management:

(A) Set out the objectives:
- Identify the key user(s) and the prime use(s)
- Pose the key questions to be answered by the monitoring program

(B) Will this be a single, ad hoc study or a monitoring system? Why?

(C) Describe how the monitoring program will be managed:
- Who will do the monitoring?
- What incentives will you use?
- What resistances do you anticipate?
- How will you evaluate performance?

(D) Propose a schedule for the monitoring effort.

(E) Designate the resources required and justify them.

8.4 (Project-suitable) Select a topic of interest (a technology or the context for one company's business interests) and conduct Level 1 (or cold start) monitoring. Follow Section 8.3.3, using suggested sources of information to fulfill the four objectives. Your brief report should include a technology description, identification of related technologies (sketch relationships as done in Figure 8.1), identification of key players, and setting out possible development pathways. In addition, note what you found to be promising hard copy sources (such as key serials, books) and on-line data bases.

8.5 (Project-suitable) Carry your project forward, conducting a Level 2 (or warm) monitoring program on the topic developed in Exercise 8.4 or on a topic with which you are already familiar. Choose among the following activities:

(A) Conduct a focused on-line search to round out what you know about a particular technology. Classify how pertinent various sources are.

(B) Interview two experts to verify the accuracy of your information. Have 5 to 10 questions ready to help you find out about unpublished sources, forecasts, other experts, who is promoting the technology, and so forth. Summarize what you learn from them.

(C) Create a historical chronology of the development of a particular technology or product line. Use this to suggest driving forces and key uncertainties. Consider a precursor analysis, qualitative or probabilistic (see Exhibit 8.3), if your data warrant. Alternatively, consider preparing a product roadmap, as discussed in Chapter 3. If neither approach appears useful in this case, explain why and offer your own analysis of the chronology.

(D) Map the competing and supporting technologies that affect this technology or product line. Use an impact matrix (Figure 8.2) to identify those in which developments could constitute milestones for your technology or product line.

(E) Map the pertinent contextual factors that affect, or are affected by, your technology or product line. Use an impact matrix to identify potential milestones.

8.6 (Project-suitable) Set up a "two-stomach system" for your technology (see Exhibit 8.2). You will need manila folders and an accessible storage site. If you have a project team, write your understanding of the topics to be covered, a deadline for information gathering, and who will be responsible for analyzing each of the topics. Agree on what those analyses will contain, their format, and their initial due date. If sensible, arrange to continue monitoring and have later topic updates scheduled.

9

BASIC TOOLS FOR QUANTITATIVE TREND EXTRAPOLATION

OVERVIEW

This chapter lays the groundwork for the extrapolation of quantitative historical data into the future. It presents basic considerations in performing a trend analysis and then introduces simple models for short-term forecasting. It also discusses least squares estimation for straight lines and associated procedures that support the models in Chapter 10.

9.1 INTRODUCTION

Extrapolation uses the past to anticipate the future. Trend extrapolation uses mathematical and statistical techniques to extend quantitative time series data into the future.

Trend extrapolation is of use to the technology manager only if the future proves to be like the past, at least in some important respects. Observations of natural phenomena have led to the recognition of scientific laws that describe growth processes. These laws also model technological growth, or innovation, processes (see Chapter 2). Unlike many natural growth processes, technological growth relies on complex, socioeconomic factors. Human beings develop patterns of behavior based upon their experiences, and these patterns allow social scientists to anticipate events in a manner analogous to the predictions of physical scientists. While social relationships are more complex and uncertain than physical ones, the past is the richest source of information about the future. As the saying goes, "I'm sorry to be so stupid about the future, but I've had so little experience with it." We do have vicarious experience with the future, however, to the degree that our past experiences condition our future responses.

A critical assumption underlies trend extrapolation of technological change: *Technical attributes often advance in a relatively orderly and predictable manner.* This is no trivial assumption; exceptions abound. Indeed, other chapters explicitly look for milestones, breakthroughs, or qualitative discontinuities. Figure 9.1 portrays decidedly irregular advances in cement production. Contrast this with the orderly trend shown in other figures in this chapter.

Fortunately, there is a second principle that suggests when one is likely to find technological change orderly and predictable: *A complex mix of influences moderates discontinuities.* This is like the rationale that statisticians use to test observations against a "normal distribution": the cumulation of many small contributions will result in a predictable pattern of deviations about a central value. With respect to technological change, this condition of multiple small influences is best met when progress reflects a series of ongoing engineering improvements that do not require scientific breakthroughs or major inventions. In terms of technological innovation (the commercialization and use of technical inventions), it is best met when adoption depends on market acceptance. That is, extrapolation of sales of a new technology that can be bought by thousands of purchasers will be more predictable than extrapolation that depends on only a few potential buyers.

Sometimes, a single decision determines the fate of a technology. For instance, development of the American supersonic transport airplane (SST) was derailed in the 1970s by the U.S. government's decision to cease major funding (a one-

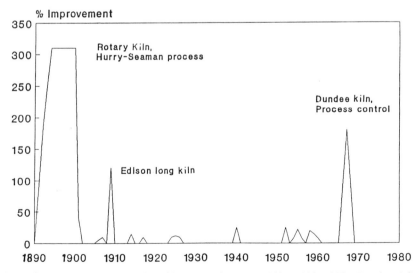

Figure 9.1 Production capacity of largest U.S. cement kiln, 1890–1980. (Reprinted from "Technological Discontinuities and Organizational Environments" by M. L. Tushman and P. Anderson published in *Administrative Science Quarterly*, Vol. 31, no. 3 by permission of *Administrative Science Quarterly.)*

vote margin in the Senate). Such technologies do not lend themselves well to trend extrapolation. (Contrast this example with cable television (CATV) covered later in this chapter. Decisions regarding CATV adoption are made by millions of households.)

Technological breakthroughs can cause fundamental discontinuities in societies. Therefore, extrapolations of technology cannot ignore the impact of technology itself. Nor is technology the only source of discontinuity. Events such as political upheavals, economic recessions, or natural disasters affect the reliability of the past as a guide to forecasting the future. These limitations in extrapolation imply that the forecaster should avoid making single, point value predictions. Any trend should be offered with confidence intervals. Furthermore, explicit sensitivity analysis should consider how a trend would be changed if critical events occurred. Such events could include advances in competing technologies, failure of a supporting technology to develop as expected, alterations in a socioeconomic trend (such as interest rate), or occurrence of a critical event (such as the Senate vote against the SST). Chapter 12 presents cross-impact models that specifically consider such interactions.

Extrapolation methods range from direct to structural techniques (see Table 4.2). The former seek to directly forecast behavior without explicitly considering effects of the broader society; the latter specifically consider cause and effect relationships. Extrapolation methods vary from very simple techniques to highly complicated statistical models that require powerful computers to apply. Sophistication of method, however, does not assure a valid forecast. Validity is seriously limited by the quality of available data.

Changing conditions and data limitations are two reasons why any forecaster should apply multiple forecasting tools. This chapter describes extrapolation tools that can contribute to technology forecasts. Trend extrapolation alone will not suffice. The chapter also describes how to select variables to forecast and presents naive models. Refinement of the basic naive model may consist of more carefully considering the past behavior of the variable, applying knowledge from analogous phenomena, or postulating causal relationships among variables. The key tool used in extrapolating various trends, least squares regression, is also presented. CATV growth provides a continuing subject for examples of trend extrapolation methods throughout the chapter.

9.2   SELECTING THE VARIABLES TO EXTRAPOLATE

In the case of CATV, the fraction of households with cable service provides an obvious variable by which to measure growth. However, in other cases in which the growth of technological capacity is to be forecast, the proper variable(s) is not always so obvious. There are at least three important criteria to apply in selecting an appropriate variable.

First, the variable chosen must measure the level of functionality of the technology. Lenz (1985, p. 14) suggests that "total output of comparable goods or services by each technology exemplifies this requirement." An understanding of the technology and its application are required for choosing the correct variable. It is not always an obvious choice. For example, in Chapter 4 it was noted that while maximum speed is an appropriate measure of fighter aircraft technology, it does not accurately portray the functionality of cargo or passenger aircraft. A measure such as seat-miles per hour or ton-miles per hour helps determine capacity, which is an important measure as airlines serve high-density markets.

Second, the variable that is chosen must be applicable to both the new technology and to any older technology it replaces. The variable lumens per watt, for example, can be applied to lighting systems ranging from the candle to the latest electrical system, because energy consumption for all lighting systems can be expressed in terms of wattage.

Third, data must be available from which to compute historical values of the variable. This last criterion often is the most limiting. If the ideal indicator is not available or is less complete than alternative measures, a compromise may be necessary. For instance, in devising indicators of high-technology development among nations, Roessner and Porter (1990) substituted technology sales for production measures. Sometimes, available data are so remote that expert opinion or other methods should be used in lieu of extrapolating inappropriate variables.

9.3 NAIVE MODELS

This section describes the basic extrapolation models. These are applicable to short-term forecasting; as one projects further ahead, they lose credibility. These naive models do not explicitly account for the effects of the surrounding environment on technology. A series of models are developed progressively to show how the use of historical data improves the forecast. Note that the simplest models are presented only to show this progression; they are not serious extrapolation models.

In the simplest naive extrapolation, it is assumed that tomorrow, next week, or next year will be exactly like today—that is, no change is expected. Mathematically,

$$X_{t+1} = X_t \tag{9.1}$$

which implies that the value of X at time $(t + 1)$ is no different than the value at time t. Such a model is applied implicitly when current sales are said to be an estimate of future sales.

The forecast represented by Equation 9.1 can be seriously misleading—too conservative in a growing industry or wildly optimistic in an era of decline. For example, consider a firm developing a device for sale to CATV suppliers. This firm might use this model to forecast potential market size as the current number of CATV subscribers. However, data in Table 9.1 show that the number of subscribers has grown each year. Thus the potential for the product is greater than forecast. On the other hand, if a new satellite dish technology is developed that is preferable to cable, the market could evaporate quickly.

TABLE 9.1   U.S. Cable TV Subscribers (In Thousands)

Year	Subscribers
1975	9,800
1976	10,800
1977	11,900
1978	13,000
1979	14,100
1980	16,000
1981	18,300
1982	21,000
1983	25,000
1984	30,000
1985	31,300
1986	37,500
1987	39,700

Source: Statistical Abstract of the United States, 1989, p. 547.

Underestimating or overestimating the potential market can be detrimental. Developing a product for a dying industry obviously courts disaster. However, being too conservative about the potential also can be dangerous. Many fledgling companies, although first to market a successful technology, cannot exploit their position because they have inadequate capital to meet growing demand. In such cases, the firm may lose the market to better-prepared competitors or may be acquired by others with the resources to exploit market position. Thus even those companies seeking a conservative posture should use all available information in the forecast.

More sophisticated extrapolation can be illustrated using data for the growth of CATV subscribers presented in Table 9.1. First, consider the prediction of market size that would be made by a forecaster using Equation 9.1, the simplest naive model. If the forecast was made only once in 1975, market size in 1976 and thereafter would be 9,800,000 subscribers. As the data in Table 9.1 shows, this forecast would be in error by an amount that grows with time.

Figure 9.2 compares actual CATV growth with that predicted by a forecaster who used Equation 9.1 to reforecast the market *at the end of each year.* This is far superior to using Equation 9.1 to project many years ahead. Notice, however, that prediction of the same number of subscribers for 1987 through 1990 looks very strange in light of the historical data available to the forecaster. It also is clear that the model underpredicts market size in every year for which there is data. This analysis suggests that historical information that accumulates during growth can be used to improve the extrapolation provided by the simplest naive model.

Figures 9.3 and 9.4 show the results of somewhat more sophisticated naive predictions that consider historical data. The forecast used to develop the first figure predicts that incremental growth during the coming year will be identical to that in the previous year. In mathematical terms this *constant growth model* is shown as:

$$X_{t+1} = X_t + (X_t - X_{t-1}) \tag{9.2}$$

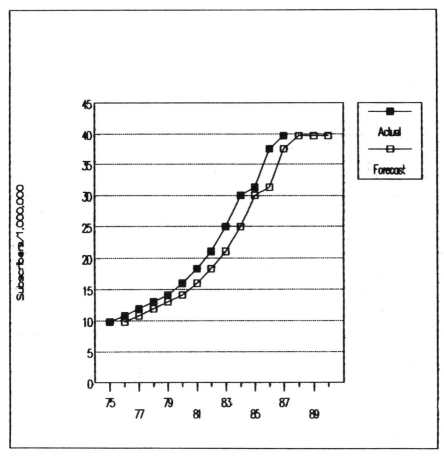

Figure 9.2 CATV subscribers in the United States: actual growth data and "No Growth" forecast.

Applying this approach at the end of 1977, a forecaster would have estimated the number of subscribers in 1978 to be

$$\frac{X_{1978}}{1,000} = 11,900 + (11,900 - 10,800) = 13,000$$

This would have been correct. Although Figure 9.3 shows that forecasts for other years are not quite as accurate, using historical information clearly improves the forecast.

Another variation on the use of historical data is provided by assuming that the fractional growth will be constant from one year to the next. That is

$$X_{t+1} = X_t(X_t/X_{t-1}) \tag{9.3}$$

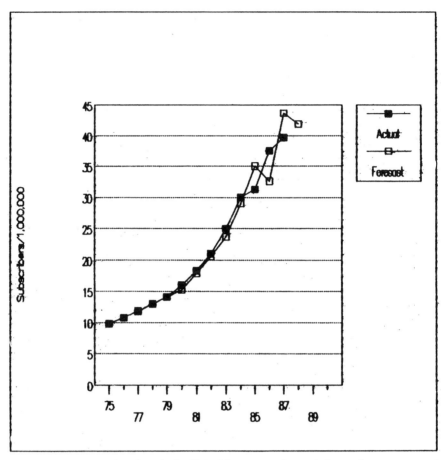

Figure 9.3   CATV subscribers in the United States: actual growth data and constant annual growth forecast.

Using this approach in 1977, a forecaster would have predicted the number of subscribers in 1978 to be 13,112 (in thousands). Figure 9.4 compares the forecast (using Equation 9.3) with the actual CATV growth. Note that the forecasts are very similar to those produced by Equation 9.2. However, as extrapolations are continued further into the future, the two forecasts will diverge due to compound growth effects.

Other models can be created as well (for example, either the increment or the fraction of growth could be averaged over the two preceding years). The important fact to be grasped from this discussion is that forecasts can be improved by using historical information about the process being forecast.

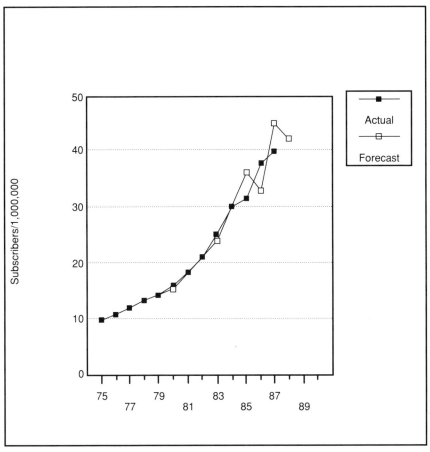

Figure 9.4   CATV subscribers in the United States: actual growth data and constant fractional growth forecast.

9.4 ADAPTIVE WEIGHTING

The techniques described in the previous section rely only on the most recent data points. As a result, valuable information that may be embodied by earlier data is lost or incompletely employed. This limitation is overcome by a class of models that forecasts future values using a weighted sum of past data. The general mathematical expression for such models is

$$X_{t+1} = \sum_{i=0}^{n} w_{t-i} X_{t-i} \tag{9.4}$$

where X_t is the value of the variable at time t and w_{t-i} is the weight that will be given to the value of the variable at time $(t - i)$.

In this formulation, the sum of the weights must be unity. This model is said to "smooth" the trend because it removes periodic fluctuations that may appear in year-to-year values. For example, one might use a three-period moving average by making

$$w_t = w_{t-1} = w_{t-2} = \frac{1}{3}$$

with $w_{t-i} = 0$ for all $i > 2$. Using this approach to project the number of CATV subscribers in 1978 gives

$$\frac{X_{1978}}{1,000} = \frac{1}{3}(11,900) + \frac{1}{3}(10,800) + \frac{1}{3}(9,800) = 10,833$$

The technique illustrated gives equal weight to information from the last three observations prior to the forecast.

Alternative approaches can be used to give the heaviest weight to the most recent data point and declining weights to earlier points. For example, *exponential smoothing* sets

$$w_{t-i} = \frac{a(1-a)^i}{d}$$

where $0 < a < 1$, and d is the sum of the numerators for all the w_i terms, as shown in Exhibit 9.1.[1]

Approaches such as exponential smoothing assume that the most recent data embody more useful information than earlier data. There are many other uses of the basic adaptive weighting approach. For example, if observations were on a quarterly rather than a yearly basis and showed seasonal fluctuations, then taking nonzero values for w only in periods $t-4$, $t-8$, $t-12$, and so on would produce a type of seasonal adjustment.

Adaptive weighting does not make use of the trend information per se. In this respect it is more like the constant model than the constant growth or constant fractional growth models. Such techniques are best used to smooth "lumpy" time series (such as daily stock prices) so that prevailing trends can be identified. They are less useful as a forecasting tool per se.

There are many other methods for improving the accuracy of naive models; the models, however, are still naive in the sense that they only use the past behavior of a variable to determine its future path.[2] The forecaster can, for example, use a

---

[1] For a more extensive discussion of exponential smoothing, see Chisholm and Whitaker (1971, pp. 22–26) or Granger (1980, pp. 88–90).

[2] Very good discussions of such methods can be found in textbooks for forecasting such as Chisholm and Whitaker (1971) and Wheelwright and Makridakis (1980).

■ **Exhibit 9.1** **Exponential Smoothing**

Forecast an exponentially smoothed value for the number of CATV subscribers in 1978 using $a = 0.5$ in a three-period model.

$$w_{t-i} = \frac{a(1 - a)^i}{d}$$

$$w_{t-1} = \frac{0.5(1 - 0.5)^1}{d} = \frac{0.25}{d}$$

$$w_{t-2} = \frac{0.5(1 - 0.5)^2}{d} = \frac{0.125}{d}$$

$$w_{t-3} = \frac{0.5(1 - 0.5)^3}{d} = \frac{0.0625}{d}$$

$$d = 0.25 + 0.125 + 0.0625 = 0.4375$$

This yields weights of $w_{t-1} = 0.571$, $w_{t-2} = 0.286$, and $w_{t-3} = 0.143$. The estimate of X for 1978 is

$$\frac{X_{1978}}{1,000} = 0.571(11,900) + 0.286(10,800) + 0.143(9,800) = 11,285$$

The value of a can be selected on the basis of the forecaster's judgment or as the value that works best when applied to past observations. Values between 0.2 and 0.4 are generally in order.   ■

---

computer to determine the weights to apply to new observations to minimize forecasting errors (see Wheelwright and Makridakis, 1980, pp. 85–100). However, the general availability of computer facilities also makes it attractive to use statistical forecasting techniques, which may be superior to techniques that rely solely on weighted sums of past observations. A discussion of linear regression analysis and its use in forecasting lays the groundwork for understanding such statistical techniques.

9.5 LINEAR REGRESSION

Linear regression provides the essential tool to determine equations for straight-line relationships. Such equations can be used to extrapolate into the future. They can also be used to fit nonlinear relationships if those relationships can be mathematically transformed into forms that are linear.

9.5.1   Regression Model Validity

The previous sections considered techniques for using past observations to project future values. These techniques were called naive because they did not use information about interactions between the forecast variable and other variables that affect it. At times, however, there is information about such relationships that can add to the forecaster's knowledge of the process—knowledge that can be used to improve forecast quality (Note the discussion in Exhibit 9.2).

For example, to forecast the development of temperature-resistant materials (such as those used in the space shuttle heat shield), it might be expected that the rates of development and diffusion are closely related to federal expenditures on the shuttle program. These expenditures fund research to improve materials characteristics, and to advance production and manufacturing processes. As a result, better materials will be developed, produced, and eventually applied to products manufactured and sold in the private sector. Thus the forecaster might hypothesize that knowledge of projected federal funding could be used to improve predictions about the development of temperature-resistant materials and about their subsequent diffusion in industry.

Knowledge of past causal relationships often can be coupled with independent projections of related variables and used with regression analysis to develop a structural model that will provide an excellent forecast. For example, past data on space expenditures and on the production of heat-resistant materials might be used to formulate a functional relationship linking these variables to material use. Then projections of expenditures on space programs prepared by the Office of Management and Budget could be used with the model to forecast material use.

As a starting point, consider an example of simple regression analysis in which the forecast variable, Y, is thought to be linearly related to one independent variable, X. The model of this causal relationship would be expressed as

$$Y = a + bX + e \qquad (9.5)$$

where a and b are the intercept and slope of the function and e is the error between the value of Y predicted by the causal relationship $(a + bX)$ and the observed value of Y. This error always exists because the causal relationship does not completely determine the value of Y. For example, Y may be slightly affected by variables W and Z, as well, or the causal relationship may not be quite linear. The goal of regression is to determine values for a and b that minimize the error.

9.5.2   Fitting the Regression Equation

Regression is the name given to a technique for estimating the values of the constants a and b in the linear model given by Equation 9.5. The least squares criterion used to estimate these constants is that they should be chosen to minimize the sum of the squares of the errors between the observed value of Y and the value predicted by the linear model $(a' + b'X)$. The primes indicate that these values are estimated

■ **Exhibit 9.2 Causality and Regression**

"Correlation does not prove causation." This well-worn phrase deserves thoughtful consideration by forecasters. Some correlations may be spurious—the stock market rises in years when the National Football Conference wins the Super Bowl (we cannot conceive a plausible causal mechanism, but who knows!). Other correlations leave one uncertain as to the direction of causal influence. For instance, imagine that government units that spend more on social services have greater unemployment rates. Does the unemployment induce social services expenditures, or do social services decrease people's incentive to find work? In other instances, correlation between two variables may reflect the influence of a third. During the 1980s, U.S. expenditures for the Army increased, as did those for the Navy. Rather than one driving the other, it is far more likely that both reflect a political decision to enhance military preparedness.

Regression among two or more variables can begin from various standpoints. In the strongest case, a solid theoretical basis is reflected in the regression equation. In the weakest, a data base is explored in order to discover correlation among certain variables. In this case, it should not be assumed that causation has been "proven" in any sense.

Four levels of regression modeling should be distinguished (Porter et al., 1981):

1. *Descriptive modeling*—Examination of a data set uncovers correlations; no basis for prediction.

2. *Simple prediction*—Even though the underlying causal processes, are not understood, continuity (stationary nature of behaviors and structures) is assumed to make usable predictions (e.g., stock market models based on previous data patterns rather than structural models).

3. *Causal models*—The underlying causal process generating the present data is known and correctly specified; however, the analysts do not know how the causal system will respond to manipulations (for instance, an ecosystem in the valley may be well described, but construction of a new highway would change it so drastically that no one can predict the results).

4. *Causal predictive models*—The underlying causal process is correctly specified and the results of deliberate interventions can be predicted. This requires knowledge that a change in the manipulable variable will have a sufficiently strong effect in the intended direction (significant regression coefficient), that this effect will not be swamped by other effects, and that the causal system will remain stable while the variable is manipulated. For example, assume production capability for a new technology-intensive product is increased. To predict sales, the system must not be perturbed by competitors increasing their capacity or developing newer products and the market's response to increased supply must be understood.

The forecaster must assess the causal strength of any regression model before using it to forecast. ■

by applying a specific criterion; usually they are not the true values. In practical terms, squaring the deviations between the estimated regression line and the data points places great weight on points far from the line. This makes it important to examine the data carefully before fitting a regression line (probably by using a computer statistical package). *Plot the data first.* Such a plot should be inspected to identify possible "outlier" data points and to decide whether they should be left as is, corrected, or excluded. Such decisions rest on knowledge of the underlying relationships and familiarity with how the data were measured. In addition, it should be determined whether a nonlinear relationship exists. For example, visual inspection of the CATV data (Figure 9.2) should warn that a linear fit will not work well.

The example of CATV growth can be used to illustrate simple regression. The simplest case would be to derive a linear relationship between CATV sales and time. This relationship can be fitted using two-variable linear regression:

$$y_i = a + bX_i + e$$

where i refers to each of the individual data points. To predict Y from X, use the following four-step calculation (in practice one is likely to use a statistical package—such as SPSS™ from SPSS Inc. in Chicago—or a programmable calculator).

First, determine the values of a and b^* that best fit the available data. These values are computed from the following equations:

$$b = \frac{\sum_{i=1}^{n} x_i y_i}{\sum_{i=1}^{n} x_i^2}$$

$$a = \overline{Y} - b\overline{X}$$

where

$$x_i = X_i - \overline{X}$$

---

*This b is the regression coefficient of Y on X. A different line, with a different slope, typically results if one were to regress X on Y—that is, to predict X values from the observed Y values. The formula for b in that equation would include the sum of the squared Y deviation scores in the denominator instead of the squared X deviation scores.

The correlation coefficient, r (the Pearson product-moment correlation), can be calculated as

$$r = \frac{\sum_{i=1}^{n} x_i y_i}{\sqrt{\sum_{i=1}^{n} x_i^2 y_i^2}}$$

or

$$r = b(s_x)/s_y$$

Conceptually, imagine a scatterplot of dots in two dimensions that represents the X, Y values. The slope of the best-fitting least squares line is given by b, and r indicates how closely the points fall to the line.

The small x (or small y) values, which are called deviation scores, are the difference between the observed X (or Y) values and the mean of the observed sample. Their use simplifies the notation and the calculation.

$$\overline{X} = \sum_{i=1}^{n} \frac{X_i}{n}$$

where $\overline{X}$ is the mean and n is the number of observations in the data sample.

$$s_y^2 = \frac{\sum_{i=1}^{n} y_i^2}{(n-1)}$$

where s_y^2 is the variance in Y, or the square of the standard deviation in Y.

$$s_e^2 = \frac{\sum_{i=1}^{n}(Y_i - \hat{Y}_i)^2}{(n-2)}$$

where s_e^2 is the standard error and $\hat{Y}$ represents values of Y predicted by using the equation

$$\hat{Y} = a + bX$$

In this sample calculation, X = time and Y = CATV subscribers (from Table 9.2). It is assumed that only the first four observations are available to calculate the least squares regression line (to simplify the numerical manipulations, let X = years − 1900):

X	x	x^2	Y	y	xy	$(Y_i - \hat{Y}_i)$	$(Y_i - \hat{Y}_i)^2$
75	−1.5	2.25	9,800	−1,575	2,362.5	30	900
76	−0.5	.25	10,800	−575	287.5	−40	1,600
77	0.5	.25	11,900	525	262.5	−10	100
78	1.5	2.25	13,000	1,625	2,437.5	20	400

$$\overline{X} = \frac{(75 + 76 + 77 + 78)}{4} = 76.5$$

$$\sum_{i=1}^{n} x_i^2 = 5.0$$

$$\overline{Y} = \frac{(9,800 + 10,800 + 11,900 + 13,000)}{4} = 11,375$$

$$\sum_{i=1}^{n} x_i y_i = 5,350$$

$$b = \frac{\sum_{i=1}^{n} x_i y_i}{\sum_{i=1}^{n} x_i^2} = \frac{5,350}{5.0} = 1,070$$

$$a = \overline{Y} - b\overline{X} = 11,375 - [1,070(76.5)] = -70,480$$

The prediction (regression) equation is thus:

$$\hat{Y}_i = -70,480 + [1,070(X_i)]$$

Next examine the residuals (differences between predicted and actual values of Y; that is, $Y_i - \hat{Y}_i$). Note that the small residuals do not guarantee a good fit for the extrapolation (see Figure 9.5). The sum of the squared residuals $(Y_i - \hat{Y}_i)^2$ from the preceding sample calculation equals 3,000.

Next extrapolate values of Y or X using the best fit equation.

a. In what year will CATV subscriptions reach a certain level—say, 30,000 (in thousands)? To answer, solve the prediction equation by substituting 30,000 for Y:

$$30,000 = -70,480 + [1,070(X)]$$

$$X = 93.9 \text{ or the year } 1993.9$$

This year, 1993.9, is far later than the actual year, 1984, in which CATV subscribers reached 30,000 (in thousands). This reflects the poor fit between this line and the true nonlinear growth pattern (see Figure 9.5).

b. What is the predicted number of CATV subscribers in a certain year—say, 1987? To answer, solve the prediction equation by substituting 87 (for the year 1987) for X:

$$\hat{Y} = -70,480 + [1,070(87)]$$

$$\hat{Y} = 22,610$$

Again, note that the predicted value falls far short of the actual. The farther out the extrapolation, the worse the prediction becomes with this misspecified model.

Finally, compute confidence intervals (C.I.) for the predicted Y. To illustrate, use the $\hat{Y}$ value for 1987 calculated in the previous step:

$$\text{C.I.} = \pm t_{\alpha/2, \text{d.f.} = n-2} \sqrt{s_e^2 \left(1 + \frac{1}{n} + \frac{(X_* - \overline{X})^2}{\sum x_i^2}\right)}$$

where t is a statistical value from a t-table and X_* is the value of X at the target extrapolation year.

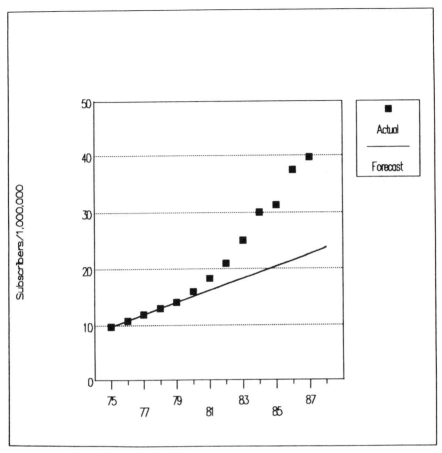

Figure 9.5 CATV subscribers in the United States: actual growth data and linear growth projection based on the first four data points.

Note the three terms, multiplied by s_e^2, within the square root. These reflect three sources of error in the extrapolated prediction: (1) error about the regression line, (2) error about the mean, and (3) error in the extended slope.

For example, assuming a 50 percent C.I.:

$$\alpha = 1 - \text{C.I.} = 0.50$$

$$\frac{\alpha}{2} = 0.25$$

$$\text{Degrees of freedom (d.f.)} = N - 2 = 2$$

Look up the corresponding t value in Appendix B.2: $t_{0.25,2} = 0.816$ (t values derive from a statistical distribution discussed in detail in most statistics books.)

$$s_e^2 = \sum_{i=1}^{n} \frac{\left(Y_i - \hat{Y}_i\right)^2}{(n-2)} = \frac{3,000}{2} = 1,500$$

$$\frac{(X_* - \hat{X})^2}{\sum x^2} = \frac{(10.5)^2}{5.0} = 22.05$$

$$\text{C.I.} = \pm 0.816 \sqrt{1,500(1 + \frac{1}{4} + 22.05)}$$

$$= \pm 0.816 \sqrt{34,950}$$

$$= \pm 153$$

So our predicted number of CATV subscribers for 1987 is

$$\hat{Y} = 22,610 \pm 153$$

$$= 22,457 \text{ to } 22,763$$

Given the limited precision of the actual data and the long-term nature of the forecast, this should be rounded off to fewer significant figures, such as

$$\hat{Y} = 22,600 \pm 150$$

$$= 22,450 \text{ to } 22,750$$

Again, note that this precise model is badly misspecified; it does not include the actual 1987 Y value (39,700) within this C.I.

The lesson to be learned from this example is that simple linear relationships between the forecast variable and time are unlikely to do a very good job as predictors of technology growth. However, regression can be a powerful tool, as later sections show.

In some cases, it may be necessary to approximate a linear regression equation for a large data set (say, in the absence of a handy computer regression program). Exhibit 9.3 illustrates a method for calculating such approximations.

■ Exhibit 9.3   Three-Point Method for Approximating a Linear Regression

Granger (1980) presents a weighted, three-point method for fitting a linear function to time series data (Y_i). This method may be easier to apply quickly for some users than the standard least squares fit. The technique assumes an odd number of data points, n. If the number is even, the forecaster should discard the earliest data point. Granger refers to the case when $n > 10$ as the "ordinary" series; when $n < 10$ it is referred to as the "short" series. Three averages are used: the weighted averages of the initial, middle, and final terms. For an ordinary series, these averages are defined as follows:

$$R = \frac{(Y_1 + 2Y_2 + 3Y_3 + 4Y_4 + 5Y_5)}{15}$$

$$S = \frac{(Y_{d-2} + 2Y_{d-1} + 3Y_d + 4Y_{d+1} + 5Y_{d+2})}{15}$$

$$T = \frac{(Y_{n-4} + 2Y_{n-3} + 3Y_{n-2} + 4Y_{n-1} + 5Y_n)}{15}$$

For a short series they are defined as follows:

$$R = \frac{(Y_1 + 2Y_2 + 3Y_3)}{6}$$

$$S = \frac{(2Y_{d-1} + 3Y_d + 4Y_{d+1})}{6}$$

$$T = \frac{(3Y_{n-2} + 4Y_{n-1} + 5Y_n)}{6}$$

For each series, the central point, d, is given by

$$d = \frac{(n + 1)}{2}$$

For either series, the curve to be fitted is

$$Y(t) = a + bt$$

For example, for the Fisher-Pry model, $Y(t) = \ln[(1 - f)/f]$ where f is the fraction of households with CATV, $a = \ln(c)$ and $b = -b$ (see Equation 10.4). For an ordinary series, the equations for a and b are as follows:

$$a = R - \frac{11b}{3}$$

$$b = \frac{T - R}{n - 5}$$

For a short series, the equations are:

$$a = R - \frac{7b}{3}$$

$$b = \frac{T - R}{n - 3}$$

The first 31 CATV data points presented in Table 10.1 (see p. 180) will be used as an example. For 31 data points, $n = 31$ and $d = 16$.

For the first five points, the following data apply:

YEAR	FRACTION WITH	$Y(t)$
1955	.005	5.293
1956	.009	4.701
1957	.009	4.701
1958	.011	4.499
1959	.013	4.330

For the middle five points, the data are as follows:

YEAR	FRACTION WITH	$Y(t)$
1968	.044	3.079
1969	.061	2.734
1970	.077	2.483
1971	.088	2.338
1972	.097	2.231

For the final five points, the data are as follows:

YEAR	FRACTION WITH	$Y(t)$
1983	.372	0.524
1984	.412	0.356
1985	.446	0.217
1986	.468	0.128
1987	.487	0.052

Thus,

$$R = \frac{[5.293 + 2(4.701) + 3(4.701) + 4(4.499) + 5(4.330)]}{15} = 4.563$$

For this same example, $S = 2.434$ and $T = 0.177$. Therefore,

$$b = \frac{0.177 - 4.563}{31 - 5} = -0.169$$

and

$$a = 4.563 - \frac{11(-0.169)}{3} = 5.179$$

The equation for the fit is

$$Y(t) = \ln[(1 - f)/f] = a + bt = 5.179 - 0.169t$$

where $\ln(c) = 5.179$ ($c = 177.5$) and $b = 0.169$. ∎

9.5.3 Using Causal Relationships

A forecaster projecting CATV growth might hypothesize that as real personal incomes increase, consumers would be more willing to pay for cable TV. Therefore, there could be a causal relationship between disposable income and CATV growth. Perhaps, one could use disposable income to predict subsequent CATV subscriber growth. This example, chosen only to exemplify the calculation, assumes that the relationship between the two variables is linear. In practice, this assumption must be carefully evaluated. It is certainly plausible for CATV subscribers to relate linearly with one variable (disposable income) and nonlinearly with another (time). It should also be noted that time can be treated as a causal variable and used in forecasts with the techniques described here. The earlier discussion of regression on time merely sought a linear trend; this section extends the search to more sophisticated relationships among variables.

To check this hypothesis, the forecaster tabulates the number of CATV subscribers versus real disposable income. The results suggest that a linear causal relationship between these two variables, rather than between CATV subscribers and time, might form a workable model for CATV growth.

TABLE 9.2 Comparisons of Actual and Predicted Data Regarding CATV Subscribers

	Actual[a]		Predicted[b] Y_i	Residuals[b] $Y_i - \hat{Y}_i$
Year	Income	Subscribers		
1975	875	9,800	5,979	3,821
1976	904	10,800	8,357	2,443
1977	940	11,900	11,309	591
1978	983	13,000	14,835	−1,835
1979	1,012	14,100	17,213	−3,113
1980	1,027	16,000	18,443	−2,443
1981	1,042	18,300	19,673	−1,373
1982	1,054	21,000	20,657	343
1983	1,088	25,000	23,445	1,555
1984	1,156	30,000	29,021	979
1985	1,190	31,300	31,800	−500
1986	1,238	37,500	35,749	1,751
1987	1,255	39,700	37,169	2,531

Source: Statistical Abstract of the United States, 1989, p. 547.

[a] Income in billions of 1972 dollars; subscribers in thousands.

[b] In thousands.

Using the linear regression approach described earlier for the full 1975–1987 data set,[3] the linear model gives

$$\hat{Y}_i = -65,771 + 82X_i$$

where $\hat{Y}_i$ is the predicted number of CATV subscribers (in thousands) in a given year and X_i is the real disposable income in billions of 1972 dollars in the same year. Comparisons of predicted and observed values are given in Table 9.2 and in Figure 9.6.

Once the forecaster is satisfied that regression has produced a useful model (see Section 9.5.4), the next step is to use that model to forecast. If the forecaster knows or can estimate a specific value of X, call it X_*, then a corresponding value for Y can be produced by the model. Let us estimate real disposable personal income in 1990 as $1,360 billion (in 1972 dollars). This estimate might be obtained from long-term economic growth forecasts made by the government, or it may be estimated by the forecaster. The forecasted number of cable subscribers (in thousands) in that year would be

$$\hat{Y} = -65,771 + 82(1,360) = 45,749$$

---

[3] The precise values of a and b will vary according to how you round off the calculations. For instance, the TOOLKIT calculates a as −66,137 and b as 82.69 for these data. The extrapolated values and statistical significance levels are very much the same from these estimates.

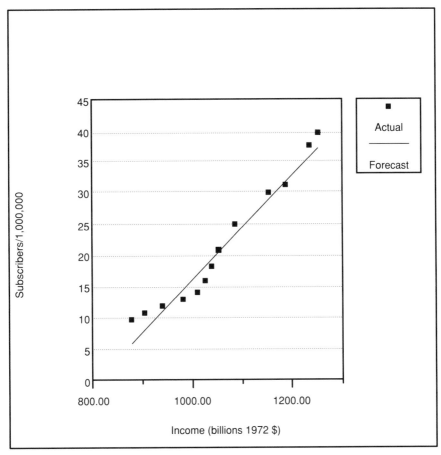

Figure 9.6   CATV subscribers in the United States vs. U.S. disposable income: actual growth data and linear regression of subscribers on income.

The statistical model makes it possible to estimate intervals, rather than simply to make point predictions. Thus it is possible to generate upper and lower bounds for the forecast that have a specific probability of trapping the true value of Y between them—that is, it is possible to determine a *confidence interval (C.I.)*. In the cable TV example, a 90 percent C.I. might be generated for the number of subscribers when real disposable personal income is $1,360 billion. A 90 percent C.I. would be calculated as follows:

$$\text{C.I.} = \pm(t_{\alpha/2, \text{d.f.} = n-2})\sqrt{s_e^2\left[1 + \frac{1}{n} + \frac{(X_* - \overline{X})^2}{\Sigma_{i=1}^{n} x_i^2}\right]}$$

$$= \pm(1.796)\sqrt{5.0 \times 10^6[1 + \frac{1}{13} + (1,360 - 1,058.8)^2/179,232.2]}$$

$$= \pm(1.796)\sqrt{5.0 \times 10^6(1.583)}$$

$$= \pm 5,053$$

or, rounding off:

$$\hat{Y}_* = 45{,}750 \pm 5{,}050 = 40{,}700 \text{ to } 50{,}800$$

(t is taken from Appendix B.2, and all other values were computed from the data in Table 9.2.) Confidence limits can also be used as the basis for best case/worst case scenario construction.

9.5.4   Evaluating the Regression Model

Section 9.5.3 developed a structural model for the dependence of CATV growth on real disposable income. Before applying it, however, investigate the validity of the model by asking two questions: (1) Is the model actually effective in explaining CATV growth? and (2) Can the hypothesis that the true values of a or b are zero be rejected? These questions are answered by applying statistical tests. An F-test is applied to answer the first question, a t-test helps answer the second.[4]

It is important that the model actually be effective in explaining something about the dependent variable, Y. This is a necessary, but not sufficient, condition to fore-cast future values of Y (recall Exhibit 9.2). One way to evaluate the effectiveness of a model is to examine how well it explains the data being used. This can be done by means of the coefficient of determination, R^2. This coefficient indicates the fraction of the total variance in the dependent variable that is explained by the model. The coefficient of determination is defined as

$$R^2 = \frac{\sum_{i=1}^{n}(\hat{Y}_i - \overline{Y})^2}{\sum_{i=1}^{n}(Y_i - \overline{Y})^2} \tag{9.6}$$

The value of R^2 is provided by the output of most statistical packages. R^2 for the CATV and disposable income example is 0.957. This implies that nearly 96 percent of the variance in the number of cable subscribers is explained by the income vari-able. Therefore, the model coupled with a good forecast of real disposable income should produce a good forecast of CATV subscribers, as long as the relationship and the pertinent contextual factors remain essentially unchanged.

While R^2 provides an indication of model effectiveness, there are statistical tests that can be used to establish the effectiveness of the model more rigorously. To use these tests, the forecaster must compute the values of the sample variance, s_y^2 and

---

[4]In this simple linear regression, the model involves a single, independent variable—in this case, disposable income. The two questions essentially are one. For example, if the hypothesis that the true value of b', the coefficient of the independent variable, is zero cannot be rejected, then one cannot conclude that CATV growth depends on real disposable income. Thus the model will fail both the t-test and the F-test. However, it is worthwhile to illustrate how these tests are applied in this simplest case. The results can then be extended to multiple regression, which is introduced in Section 9.7.

its square root, the sample standard deviation, s_y. The sample variance and sample standard deviation are used in the F-test and t-test described in Exhibit 9.4.

A statistical test used to determine the adequacy of model fit is the F-test. This test is the equivalent of testing whether all coefficients of the regression equation could be zero. Two steps are required to apply the F-test. First, determine a critical value for the F statistic. This value depends on the degrees of freedom of the data set used to develop the model and on the confidence level to be placed on the question of model fit. Next, determine the value of the F statistic for the model. These steps are illustrated for the CATV model in Exhibit 9.4.

■ **Exhibit 9.4 F-Test of the Regression Model**

Critical F values are determined from standard tables such as those included as Appendix B.1. For this example, a portion of such a table is presented in Table 9.3. To use this table, determine the degrees of freedom (d.f.) of the numerator and denominator of the F statistic, defined as

$$ F = \frac{\sum_{i=1}^{n} \frac{(\hat{Y}_i - \bar{Y})^2}{k}}{s_e^2} $$

The d.f. of the numerator are the number of independent variables (k)—in this example, the number is one. The d.f. of the denominator equal the number of data points used to construct it (n), less the number of independent variables in the model, minus 1—in this example, the d.f. of the denominator (k) equal $13 - 1 - 1$ or 11. Using 1 as the column entry and 11 as the row entry into Table 9.3, there are several critical values depending on the probability selected. For example, if a

TABLE 9.3   Portion of a Standard F Distribution Table[a]

d.f. of Denom- inator	Prob- ability	d.f. of Numerator					
		1	2	3	4	5	...
10	0.25	1.49	1.60	1.60	1.59	1.59	...
	0.10	3.29	2.92	2.73	2.61	2.52	...
	0.05	4.96	4.10	3.71	3.48	3.33	...
	0.01	10.00	7.56	6.55	5.99	5.64	...
11	0.25	1.47	1.58	1.58	1.57	1.56	...
	0.10	3.23	2.86	2.66	2.54	2.45	...
	0.05	4.84	3.98	3.59	3.36	3.20	...
	0.01	9.65	7.21	6.22	5.67	5.32	...
12	0.25	1.46	1.56	1.56	1.55	1.54	...
	0.10	3.18	2.81	2.61	2.48	2.39	...
	0.05	4.75	3.89	3.49	3.26	3.11	...
	0.01	9.33	6.93	5.95	5.41	5.06	...

[a] The full F-table appears in Appendix B.1.

probability of 5 percent (that is, 0.05) is selected, the critical value of F is 4.84. This implies that in constructing many different models with a computed F value of 4.84, the forecaster would err in inferring that a real relationship exists between the dependent and independent variables 5 percent of the time. If the value of F were 9.65, the inference would be incorrect 1 percent of the time.

The next step in applying the F-test is to compute the F statistic for the model. The statistical information necessary to compute F is available from information developed to calculate a' and b'. For the model of disposable income with CATV subscribers, the mean number of calculated subscribers (in thousands) is 21,050 (the sum of column 3 in Table 9.2 divided by 13 years). Computing the numerator using the predicted values of Y in column 4 of Table 9.2, the mean value of Y, and the sample variance gives

$$F = \frac{1.206 \times 10^9/1}{5.0 \times 10^6} = 241$$

Since this value is far greater than the critical value for a 1 percent probability, reject the hypothesis that there is no relationship between CATV growth and real disposable income with much less than a 1 percent probability of error. ∎

The t-test is performed to assess the probability with which the forecaster can reject the hypothesis that the coefficient b' is different from zero. This test is performed in much the same way as the F-test. That is, a critical value of t is determined from standard tables and then it is compared to the value of t computed for the data used in determining b'. A portion of a standard t-distribution table is shown in Table 9.4 (a more complete table is provided in Appendix B.2.) As before, the row entry is the number of degrees of freedom of the data computed as $(n - 1 - k)$ where n is the number of data points (13) and k is the number of independent variables in the model (1). From Table 9.4, the critical value of t for a 5 percent probability is 1.796, or 2.718 for a 1 percent probability. As before, these probabilities indicate the probability that rejecting the hypothesis that b' is different from zero will be incorrect (see Exhibit 9.5).

■ Exhibit 9.5 *t*-Test of the Regression Coefficients

As with the F-test, once the appropriate significance level has been set, one is ready to compute the value of the t statistic for the regression coefficient. This statistic is defined as

$$t = \frac{b'}{\sqrt{s_e^2/\sum_{i=1}^n (x_i)^2}}$$

Recall that x is the deviation from the mean. The mean of X from column 2 of Table 9.2 is $1,058.8 billion. Squaring the difference between each value in column 2

TABLE 9.4 Portion of a *t*-Distribution Table[a]

	Probability					
	0.25^b		0.05^b	0.025^b	0.01^b	
d.f.	0.50^c	. . .	0.10^c	0.05^c	0.02^c	. . .
. . .						
. . .						
10	0.700	. . .	1.812	2.228	2.764	
11	0.697	. . .	1.796	2.201	2.718	
12	0.695	. . .	1.782	2.179	2.681	
. . .						

[a]The full *t*-distribution table appears in Appendix B.2.

[b]Probability for one-tailed tests $= \alpha/2$.

[c]Probability for two-tailed tests $= \alpha$.

Note: The inferential test alluded to compares a null hypothesis with an alternative hypothesis. When that alternative hypothesis only asserts that the parameter in question is different from the one hypothesized, it is nondirectional, or two-tailed. Sometimes the alternative hypothesis is directional (that is, it might specify that the parameter in question is larger than some value); this sets up a one-tailed test. The tails pertain to the expected distribution of possible parameter values, as in a bell-shaped curve with a bulge in the middle (highly likely values) and narrowing as one moves toward the extremes (the lower and upper tails.)

and the mean, in turn, and summing the squares gives \$179,232.3 billion. Thus:

$$t = \frac{82}{\sqrt{5.0 \times 10^6 / 1.79 \times 10^5}} = 15.52$$

Comparing to the critical values of t, reject the hypothesis that b' is zero with much less than a 1 percent probability of error. ∎

9.6 TRANSFORMATIONS

Many relationships are not linear. Most technological change processes follow non-linear growth models—in particular, the S-shaped curves. The most effective way to handle these nonlinear models is to find an appropriate way to transform them to a linear relationship. Straight lines are best for graphic extrapolation. For instance, if a technology, Y, is growing exponentially, plotting log Y versus time will give a straight-line relationship. A number of information technology parameters (such as computer memory per unit volume) have maintained roughly exponential growth for several decades. These are typically plotted with the Y axis evenly divided into powers of 10. That is, the interval from 1 to 2 represents growth from 10 to 100 units; the interval from 2 to 3 represents growth from 100 to 1,000 units—decidedly nonlinear!

Transformation to a linear function can at times fit an equation to a nonlinear relationship. Exhibit 9.6 presents the most common transformations and illustrates how to perform the required calculations. These are elaborated further in Chapter 10.

Other functional forms can also be made linear. The power expression

$$Y = aX^b$$

can be transformed into a linear equation simply by taking the log of both sides to produce

$$\log Y = \log a + b \log X$$

which is linear if the dependent variable is taken to equal log Y and the independent variable as log X. As further examples, the hyperbolic relationship

$$Y = a - b(1/X)$$

can be dealt with by treating $1/X$ as a variable, say X'. And, the parabolic relationship

$$Y = a + bX + cX^2$$

can be handled by treating X^2 as just another variable, say Z, in the linear relationship.

The approach detailed can be easily adapted to calculate a regression equation for one of the models listed in Exhibit 9.6 This is illustrated in Exhibit 9.7.

■ **Exhibit 9.6   Common Transformations for Technological Growth Curves**

GROWTH MODEL	TRANSFORMATION
Exponential	$Z = \log_{10} Y$ (or $\ln Y$)
Pearl	$Z = \ln[(L - Y)/Y]$
(Fisher-Pry, single variable)	where L = upper growth limit
Gompertz	$Z = \ln[\ln(L/Y)]$
Substitution	
(Fisher-Pry, two variables)	$Z = \ln[f/(1 - f)]$
	where L = upper growth limit and
	f = fraction of the market held by the new technology ■

■ **Exhibit 9.7   Least Squares Calculations with a Transformed Variable**

Consider the following imaginary growth statistics for unit sales by a small manufacturing company, Y, (in thousands). Assume that $L = 100$ (in thousands).

YEAR	Y	$Z = ln[(L - Y)/Y]$
1985	9	2.314
1987	21	1.325
1989	41	0.364
1991	65	-0.619

First, the appropriate growth model must be chosen. (Growth models are discussed in Chapter 10.) This example assumes that a Pearl (or identically, a Fisher-
Pry model for one variable changing over time) model is appropriate. Then calculate
Z as shown. From this point, follow the same procedures as shown in Section 9.5.2
for Z instead of Y. For instance, calculate the deviation scores, z, and multiply these
by the deviation scores, x. Sum these and divide by the sum of x_i^2 to determine b.
In this illustration, the linear regression is:

$$\hat{Z} = 43.79 - [0.488(\text{year} - 1900)]$$

Extrapolating to 1997, for example:

$$\hat{Z} = -3.546$$

Next, take the antilog and solve for Y:

$$\frac{(100 - Y)}{Y} = .0288$$

$$Y = 97,200 \qquad ■$$

9.7   REGRESSION AND FORECASTING—SOME EXTENSIONS

This chapter has introduced simple regression as a forecasting tool. Such an
overview is necessarily brief and omits many important assumptions and qualifications. Additional regression topics can support management decisions that depend
on forecasts of technology and its effects. This section introduces some useful
extensions to simple regression methods.

Perhaps the most important extension is to expand simple regression analysis
to consider simultaneously the causal influence of more than a single explanatory
(independent) variable. Multiple regression often can provide a better explanation

of the past behavior of the forecast (dependent) variable and a better basis for predicting its future levels. In the example of CATV growth, the simple regression considered the effects of real disposable personal income on the number of subscribers and the R^2 indicated that much of the variance was captured. However, including the number of households composing the potential market or the hours of nonnetwork programming provided by CATV as variables might improve the fit of the regression and enhance its predictive power.

A multiple regression with two explanatory variables has the form

$$Y = a + bX + cZ + e \tag{9.7}$$

The estimates of a, b, and c once again are determined by the regression procedure so as to minimize the sum of squared errors. In the case of two independent variables, these estimates will specify a two-dimensional plane in a three-dimensional space (Y, X, and Z). However, the procedures can be generalized to provide for any number of independent variables as long as there are adequate numbers of observations available.

There are many computer packages available that will generate results similar to those given earlier for simple regression. Before extensive statistical inference or forecasting is done, the reader should consult expanded discussions on multiple regression because there are many pitfalls underlying the assumptions on which these models rest. Once results are available from the multiple regression, forecasting procedures are similar to those used in simple regression. First, the F-test will determine whether the equation has really explained anything about the dependent variable, and the R^2 will indicate the goodness of the fit of calculated values to actual data. Then, t-tests can be applied to determine whether specific coefficients are significantly different from zero (that is, whether the associated variable is important to explaining and predicting the dependent variable).

The generation of the C.I. for the forecast is somewhat more complicated because of the additional explanatory variables, but the typical statistical package printout provides all the necessary information.[5]

Another way time can be included in regression is by using dummy variables (a variable with a value of either 0 or 1) to seasonally adjust data and forecasts. For example, summer (S), fall (F), and winter (W) variables could be included in the equation.

$$Y = a + b_1 X + b_2 S + b_3 F + b_4 W + e$$

For this approach, the data presented in Table 9.1 would be expanded by adding three columns (summer, fall, and winter) following the year. Observations of the

---

[5]Econometrics textbooks such as those by Gujurati (1978), Johnston (1972), Wonnacott and Wonnacott (1979), and Goldberger (1964) are excellent sources for the techniques and problems of multiple regression. Forecasting methods books, such as those by Wheelwright and Makridakis (1980) or Granger (1980), also discuss the major issues, although they are not as extensive as those on econometrics.

number of subscribers made in the fall quarter, for example, would have a 1 in the fall column and a 0 in each of the winter and summer columns. Using this approach, the estimated coefficients of the regression will change the intercept of the regression line depending on the quarter. Thus, a forecast for a Fall Quarter value would be adjusted up or down depending upon how the number of subscribers changed during that season. In this example, the spring quarter provides the base from which the adjustments for the other quarters are made. This technique is analogous to seasonally adjusting data for projections, and most regression computer packages make it very easy to do.

Analysis of time series data seriously compromises a major assumption of the statistical analysis of regression equations—namely that observations, and errors in those observations, are independent of each other. Econometricians and others who work with time series data have developed sophisticated procedures to cope with this problem. Two-stage least squares estimation procedures and Box-Jenkins time series analyses are two notable approaches. These are beyond the scope of this book, and are not viable for most technological trend analyses. For instance, the Box-Jenkins approach first estimates the autocorrelation process in the series and then makes suitable adjustments so that the resulting series is well behaved. However, for this estimation to be credible, 50 data points or more are required. Rarely would a technology forecaster have the luxury of such extended series. Furthermore, the forecaster usually seeks to predict long-term change (five or more years in the future). The uncertainties of fast-paced technological change and complex socioeconomic influences dwarf the concerns of biased time series statistical tests.

EXERCISES

9.1 What type of additional assumptions are made when causal models are used instead of simple extrapolation? If they require more assumptions, how can causal models make the forecaster more confident about a projection?

9.2 Adaptive weighting schemes allow for flexibility in extrapolation. Identify three cases in which data may fluctuate in a cyclical fashion (for instance, snowmobile sales are probably seasonal). How might you use adaptive weighting to capture these cyclical characteristics to improve extrapolation in each of the three cases?

9.3 The TOOLKIT includes the even-numbered years of the time series of data for U.S. CATV subscribers (extended series compared to Table 9.1) under the file name CATV2.dat. The percent of households with CATV (Y) with time (X) is distinctly nonlinear (examine a plot of the series using TOOLKIT). However, consider the subset of the data for the even years 1966 through 1972. Use exponential smoothing for three periods—1966, 1968, and 1970—to predict a value for 1972, using $a = 0.5$. Repeat the calculation using $a = 0.4$. Repeat it once more using a different a value of your choice. Comment on what difference the value of *a* makes. Comment on how well the exponentially smoothed values fit the actual data for 1972.

9.4 Using the same four data points as in Exercise 9.3—that is, percent of households with CATV for 1966, 1968, 1970, and 1972—calculate a linear regression equation by hand. Follow the method of Exhibit 9.3. Compare your coefficients *a* and *b* with those obtained by a linear model of the full series (use the TOOLKIT). Comment. Calculate R^2 for your four-point linear model.

9.5 Use the equation calculated in Exercise 9.4 to extrapolate the percentage of households with CATV through 1978. Compute 50 percent and 90 percent confidence intervals. Do the actual data fall within these confidence intervals? Comment.

9.6 Use the first four data points for disposable income versus CATV subscribers (see Table 9.2). Calculate a leading indicator model that predicts subscribers from income. Extrapolate subscribers using $1,255 (in billions) as the income figure. Compare your prediction accuracy with the fit of the whole series model (available in the text or by manipulating file CATV.dat in the TOOLKIT). Calculate an *F*-test and a *t*-test for your four-point regression. Comment.

9.7 Repeat Exercise 9.6, but let subscribers be the *X* variable and income the *Y* variable. Again, predict subscribers using $1,255 (in billions) as the income figure. Explain any difference from the previous result.

9.8 Give one technology forecasting example in which two or more variables appear to be related in a nonlinear way. Describe how you could still make use of linear regression to derive an equation for this example.

____10
EXTRAPOLATING TECHNOLOGICAL TRENDS

OBJECTIVES

This chapter explains and illustrates the use of the most applicable forms of trend extrapolation for technology forecasting and presents trend analysis as a four-step process. The chapter emphasizes two key growth models, Fisher-Pry and Gompertz. Finally, it illustrates how the Lotka-Volterra equations offer a promising general framework for trend modeling.

10.1 TREND ANALYSIS IN TECHNOLOGY FORECASTING

Technology forecasting relies largely on naive (direct) time series analysis. This implies major assumptions about the nature and permanency of both context and structure, as discussed in Chapter 9. Trend analysis methods can yield valid forecasts when supporting and competing mechanisms in the larger environment remain constant over the time horizon of the forecast or when changes in these mechanisms cancel one another. Even under those appropriate conditions, trend extrapolations should be used in conjunction with complementary technology forecasting methods, especially expert opinion and monitoring. This chapter stresses extrapolation over time. It applies regression techniques to fit selected nonlinear relationships that are especially suited for technology forecasting.

10.2 STEPS IN TREND ANALYSIS

Once the variable(s) have been chosen and the necessary data have been obtained (see Chapter 9), trend analysis can begin. Exhibit 10.1 outlines the basic steps.

Step 1, Exhibit 10.1, model identification, draws upon insight into typical patterns of technological innovation (see Chapter 2), solid knowledge about what is driving the change and empirical evidence. *S-shaped growth* should be considered the most likely form (as discussed in Chapter 4). Section 10.3 contrasts the Fisher-Pry and Gompertz models—two approaches that produce S-shaped curves. However, other models sometimes merit serious consideration as well.

■ **Exhibit 10.1   Steps in Trend Analysis**

1. Identify the proper model. Prominent alternatives include:
   a. S-shaped growth curves
      1. Fisher-Pry (or equivalently, Pearl)
      2. Gompertz
      3. Technical progress function
   b. Learning curve
   c. Exponential growth
   d. Linear
2. Fit the model to the data
   a. Graphically
   b. Solve for the constants in the equation
3. Use the model to project
   a. Graphically
   b. Mathematically
4. Perform sensitivity analysis and interpret the projections.
   a. Compute confidence intervals
   b. Consider outside factors.   ■

Exponential growth often holds over certain periods, or epochs (Hamblin, Jacobsen, and Miller, *et al.*, 1973). The growth rate then shifts and another epoch emerges. Over multiple epochs, continuing exponential growth becomes apparent, but succeeds at different rates. Exponential growth may hold for the time period of interest, but possible physical or social limits that could slow or stop growth must be anticipated.

An *envelope curve* can be constructed by stacking S-shaped curves, one after and over another. Figure 10.1 shows a classic example of how a series of technological developments, each S-shaped, can combine to drive a parameter forward, possibly along an exponential frontier. As one technology approaches a peak, R&D provides a successor that fulfills the appetite of a still-hungry market. An envelope curve could be plotted through the peaks and the valleys of the successive technologies to depict the general trajectory of the development pattern.

The *technical progress function* is another important model. It measures growth as a function of effort instead of time. As depicted in Figure 10.2A, the notion is

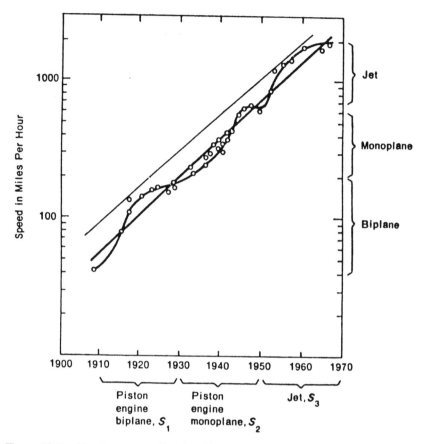

Figure 10.1   Envelope curve. (Reprinted by permission of the publisher from *Technological Forecasting for Decision Making* by J. P. Martino. Copyright 1983 by Elsevier Science Publishing, Co. Inc.)

that progress in developing a technology starts slowly as many impediments must initially be overcome, advances rapidly for a period, and then slows as the easy improvements are "mined out." This is, of course, the S-shaped growth curve in another guise. The tapering off of technical progress for constant increments of additional effort (Figure 10.2A) implies dwindling productivity (Figure 10.2B). The R&D organization receives the greatest payoff from its effort during the steep portion of the technical progress curve (Figure 10.2A). After that, marginal returns per unit of effort diminish (later portion of Figure 10.2B). The technical progress function provides a vital signal to those who would manage R&D for a technology—that is, at some point continued investment in R&D will deliver less and less. This means that the natural momentum of continuing to do what has delivered good results in the past, must be stopped—that large, successful research group must be reassigned to fresh tasks or R&D productivity will drop.

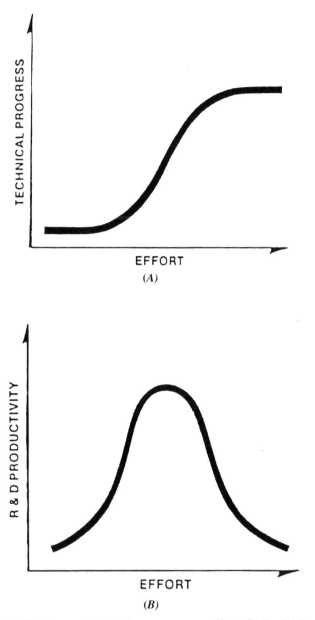

Figure 10.2   (*A*) The S-curve techinical progress versus effort. (*B*) The R&D productivity curve. (*Source:* Becker and Speltz, 1986. Reprinted by permission. ©Industrial Research Institute.)

A related notion is the *learning curve*, which addresses the improvement in productivity often seen as a technology production process matures. This is conventionally depicted as a power function:

$$Y = aX^{-b}$$

or equivalently,

$$\log Y = \log a - b \log X \qquad (10.1)$$

where Y is the number of direct labor hours required to produce the Xth unit; a is the number of direct labor hours required to produce the first unit; X is the cumulative number of units produced (not time); and b is a parameter that measures the rate that labor hours are reduced as cumulative output increases (Argote and Epple, 1990).

Analysis of the learning curve for a technology can help predict declining production costs. The decline is attributable to organizational learning as personnel gain experience in refining the production process—in other words, the learning curve is due to both technology and people factors. Each increase in cumulative output leads to a reduction in unit production cost. While this "progress ratio" varies greatly from technology to technology, the modal value is about 80 percent—that is, each doubling of cumulative output leads to a 20 percent reduction in unit production cost (Argote and Epple, 1990). Figure 10.3 shows the learning curve for production of an advanced military jet built in the 1970s and 1980s. It shows the reduction in the number of hours needed to assemble the aircraft as workers learned the intricacies of construction, and less labor means a lower cost per unit. Projection of a learning curve can help a technology manager monitor his or her own production processes to ensure that learning is progressing reasonably. Projection of a competitor's learning curve could help gauge whether to enter a market.

Having identified a promising model, the second step is to fit the model to the data. Should that fit prove poor, the forecaster may wish to reconsider the choice of model. The choice of a model should never be made by fishing through a grab bag of models and picking the one that fits the data best. Noisy data can mask true relationships; there is no substitute for solid conceptual underpinning.

Begin by graphing the data. Remember that individual data points may be problematic. This is important as outlier points can exert great influence over mathematical curve fits. Outliers may result from special circumstances, mistaken measurement, transcription errors, or just a divergent value that cannot be ignored. As discussed in Chapter 9, the forecaster may want to examine various transformations. For many forecasting purposes, fitting a line to properly transformed data will provide a satisfactory basis for extrapolation. Even if the equation for the trend is eventually calculated, the graph will provide an excellent check. It is easier to detect a bad extrapolation on a graph than from an equation.

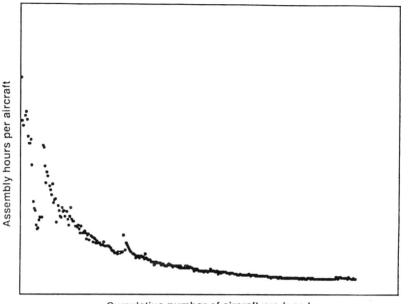

Figure 10.3 Learning curve—assembly hours per aircraft and cumulative number produced. (*Source:* Argote and Epple, 1990. Reprinted by permission of the author and publisher from *Science*, Vol. 247, pp. 920–924. Copyright 1990 by the AAAS.)

Details about how to fit an equation for a model based on a sample of data are provided in Chapter 9. Once the equation is determined, an extrapolation can be made graphically and/or mathematically (Step 3). Chapter 9 provides examples.

Next, perform a sensitivity analysis for the extrapolation (Step 4). Calculation of confidence intervals (see Chapter 9) provides vital information on the range of future values to be expected. Consider also what outside influences are important—that is, which ones could alter the trend substantially and that are reasonably likely to occur. This sensitivity analysis may be quantitative (see Appendix C) and/or qualitative. Cross-impact methods (Chapter 12) can be of use, and expert opinion (Chapter 11) can be quite helpful at this juncture. A strategy that often proves highly effective includes these steps:

- Show your trend extrapolation to selected experts for their reactions,
- Raise specific questions about the external influences identified to see if the experts agree that they are likely and how they would alter the trend,
- Ask the experts to identify other factors likely to alter the trend.

After the sensitivity analysis has been performed, the projections should be interpreted. Forecasters often consider their job to be done when they provide trend(s), but this is not so. The implicit knowledge that has been gained in determining that trend should be made explicit. The forecaster needs to indicate

- Why a certain model has been selected
- How strong or weak the data are
- What factors are likely to interact with the trend and how probable they are
- How much confidence he or she has in the trend

Wherever possible, an open dialogue with the intended users will add significant value to the trend analysis.

10.3 THE S-SHAPED CURVES

The familiar S-shape (more formally, sigmoidal) curves describe many natural phenomena and also fit the technological growth processes. This could be stated as the "pumpkin theory" of technological innovation (see Figure 10.4), the yeast theory, or the embryo theory. These natural growth processes share the properties of relatively slow early change, followed by steep growth, then a turnover as size asymptotically approaches the limit. Various mathematical formulations can depict such patterns. Two such models are presented here.

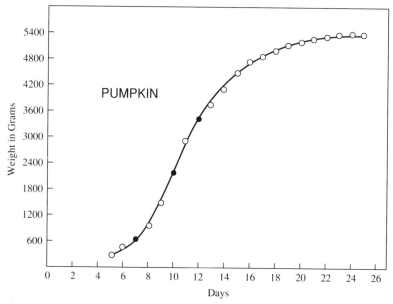

Figure 10.4   The pumpkin curve. (*Source:* Martino, 1972. Reprinted by permission of the publisher from *Technological Forecasting for Decision Making* by J. P. Martino, p. 184. Copyright 1972 by Elsevier Science Publishing Co., Inc.)

10.3.1   The Fisher-Pry Model

The Fisher-Pry model predicts characteristics loosely analogous to those of biological system growth. It also is referred to as a substitution model because it is often applied to forecast the rate at which one technology will replace (or be substituted for) another.

It is convenient to take the fraction of potential market penetration, f, as the dependent variable where $f = Y/L$ and L is the upper bound for growth of the variable Y (such as the number of units sold). The fundamental assumption of this approach is that the rate of change of f is proportional to both f and to the fraction of the market remaining, $1 - f$.[1] That is,

$$\frac{df}{dt} = b[f/(1 - f)] \tag{10.1}$$

This is a good assumption when initial sales make the sale of subsequent units easier. For example, when sales lead to growth in distribution, service, and/or repair networks that, in turn, encourages additional buyers.

Integrating Equation 10.1 between t_o, the time at which substitution is 50% complete, and t is displayed as

$$\int_{t_0}^{t} b[f/(1 - f)]dt$$

Solving for f yields

$$f = \frac{1}{1 + \exp[-b(t - t_0)]}$$

or

$$f = \frac{1}{1 + c\, \exp(-bt)} \tag{10.2}$$

The expression given by Equation 10.2 was suggested years ago and was utilized in demographic forecasting by Raymond Pearl. It is often referred to as the *Pearl curve*, particularly when used to examine the absolute growth of a technology rather than the fraction of a market taken over by a technology. The form developed by Fisher and Pry,

$$f = 0.5[1 + \tanh a\,(t - t_0)]$$

can be reduced to Equation 10.2 with a bit of straightforward manipulation. However, since the name "Fisher-Pry curve" is commonly used for this substitution function, we will refer to all forms of the equation as such here.

---

[1] Such models are also known as logistic curves.

Fisher and Pry actually use a special version of the more general Pearl equation; the linear form given by

$$\ln[f/(1 - f)] = 2\alpha(t - t_0) \tag{10.3}$$

where f = the fraction of that year's market held by the new technology; $1 - f$ = the fraction held by the old technology; α = half the annual fractional growth in the early years; and t_0 = the time at which f = 1/2. Note the different interpretation of f in this case of competition from that in which f is the fraction of the ultimate market limit, L. In sum, this version offers an alternative formulation for an S-shaped curve with the same functional form.

In their original article, John Fisher and Robert Pry (1971) presented 17 cases of substitution that fit their model. Subsequent investigators have expanded the catalog so that nearly 100 cases of technology substitution have been found to follow the Fisher-Pry forecast (Lenz, 1985). Figure 10.5 illustrates the curves for various products and processes. These curves follow Equation 10.3. They appear

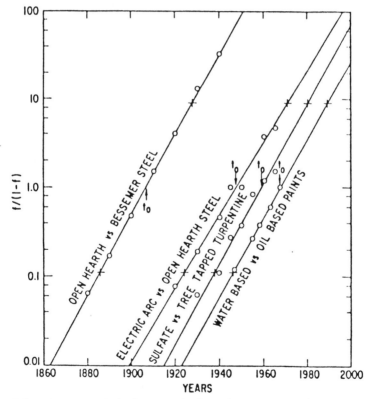

Figure 10.5   Fisher-Pry substitution curve fits for various products and processes. (*Source:* Fisher and Pry, 1971. Reprinted by permission of the publisher from "A Simple Substitution Model of Technological Change" by Fisher and Pry, *Technological Forecasting and Social Change*, pp. 75–88. Copyright 1971 by Elsevier Science Publishing Co., Inc.)

linear in the plot because the transformation ln [f/(1-f)] is a linear function of time for the model. The issue of modeling change in a technology over time, in conjunction with competition from other technologies, will be addressed further in Section 10.4.

The Fisher-Pry curves described by Equation 10.2 (equivalently, Equation 10.4) or by Equation 10.3 are both S-shaped, although not identical. The remainder of this section concentrates on the more general form—described by Equation 10.2 (or Equation 10.4).

From Equation 10.2, there are inflection points at $f = 0$, $f = 0.5$, and $f = 1$, which occur at times $t = -\infty$, $\ln(c)/b$ and $+\infty$, respectively. The first corresponds to the minimum value of f, the last to the maximum, while $[\ln(c)/b, 0.5]$ is the point about which the curve is symmetrical. As demonstrated in Figure 10.6A, changes in the value of b alter the shape of the curve; Figure 10.6B illustrates how

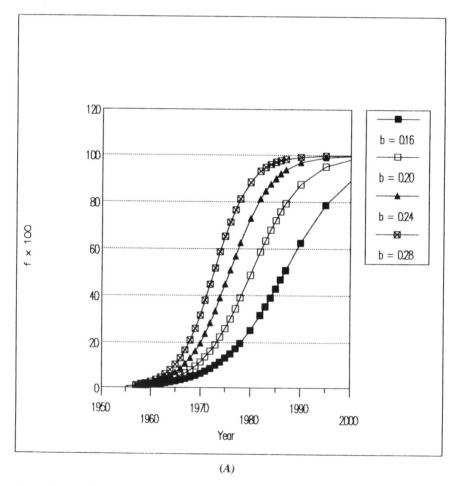

(A)

Figure 10.6 Fisher-Pry model. (*A*) Variation with values of b. (*B*) Variation with values of c.

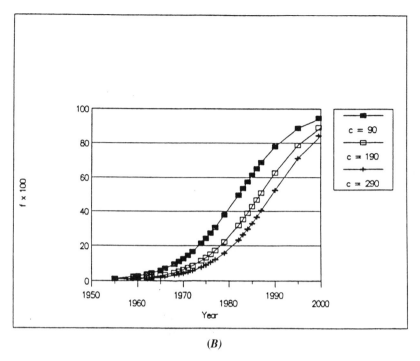

(B)

Figure 10.6   *(Continued)*.

changes in the value of c alter the location of the curve. Thus the model offers considerable flexibility to fit time series data.

To create a Fisher-Pry forecast, the analyst must determine the values of b and c that produce the best fit to the time series data. Although this can be done using the form given by Equation 10.2, it is much easier to transform the equation as follows:

$$Z = \ln[(1 - f)/f] = \ln(c) - bt \qquad (10.4)$$

In this form, Z is a linear function of time. Existing time series data are used to compute values of Z. These are plotted versus time, and a straight line is fitted to the data to determine the values of b and c. Alternately, values of $(1 - f)/f$ can be plotted versus time using semilog graph paper. Five-cycle paper will allow the forecaster to cover a range from 0.001 to 100, which will accommodate values of f from 1 percent to 99 percent. The forecaster can fit the resulting plot with a straight edge or by using the least squares straight-line expressions presented in Chapter 9. Figure 10.7 shows the results of a least squares fit used to determine b and c for CATV data presented in the first columns of Table 10.1.

The accuracy of the growth portrayed by the resulting expression, of course, depends on the validity of the fundamental assumption about the invariance of the external environment. This assumption is inherent in all direct trend extrapolation

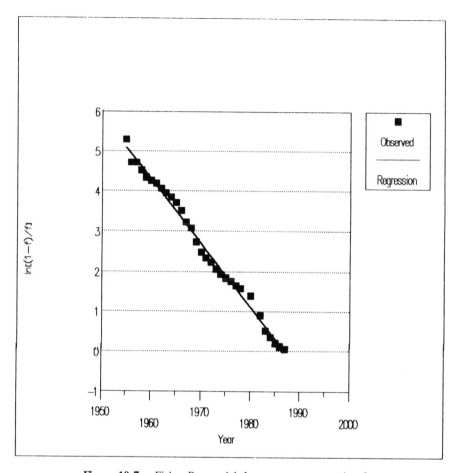

Figure 10.7   Fisher-Pry model: Least squares regression fit.

methods. Validity also depends on the extent and accuracy of the data the forecaster has in hand when b and c are determined. Figure 10.8 illustrates this by showing the different forecasts that would have been prepared for CATV growth depending on the number of years of available data. This variation in forecasts has led some to suggest that the Fisher-Pry model should be employed only after substitution has reached a level of 10 percent (that is, $f = 0.1$). For the CATV example, that point would be reached in 1972 (point number $n = 18$).

Figure 10.8 shows that waiting until substitution is substantial does indeed produce a more accurate forecast of CATV growth. Unfortunately, the technology manager frequently must make critical decisions at a much earlier time to capitalize on technological or market advantages (Lenz, 1985). The only answer to this dilemma is that forecasts should be made early and often—that is, the forecast should be made as soon as data are judged to indicate a trend and revised continually as new information becomes available.

TABLE 10.1   Fisher-Pry Forecast of CATV Penetration Including 90 Percent Confidence Limits

n	Year	Actual %	Forecast %	90% C.I. Lower	90% C.I. Upper
1	1955	0.5	0.61	0.47	0.79
2	1956	0.9	0.72	0.56	0.93
3	1957	0.9	0.84	0.65	1.09
4	1958	1.1	0.99	0.76	1.27
5	1959	1.3	1.15	0.90	1.48
6	1960	1.4	1.35	1.05	1.74
7	1961	1.5	1.58	1.23	2.03
8	1962	1.7	1.85	1.44	2.37
9	1963	1.9	2.16	1.69	2.76
10	1964	2.1	2.53	1.98	3.23
11	1965	2.4	2.95	2.31	3.76
12	1966	2.9	3.44	2.70	4.38
13	1967	3.8	4.02	3.16	5.10
14	1968	4.4	4.68	3.68	5.92
15	1969	6.1	5.44	4.29	6.88
16	1970	7.7	6.32	5.00	7.97
17	1971	8.8	7.34	5.81	9.22
18	1972	9.7	8.50	6.75	10.65
19	1973	11.3	9.82	7.82	12.27
20	1974	12.7	11.33	9.05	14.09
21	1975	13.8	13.03	10.45	16.14
22	1976	14.8	14.95	12.03	18.42
23	1977	16.1	17.09	13.82	20.95
24	1978	17.1	19.47	15.82	23.72
25	1979	NA	22.09	18.04	26.76
26	1980	19.8	24.96	20.49	30.03
27	1981	NA	28.06	23.19	33.51
28	1982	29.0	31.39	26.13	37.18
29	1983	37.2	34.93	29.30	41.00
30	1984	41.2	38.63	32.69	44.94
31	1985	44.6	42.47	36.26	48.94
32	1986	46.8	46.41	40.05	52.89
33	1987	48.7	50.39	43.90	56.87
36	1990		62.11	55.69	68.12
41	1995		78.44	73.45	82.72
46	2000		88.98	85.87	91.48

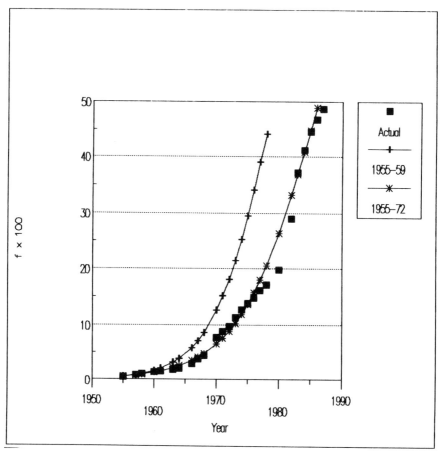

Figure 10.8 Fisher-Pry model: Variation with number of years of available data.

Figure 10.9 compares the Fisher-Pry model using coefficients from a least squares fit of the 31 CATV data points (see Table 10.1) to measured values of f. Also shown in the figure are the 90 percent confidence interval values for the forecast. These confidence limits were computed by the methods presented in Chapter 9 applied to the forecast in the form suggested by Equation 10.4. Values of the 90 percent confidence limits of the forecast transformed into the f versus time domain are given in Table 10.1. Note that the confidence intervals are not perfectly symmetrical about any point (reflecting the nonlinear Fisher-Pry equation). Notice also that, unlike the case for unbounded linear extrapolation, the confidence intervals do not expand in a cone shape as the extrapolation period lengthens. Instead, the confidence intervals, too, are constrained as the curve approaches the limit.

10.3.2 The Gompertz Model

The underlying dynamics of the Gompertz model are quite different from those of the Fisher-Pry model. In fact, Gompertz is sometimes referred to as a mortality

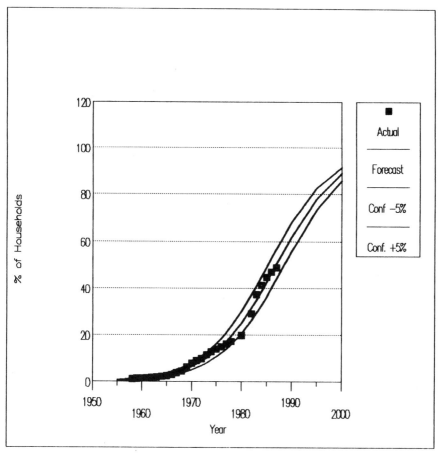

Figure 10.9   CATV subscribers in the United States: Fisher-Pry fit, 90 percent confidence intervals, and actual data.

model, whereas Fisher-Pry is characterized as a growth model. Benjamin Gompertz, an English demographer, found that the mortality rate of a population grew exponentially as it aged. With respect to technology forecasting, the Gompertz model is most appropriate in cases in which equipment replacement is driven by equipment deterioration rather than by technological innovation.

The Gompertz model is defined by this equation:

$$f = \exp[-b\exp(-kt)] \qquad (10.5)$$

Once again, $f = Y/L$, and L is the upper limit for the number of units, Y. While equation 10.5 produces an S-shaped variation, as does the Fisher-Pry model, Figure 10.10 shows that the shapes predicted by the two models for a given set of data are quite different. As with the Fisher-Pry expression, Equation 10.5 has three inflection points: $f = 0$; $f = 0.368$, and $f = 1$. These correspond to $t =$

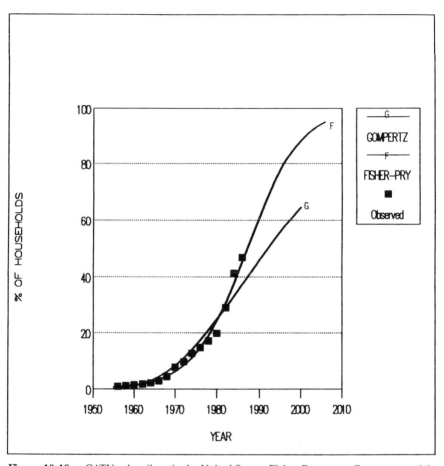

Figure 10.10 CATV subscribers in the United States: Fisher-Pry versus Gompertz models.

$-\infty$, $\pm \ln(b)/k$, and $\pm = +\infty$. However, the Gompertz curve is not symmetrical about its central inflection point.

The position and shape of the Gompertz curve can be adjusted independently — the position by changing b; the shape by altering k. Thus, as with the Fisher-Pry model, it offers considerable flexibility in obtaining a satisfactory fit of time series data. The process of fitting the Gompertz curve is analogous to fitting the Fisher-Pry curve. However, the log of Equation 10.5 must be taken twice to produce this linear expression:

$$Z = \ln[\ln(L/f)] = \ln(b) - kt \qquad (10.6)$$

The forecaster can find values of b and k by fitting a straight line to values of Z computed from the data by using a straight edge on a plot of Z or by least squares regression fit.

10.3.3 Choosing Fisher-Pry or Gompertz

The model chosen by the forecaster must embody underlying characteristics that reflect those of the process to be forecast. This is not always easy: the process often is not well understood. However, the Fisher-Pry and Gompertz models offer an opportunity to illustrate model selection. The following discussion reflects the general concerns of model selection and of differention between growth and mortality models.

Equation 10.1 shows that Fisher-Pry, a growth model, assumes that the rate of change of the process is proportional to *both* the fraction of the market penetrated by the technology and the fraction that remains to be penetrated. It depends on the number of uses for which the new technology has been applied and on the number for which it is yet to be applied. This is analogous to the process followed by the diffusion of a new, technically advanced product. In such cases, knowing someone who owns one helps prospective buyers assess the technology's potential for them. Since the availability of the technology and of spares, repair facilities, and advice normally grows with the number of units in the field, diffusion is further enhanced by the sale of additional units. Thus penetration of the market is driven, not only by potential sales, but also by the sales that have been achieved.

In the diffusion process modeled by Fisher-Pry, initial sales of a new technology are difficult despite its promise and the size of the potential market. This is true largely because the technology is unknown and unproven. Moreover, in the early stages, field support is likely to be poor, there may be institutional barriers to overcome (such as, licensing), and survival of the technology and its supplier is uncertain. Thus adoption implies considerable risk. As applications grow, so does general knowledge of the technology and its advantages. The support infrastructure improves, as does confidence that the purchaser will not suffer either from inadequate support or from the financial failure of the manufacturer. From this point, penetration grows rapidly. At some stage (for example, after 50 percent penetration), further penetration becomes increasingly more difficult because it often involves sales to companies that may not benefit as greatly from the technology or that have marginal capability to finance adoption. Thus the rate of penetration slows.

The Gompertz model embodies quite different dynamics. For example, although not readily apparent from Equation 10.5, for penetrations greater than 50 percent, the rate of penetration depends primarily on the fraction of the market remaining. Thus the Gompertz model is appropriate for forecasting market penetration by technologies for which initial sales do not make subsequent sales easier. This dynamic usually is found when a new technology offers no clear-cut advantages over an old. In such instances, an older technology is replaced by a newer technology that performs the same tasks with essentially the same financial and/or functional efficiency. Purchases simply replace equipment that has worn out or has been destroyed. It is this characteristic of the Gompertz model that leads to its classification as a mortality model.

Since the dynamics of the two models are different, it is reasonable to expect significant differences in the forecasts they produce. For example, the Fisher-Pry model forecasts a more rapid penetration than does the Gompertz. However, either model can be made to fit a handful of initial data points. Thus goodness of fit cannot be used as a criterion to determine which model to use; instead, the dynamics of the model must be matched to the dynamics of the process being forecast.

The process of CATV growth considered in this chapter is an example of new technology diffusion. CATV offers users significant benefits over older telecast delivery technologies. In this case, older technology is not merely replaced when it wears out. In fact, existing TV sets are converted to CATV, and the new and old technologies often function side by side in the same neighborhood. In the initial stages, diffusion of CATV was slowed by uncertainties about the potential of the technology and the future of the cable companies themselves. These uncertainties were mitigated, however, because conversion to CATV did not prevent TV sets from functioning in their original mode. Initial diffusion was more seriously inhibited by the need to build a cable infrastructure and by licensing requirements that usually must be satisfied location by location. In short, the dynamics of CATV growth provide a compelling case for selecting the Fisher-Pry model. Data show that diffusion has now proceeded to nearly the 50 percent level. The forecast (see Figure 10.9 and Table 10.1) suggests that it will proceed to 90 percent completion in the next decade.

Not all substitution processes are as clear-cut as the example of CATV. Some, for example, appear to be driven by a mix of new technology diffusion and old technology mortality. For example, some owners of prestige automobiles may replace them with newer models that offer significant technological improvements (such as anti-lock braking systems or fuel injection). Others may replace to "keep up with the Joneses" even when no clear technological superiority is offered. Still others may be driven (no pun intended) to replace worn-out units by nearly identical ones because of the prestige and tradition of an established design ("only the very best people own a ... "). In cases in which the underlying dynamics of the penetration are unclear, the best guide is to consult those whose job it is to understand the dynamics of the market. The overly optimistic forecasts of demand for innovative technology products, such as home computers and expert systems, for example, can be partly attributed to the failure of forecasters to consult such experts (Wheeler and Shelley, 1987). However, Schnaars (1989) notes that such experts often get caught up in the "zeitgeist" (prevailing wisdom) of the times and tend to be overly optimistic. Historical analogy to similar situations also can be helpful.

The forecaster must match the dynamics of the model to those of the process. To reiterate, the dynamics of the Fisher-Pry model are appropriate to cases of technology diffusion; the dynamics of the Gompertz model are appropriate to cases of replacement driven by equipment deterioration rather than technological advantages. Occasionally, as Lenz (1985) notes, industries have been able to match equipment deterioration to new technology innovation; however, the difference in the rates associated with each mitigate against achieving that balance often. When the dynamics of the process are unclear, consult the experts, compare alternative models, and pray.

10.3.4 Selecting an Upper Bound for the Forecast

The preceding discussions, have emphasized forecasting the market fraction, $f = Y/L$, where L is the upper bound for Y. This approach simplifies the presentation; however, it also disguises the fact that L is a third parameter that must be estimated to employ forecasts made with the Fisher-Pry and Gompertz models. An accurate estimate of L is important for a number of reasons. First, L must be known to formulate the time series data for forecasting. Second, an incorrect upper bound can seriously distort values of the fitting coefficients (b and c in the Fisher-Pry model) and hence the subsequent forecast. Further, the technology manager generally needs a forecast, not only of the fraction of the market that will be penetrated, but also of the number of units that will be sold.

In the case of CATV, the upper limit was set as the number of households with TV sets—a number that grows with time and, therefore, is itself a subject for forecast (as considered in Chapter 9). The range of f for CATV could be established as zero to one with some confidence. In other instances, this cannot be done so easily (for example, in forecasting the upper limit for the sales of subcompact automobiles, the total number of automobiles sold would not be an accurate upper bound). Nor is the limit for the functional capacity of a technology (for example, the precision of a manufacturing method or the level of concentration of a chemical compound that can be detected) so easily set.

In some instances, forecasters have used the data to establish the upper limit (for example, in the Fisher-Pry method, data fits would be performed for b, c, and L). As Martino (1983) notes, this is bad practice because initial data are relatively insensitive to the upper bound and thus do not provide reliable guidance. Rather, L should be set by natural or fundamental limits to the process. This, of course, requires knowledge of the technology and the market place that the forecaster may not possess. Therefore, in cases where the upper bound is unclear, the forecaster should work closely with experts in the field to determine a reliable estimate.

10.4 TOWARD A GENERAL MODEL OF TECHNOLOGICAL CHANGE: THE LOTKA-VOLTERRA EQUATIONS[2]

10.4.1 Theory

The Italian mathematician Vito Volterra proposed the Lotka-Volterra equations at the turn of the century to model population changes of sharks and food fish in the Adriatic Sea. Since then, the model has been expanded and applied in the fields of demography and ecology. It successfully represents many biological populations based upon data from field observations in the wild. Although simple, the model is one of the theoretical cornerstones of the discipline of mathematical ecology.

As per earlier discussion, technology forecasting has frequently and successfully borrowed from demographic models for the purpose of trend extrapolation.

---

[2]This section was contributed primarily by Scott Cunningham at Georgia Tech. The authors greatly appreciate his contribution.

Both the Pearl (Fisher-Pry for a single technology) and the Gompertz curves were originally developed in demography and later applied to technology. In using these models, the technology forecaster draws upon an analogy between populations of technologies and populations of organisms. Demographic models often produce highly accurate forecasts for technological growth.

The Lotka-Volterra equations have helped to frame a number of long-range technology forecasts. The accuracy of such extrapolations and the strong biological bases for the Lotka-Volterra model suggest that these mathematical formulations may relate to the underlying sociotechnical processes that lead to the growth and diffusion of technologies.

A system of Lotka-Volterra equations (describing interactions between two technologies) is shown below:[3]

$$\frac{dX}{dt} = X(\alpha_1 - \beta_1 X - \gamma_1 Y) \tag{10.7}$$

$$\frac{dY}{dt} = Y(\alpha_2 - \beta_2 Y - \gamma_2 X)$$

More elaborate systems may be created by adding additional variables (technologies) and by introducing new competition terms.

Consider the basic dynamic assumptions made by this model. First and foremost, the Lotka-Volterra model assumes that technology develops within a larger market system in which technologies may compete; the model expressly includes variables to represent these features. Second, such systems tend to be open to many forces not included in the model (such as governmental regulations), but the Lotka-Volterra model assumes that Equation 10.7 captures the essential dynamics.

Specifically, the Lotka-Volterra model assumes that technological growth depends on two factors: existing technology populations and market potential. The resultant growth depends on multiplication of these two factors, as shown in Table 10.2.

Consider next the purpose and meaning of the coefficients. Note in the table that α, β, and γ all occur within the market potential portion of the equations. Let X and Y constitute two competing technologies (Equation 10.7), measured either by units sold or by profits earned. α pertains to the production or sales capacity of the market and has units of 1/time. α_1 and α_2 will be equal when both technologies have equal market capacity. β is the inhibition coefficient. It gauges the loss of potential market caused by increases in the technology. When forecasting sales it has units of 1/[money $\times$ time], and when forecasting units of technology it has units of 1/[unit $\times$ time]. γ is the competition coefficient. It reflects the share of market

---

[3]This book refers to Equation 10.7 as the Lotka-Volterra equations—plural because more than a single technology can be modeled. The original Lotka-Volterra equation lacked the second term, $-bX^2$ (when multiplied out), so that in the absence of competition, growth would be exponential. Equation 10.7 is known as the predator-prey version of the equations; it provides a quite general depiction of technological growth.

TABLE 10.2   Dynamic Origins of the Lotka-Volterra Equation

Growth in technology is the result of current levels of technology times current market potential:

$$\frac{dX}{dt} = X \cdot L(X) \qquad \text{where}$$

Symbol	Description	Used in system as:
$\dfrac{dX}{dt}$	Growth in technology	$\dfrac{dX}{dt}, \dfrac{dY}{dt}$, etc.
X	Existing levels of technology	X, Y, etc.
$L(X)$	Market potential	$(\alpha_1 - \beta_1 X - \gamma_1 Y)$, $(\alpha_2 - \beta_2 Y - \gamma_2 X)$, etc.

capacity taken by the competing technologies and has the same units as β. β_1 and β_2 need not be equal, nor must γ_1 equal γ_2. These three sets of coefficients are necessary for a complete solution to the Lotka-Volterra equations and will vary according to the characteristics of the technologies in question. The value of the coefficients will also vary depending upon the units chosen for study (such as years and thousands of items shipped per year.)

The form of Equation 10.7 may be unfamiliar to forecasters who usually address changes in amounts of technology over time (e.g., annual sales) rather than changes in degree of technological growth over time (e.g., rate of sales gains). Most trend extrapolation models employ a single mathematical equation that describes amount of technology over all time periods. These equations are known as *closed form solutions* by mathematicians. Equation 10.7, a system of differential equations, decribes changes in growth. Although it will give a definite value of growth for any instant at which all the conditions of the equation are known, it is impossible to solve these equations in a single closed form. The growth formulation of the Lotka-Volterra equations is sufficiently general to embrace a very large family of growth curves. The Lotka-Volterra equations can approximate each of the closed form forecasting models discussed earlier in this chapter.

Given the dynamic form of the Lotka-Volterra equations and the general description of its variables, it is now appropriate to discuss several relevant concepts borrowed from biology. The first is the definition of an ecology, defined here in a technological context as a "system." The properties of a system, for the purposes of the Lotka-Volterra equations, are as follows:

- A system is limited to a few closely interrelated technologies and markets.
- Each technology and market within the system is at the same level of generality.
- Systems are hierarchial and may be refined at progressively higher or lower levels of detail.

- The system is abstracted so as to be relatively closed (that is, relatively immune to external influences).
- Systems are simplified models of more complex sociotechnical environments — models that may be manipulated mathematically.

Another important concept in population biology is that of the niche. For technology growth modeling, the niche can be considered to be the prospective market for which the technology is suited. Considering this definition, technological competition is the process by which two technologies vie for a niche, or more broadly, the ways in which two technologies interact with each other within a sociotechnical system.

Ecologists speak of the concept of *biopotential*. Species vie for biopotential; technologies compete for market potential. The actual potential of the market might consist of human, economic, and/or technological resources. However, for the purposes of the Lotka-Volterra equations, this market potential is given in terms of the potential production or sales capacity for the technology. Potential is shown in Table 10.2 as $L(X)$.

There are two mutually exclusive types of biological systems that are of particular interest for technology forecasting: competitive and inhibitive systems. Inhibitive systems are those in which market size primarily limits growth. They are equivalent to the growth models discussed earlier in the chapter. In contrast, competitive systems are primarily limited in growth by the nature and extent of competitive behaviors with other technologies. In competitive systems, one technology steadily drives the other into extinction or disuse. Competitive systems relate closely to the mortality models discussed earlier in this chapter. Competitive systems occur when $\beta_1 \times \beta_2 < \gamma_1 \times \gamma_2$. Inhibitive systems predominate when $\beta_1 \times \beta_2 > \gamma_1 \times \gamma_2$.

10.4.2 Functional Equivalences of the Model

Under what conditions do the Lotka-Volterra equations produce the special case trend extrapolation models discussed earlier in the chapter? Although the equations may not be reduced to a closed form solution for the vast majority of potential systems, they can produce solutions that duplicate any of the standard forecasting curves to any arbitrarily high level of precision. The Lotka-Volterra equations are of interest because they can depict such a wide range of dynamic behaviors. They are sufficiently general to encompass many different kinds of trend extrapolation models. Linear, exponential, Pearl, Gompertz, substitution models, and oscillatory behaviors can all be matched by the Lotka-Volterra model. Table 10.3 lists specific conditions under which these solutions will be evidenced by the Lotka-Volterra equations.

Linear behaviors will be evidenced by slow-growing technologies or for any technology examined over a sufficiently short period of time. If examined for greater lengths of time, exponential or S-curve characteristics will likely appear.

TABLE 10.3 Summary of Relationships between Lotka-Volterra Equation and Other Forecasting Models

Trend Model	Closed-form Solution	Growth Description	The Lotka-Volterra Equations $$\begin{bmatrix} \dot{X} & X(\alpha_1 - \beta_1 X - \gamma_1 Y) \\ \dot{Y} & Y(\alpha_2 - \beta_2 X - \gamma_2 Y) \end{bmatrix}$$ Approximate this trend Model when . . .
Linear	$X = At + B$	$\dfrac{dx}{dt} = C$	Slow growth or short periods of time, competition, and/or inhibition are significant ($\gamma, B \rightarrow \infty$)
Exponential	$X = \exp(At)$	$\dfrac{dx}{dt} = AX$	Occurs for low saturation ($\alpha > X$)
Decaying exponential	$X = A - A\exp(-Bt)$	$\dfrac{dx}{dt} = (A - CX)$	Occurs for high saturation ($X \rightarrow \alpha$)
Pearl	$X = \dfrac{A}{1 + B\exp(-Ct)}$	$\dfrac{dx}{dt} = X(-A - CX)$	Usually inhibitive systems, but can model competition for technologies that have similar growth rates ($\gamma \rightarrow 0$ or $\alpha_1 \sim \alpha_2$)
Gompertz	$X = A\exp(-C\exp(-Dt))$	$\dfrac{dx}{dt} = CXe^{-At}$	Competitive systems; models older technology growing obsolescent ($\beta \rightarrow 0$)
Substitution models	No general closed-form solution	Various	Significant competition between growth and equilibrium technologies or between two technologies with similar growth rates

The Lotka-Volterra equations approximate such behaviors when the coefficients β and/or γ are very large.

Exponential behaviors often occur at the beginning of growth regimes. Many exponential growth patterns will in fact begin to level out and produce curves more characteristic of the Pearl or Gompertz models, when observed for a sufficient length of time. The Lotka-Volterra equations will produce solutions that approximate the exponential models when the market is not saturated. It should be noted again that the Lotka-Volterra equations will not produce a closed-form exponential curve. They will, however, produce a curve that is equivalent to the exponential model for any arbitarily defined level of accuracy and any desired length of time.

The Lotka-Volterra equations reduce in closed form to the Pearl (or one-variable Fisher-Pry) model when there is no competing technology. The Pearl curve can otherwise be approximated by the Lotka-Volterra equations whenever competition is insignificant, such as when the competing technology Y is not near saturation, or whenever the competition coefficient γ is relatively small. Pearl curves are also produced in competive situations between technologies that have similar growth rates. The Pearl curve is basically an inhibitive model of technology growth.

In contrast, the Gompertz curve occurs in situations where there is intense competition between technologies. The competition equation will produce behaviors that are functionally equivalent to the Gompertz model when there are competing technologies that take prospective markets away from the technology X of interest. The Gompertz equation, then, is chiefly a model useful for competitive systems.

Substitution is an important class of behavior often evidenced by growing technologies. The Fisher-Pry model represents substitution of a single, saturated market by a growth technology. Competition between more than two technologies, or competition between two growing technologies can be forecasted using Marchetti's multiple substitution analysis. Details of this analysis are available (Marchetti, 1983; 1988). Substitution analyses that rely on logistic growth and logistic decay as descriptors of technological takeover constitute a specific set of solutions to the Lotka-Volterra equations. The Lotka-Volterra equations will produce this kind of growth pattern whenever competitive systems occur that include technologies of similar growth and takeover rates.

Figure 10.11b shows the forecasting potential of multiple substitution analyses. Marchetti and Nakicenovic (1979) show that equations fit from 20 years of data for four competing energy technologies provide good fits with historical developments over the next 60 years. Marchetti (1988) attributes the relatively poorer fit for natural gas to the fact that its market penetration was still small (about 2 percent) at the end of the base 20-year period. Note that this model anticipates a window of opportunity for a new energy source beginning in the 1970s. Nuclear power seemed to be growing to fill this niche, but external influences may well limit it. Extension of these energy substitution models at the International Institute for Applied Systems Analysis (IIASA) suggests another substitution due to begin about 2025 (possibly solar or fusion energy?).

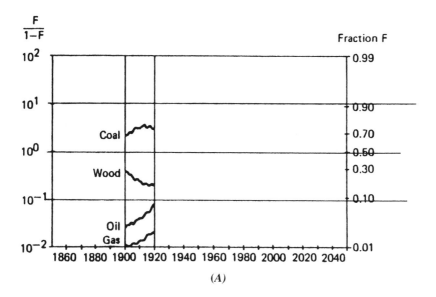

(A)

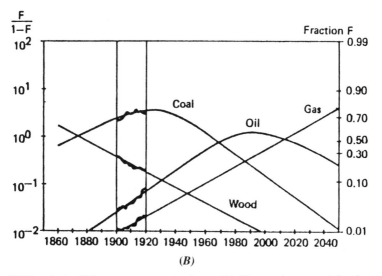

(B)

Figure 10.11   Lotka-Volterra energy projections. (*A*) 20-year data record for four comparing energy technologies. (*B*) Resulting projections. (*C*) Fit with actual data. (*Source*: Marchetti and Nakicenovic, 1979. Reprinted by permission.)

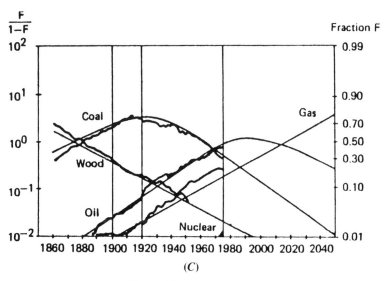

(C)

Figure 10.11 *(Continued)*

Note the left side of Figure 10.11 is scaled as $f/(1 - f)$ in powers of 10 (a log scale), as used previously in the book. The alternative scale on the right results from setting

$$f/(1 - f) = P$$

then solving for f

$$f = P/(1 + P)$$

The fraction of the total market, f, can be solved for various values of P; Marchetti emphasized the time constant—the time to go from 10 percent to 90 percent of the potential market. He notes that U.S. transportation technologies—canals, railroads, paved roads, and airways—show a period of about 55 years between their respective $f = 0.5$ (50 percent) market penetrations. This underlies his forecasts of windows of opportunity for successor technologies. It also lends support to the hypothesis of bursts of technological innovation in long waves (Volland, 1987).

Of his substitution analysis and its relationship to the Lotka-Volterra equations, Marchetti (1983, p. 3) comments:

> I am fairly convinced that the equations Volterra developed for ecological systems are very good descriptors of human affairs. In a nutshell, I suppose that the social system can be reduced to structures that compete in a Darwinian way, their flow and ebb being described by the Volterra equations, the simplest solution of which is a logistic.

Marchetti's multiple substitution analysis is strongly suggestive of the importance and power of successive families of logistic curves. While Marchetti normally examines single families of competing technologies, he notes that such families often grow within a larger market whose growth also can be described as logistic. This conception of successive layers that describe hierarchies of logistic functions has interesting implications for the technology forecaster.

It can be illustrated that logistic growth that occurs within a larger logistic market will indeed produce another Pearl or Fisher-Pry curve. It is clear, however, that the forecaster makes choices about the market and level of detail he or she wishes to study. Figure 10.12 illustrates a hierarchy of transportation technologies. At each successive level logisitic growth and competition functions may be successfully used to describe growth. In technology, as in nature, the exact taxonomy used for classification is not as important as representing the correct functional form. Substitution analysis, such as Marchetti's or Fisher-Pry, cannot be successfully used to forecast growth when addressing two technologies of radically different growth rates or from different market niches.

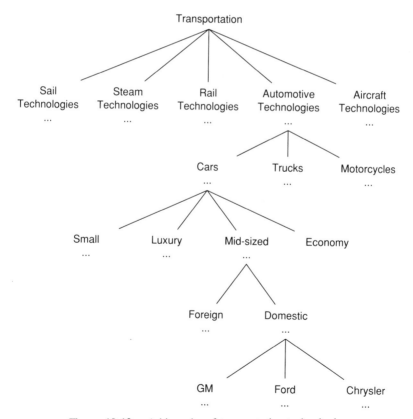

Figure 10.12   A hierarchy of transportation technologies.

Oscillatory models are a final class of models accommadated by the Lotka-Volterra equations. Periodic behaviors are commonly found in natural populations, and they can be successfully modeled using the Lotka-Volterra equations. Oscillatory behaviors have been observed in consumption and mining patterns in the United States and in car and transportation systems in Europe. It is suspected that the growth of certain technologies (such as robotics in the United States) follow an oscillatory rather than S-shaped growth pattern. Oscillatory behaviors present a special challenge to the forecaster using trend extrapolation methods. These growths often show a logistic start followed by an overshoot and then oscillation around a supposed limit. The unwary forecaster who naively applies a Pearl or Gompertz curve to start-up data, to data which are in fact oscillatory, will produce forecasts that are in serious error. The more complex population models, such as Lotka-Volterra, can represent such behaviors if the forecaster has correctly surmised their form.

Mathematical modeling using the Lotka-Volterra equations shows that oscillatory behaviors occur in certain specially defined systems that are known to biologists as preditor-prey systems. In the equivalent technological systems, the preditor technology benefits from interaction in a marketplace, while the prey technology loses from interaction. A gain from interaction is modeled with a negative competition coefficient (which subtracts from the available market). Figure 10.13 shows such oscillatory behavior seen in copper mining in the United States. Note the logisitic start-up in the early 1900s and the subsequent behavior not accounted for by Pearl or Gompertz models.

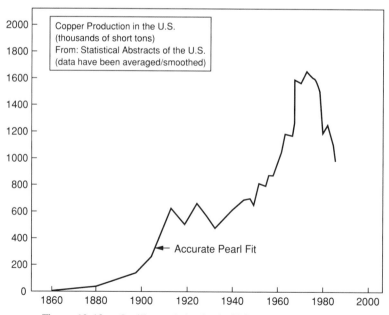

Figure 10.13   Oscillatory behavior in U.S. copper production.

10.4.3 Issues

This section has discussed the theoretical derivation of the Lotka-Volterra equations, their potential as an effective model of competing technologies, and the ways in which the model provides an overarching rationale for many conventional curve-fitting routines. Growth data, applied in the basic formulation of the Lotka-Volterra equations, are noiser than population data (for instance, it is easier to specify precisely the location of an automobile than its rate of change of location—velocity). Fitting change data to systems of equations can be done using nonlinear multiple regression, a procedure that grows more complicated as the number of competing technologies increases. Extensive data sets, often not available to the forecaster, are needed.

Until a technique is developed that allows a general solution to the Lotka-Volterra equations based upon time-series data, they cannot be used as a completely general trend extrapolation tool. Marchetti (1988), however, refers to excellent results in fitting some 100 empirical cases of technological growth using a numerical approximation program that constitutes a significant subset of potential solutions to the Lotka-Volterra equations (Marchetti and Nakicenovic, 1979).

The Lotka-Volterra equations offer a number of possible aids to the manager of technology:

1. An overarching model that is sufficiently general to describe most technological growth situations
2. A convincing rationale for trend extrapolation that provides clearly defined assumptions about the nature of technological growth (this rationale may be applied no matter what specific curve-fitting model is being used)
3. A rigorous and analytical treatment of competition within technological systems
4. A host of useful biological and population dynamics concepts and analogies that may be used when considering the interrelationships of technology and society
5. A solution set of curves that are richer than all current trend extrapolation techniques; such curves, once refined, might well prove as useful as current models have in forecasting technology growth (the Lotka-Volterra equations can model oscillatory as well as inhibitory systems not represented by current forecasting tools)
6. Links between trend extrapolation and simulation, modeling, and systems dynamics (the Lotka-Volterra equations offer an analytical technique by which alternative cases and scenarios about new and emerging technologies can be explored and refined.

10.5 CONCLUSION

Chapters 9 and 10 provide the essential tools for trend extrapolation. Chapter 9 presents the basics, including computational procedures to fit linear equations. By using appropriate transformations of nonlinear functions, these procedures can be readily adapted to all the commonly used technology trend models.

Chapter 10 provides a four-step approach to analyze technological trends. It emphasizes the Fisher-Pry (Pearl) and Gompertz curves—the S-shaped curves—and introduces the Lotka-Volterra equations as a promising general model of technological growth and competition.

The validity of trend analysis depends critically on a number of assumptions. Extrapolation assumes that the underlying structures and interacting forces remain the same throughout the period of extrapolation. This assumption always needs to be carefully considered. Sometimes, it is determined that trend extrapolation would not be justified. Even when it is justified, trend analyses should include sensitivity analyses, seeking to identify factors likely to disrupt the trend and estimating how the trend would be altered. Furthermore, trend analyses need to be combined with the use of complementary methods, especially monitoring and expert opinion. Taken together, these can provide the bases for a sound forecast.

EXERCISES

10.1 Examine the data for 1978 to 1988 in Table 10.4 for tapes and CDs. Try to establish a leading indicator relationship between the two series (recall the presentation on leading indicators and regression in Chapter 9). Use the relationship to project CD shipments for 1990. Comment on your results.

TABLE 10.4   Shipments of Audio Media (millions of units)

Year	Long Play/ Extended Play Records (LPs)	Cassette Tapes (Tapes)	Compact Discs (CDs)
1978	340	65	0
1979	315	80	0
1980	320	110	0
1981	295	140	0
1982	245	185	0
1983	210	240	2
1984	205	335	10
1985	170	342	22
1986	130	345	53
1987	110	412	110
1988	70	452	145

Source: Recording Industry Association

10.2 Develop a Fisher-Pry substitution model for CDs. Figure f as the share of audio shipments held by CDs in a given year (lump records and tapes as the old technology). Project f for 1990. Calculate a 90 percent confidence interval for 1990. Comment on the strengths and weaknesses of your Fisher-Pry substitution model.

10.3 Try the alternative procedure given in Equation 10.3 to calculate Fisher-Pry substitution for CDs (Table 10.4). Which year-to-year increase would you choose to figure α ? Note that α is defined in terms of annual fractional growth. You cannot simply compute this on the basis of growth over, say, a five-year period, divided by 5. Instead, much like compound interest, you must compute it as

$$(1 + \alpha)^n$$

where n is the number of years spanned.

Using this alternative procedure, you should:

1. Calculate a value for α;
2. Then, using the f values for one known year, calculate t_O;
3. Use this equation to project f in a certain year (or, conversely, to determine the year f will reach a certain level).

Project f for 1990. Compare with the results of Exercise 10.2.

10.4 Enter data for tapes in the TOOLKIT. Compute both Fisher-Pry and Gompertz models. Comment on their strengths and weaknesses.

10.5 Show that the Fisher-Pry equation given by Equation 10.4 and the linear form of the Pearl equation (Equation 10.3) are the same equation when b and $\ln(c)$ are expressed in terms of α and t_0.

___11
EXPERT OPINION

OVERVIEW

Gathering and analyzing the opinions of experts are very important activities for the technology manager. This chapter describes approaches for identifying experts and selecting a technique. Common methods for gathering opinion are introduced and contrasted: interviews, meetings, surveys, the Nominal Group Process, the Delphi technique (and variations on it), and the POSTURE method. The example of building a computing environment for a small engineering college (Commission for Computer Integration, 1989) is used as a unifying theme throughout the chapter.

11.1   INTRODUCTION

Throughout his or her career, the technology manager will be faced with many decisions for which he or she lacks specific expertise. In some instances, the need for decisions based on this expertise will be immediate. In others, the manager will need to make preliminary decisions to position the firm advantageously vis-à-vis longer-term projections of technology growth. Regardless of which type of decision is faced, expert input is often crucial. However, it is important that the manager not forfeit responsibility for any decision to experts. Experts seldom command the whole range of conditions that the manager must consider. Nor do they have the ultimate responsibility to sell the decision to management, to implement programs, and to reap the reward or shoulder the blame for failure. The manager must synthesize expert input with a personal assessment of risks and potential benefits and add an understanding of the corporate environment that only he or she can provide. The manager must devise methods to secure expert input that can be used to reach a decision.

Two fundamentally different kinds of expert input may be required: *fact* and *opinion*. The former deals with existing facts, the latter with future facts. The search for facts generally is the more tractable situation. Experts may be biased or may have differing interpretations of the facts, but their inputs can be cross-checked against each other and against those of other sources. The search for future facts is more uncertain. While individuals in the field or related fields can be expected to give informed opinions, bias often is difficult to isolate, and even a group of experts can be caught up in prevailing fashions (Schnaars, 1989). While expert input about facts *may* differ, expert opinions *will* differ. This chapter emphasizes special problems inherent in obtaining opinion.

To gather expert input, a manager must accomplish two fundamental tasks: (1) the expertise required and the individuals who possess it must be identified, and (2) method(s) to obtain their input must be chosen. The choice of method often will influence the selection of experts and vice versa. Different methods will be dictated by the location, availability, and schedules of experts, and by their individual and group characteristics. For example, the experts identified may not be employed by the firm or may be geographically scattered. They may hold vastly different world views, speak different technical languages, be mutually antagonistic, or simply have no history of effective communication and cooperation. Under such circumstances, a traditional, interactive, face-to-face meeting may be inadvisable or impossible.

Before selecting a method, the manager also must assess the type of information needed, the detail and accuracy required, the timing, and its potential importance to achieving corporate goals. Each method has different strengths and weaknesses, produces different forms of input, and implies a wide range of costs. Often, more than one method will be required. The tool must be adapted to the task—a sledge-hammer is of little use to attack a fly on a window.

11.2 SELECTING EXPERTS

Before selecting experts, the technology manager must determine the range of expertise he or she requires to provide input for the decision. This is not a simple or an error-free task, and it must be carried out iteratively. Perhaps the best place to start is by probing the dimensions of the decision. Only when these are understood can one determine the input required. Decision dimensions often are controlled by factors internal to the manager's organization. Therefore, he or she should turn first to individuals within the firm; they compose the first group of experts to be consulted. While they may not be experts in the technology being considered, they are expert in those areas in which the technology will be used, maintained, administered, sold, and/or manufactured. Thus their input is vital as is their commitment to and support of the decision.

To identify those within the company that the decision will affect, the manager might begin with a series of questions.

- Who will implement the new technology?
- Who will maintain it?
- For what tasks will it be used?
- What departments or activities will be influenced by the results of the application?
- What resources are available?
- What potential costs and benefits might be expected?

Answers to these questions, no matter how tentative, help the manager map information needs.

For example, suppose a small engineering college (SmEC) with a limited computer base wishes to expand and modernize its computer system to provide a platform for modern educational and engineering applications. The data systems manager is to determine the computing system required and to make decisions about the way in which its use will be integrated into the engineering curricula at SmEC. The data systems manager might begin by soliciting input from management (the president, business manager, development director) about resources that will be available. Next he or she might turn to the faculty to determine academic uses that might be made of the system, to students who will use it, and to the computing center director who will implement, maintain, and operate the system. After consulting with these individuals, the data systems manager might form a committee to probe more deeply.

Next the data systems manager should determine if others within the organization have faced similar decisions. Who are they? How did the decision differ? What information did they seek? From whom? What information do they wish they had possessed?

The manager should then determine if there are individuals in other organizations who recently have made similar decisions. Articles and ads in trade literature can be useful sources for such contacts, as can trade and professional organizations. In this step, the manager broadens the range of experts. The individuals identified may have experienced the changes produced by decisions similar to that which the data systems manager must make.

Having completed these steps, the data systems manager can make a preliminary estimate of the expertise that will be required to provide an information base. For instance, the data systems manager at SmEC might decide that he or she needs to know what hardware and software systems are available and what the requirements and costs (factual information) of such systems are. Other factual information needs could include how the systems presently are used in other engineering colleges, faculty resource requirements, and measures of effectiveness and acceptance by faculty and students. Opinions might also be needed to acquire other necessary information: projections of improvements in computer workstations; software and networking technologies; probable changes in industry and computer standards; assessments of potential educational applications and changes in faculty and student productivity; and estimates of probable accrediting agency positions on computer-based learning.

Vendors, technology developers, and established users are good sources for factual information. After confidentiality agreements are executed, they are knowledgeable about the directions of their company's product development and research. They also are good sources for opinions about industry directions and new product introduction and timing. While vendors are biased, their bias is clearly indicated by the logos on their business cards; the bias of experts from other areas is not so easily determined.

For opinions about issues farther removed in time and/or content from production and sales or about issues that span multiple technical, social, or economic fields, the problem of identifying experts may require trial and error. Lipinski and Loveridge (1982) note that in nearly any field there is a peer group who knows the current pecking order of the group. These various peer group networks should be searched until individuals emerge who are recommended repeatedly. Individuals within the firm, government, or university, and "think tank" personnel often can supply starting points for network searches, as can professional organizations, journals, and periodicals.

Merely identifying individuals is not the end of the process. Although various classifications of the type and degree of expertise are possible, Lipinski and Loveridge (1982) suggest considering the following types of experts:

- Generalists with a spread of interests and perceptions that give them a high level of awareness of the broad context
- Persons of thought who have particular and deep knowledge of specific fields
- Persons of present or future action whose present or likely future positions make it possible for them to affect the technology (p. 214)

They conclude that experts possess three important characteristics: substantive knowledge in a particular field, ability to cope when faced with uncertain extensions of that knowledge, and imagination. In most cases, the technology manager will not have the time or resources to determine all three characteristics of each expert. However, the scale for self-ranking of expertise presented in Table 11.1 may be useful in assessing substantive knowledge.

Finally, the technology manager must ascertain that the range of expertise represented by the experts selected covers that previously determined to be necessary to support the decision. It should be realized that most experts will have subsidiary knowledge in related fields as well as deeper expertise in their primary field. This subsidiary knowledge may be sufficient to cover less-critical categories of information. Galanc and Mikus (1986) describe means to determine the minimum number of persons from each area necessary to cover the field of interest.

TABLE 11.1   Self-evaluation Criteria: Guidance to Self-ranking of Expertise

1.	You are unfamiliar with the subject matter if the mention of it encounters a veritable blank in your memory or if you have heard of the subject yet are unable to say anything meaningful about it.
2.	You are casually acquainted with the subject matter if you know at least what the issue is about, have read something on the subject, and/or have heard a debate about it on either a major television or radio network or an educational channel.
3.	You are familiar with the subject matter if you know most of the arguments advanced for and against some of the controversial issues surrounding this subject, have read a substantial amount about it, and have formed some opinion about it. However, if someone tried to pin you down . . . you would soon have to admit that your knowledge is inadequate.
4.	You were an expert [with the subject matter] some time ago but feel somewhat rusty now because other assignments have intervened, [However,] you have kept reasonably abreast of current developments in the area, thus trading breadth of understanding for depth of specialization.
5.	You should consider yourself an expert if you belong to that small community of people who currently study, work on, and dedicate themselves to the subject. Typically, you know who else works in this area; you know the literature of your country and probably the foreign literature; you attend conferences and seminars on the subject, sometimes reading a paper and sometimes chairing the sessions; you are most likely to have written and/or published the results of your work. . . . Other experts in this field may disagree with your views but invariably respect your judgment; comments such as "this is an excellent person on this subject" would be typical when inquiring about you.

Source: Lipinski and Loveridge, 1982. Reprinted by permission from *Futures,* vol. 14, no. 2, 1982.

11.3   SELECTING THE TECHNIQUE

Nelms and Porter (1985) provide a useful scheme for classifying the various expert opinion techniques and, more importantly, for choosing among them. They note that Gustafson and colleagues (1973) identify three distinct processes in expert opinion capture: *talk* (interaction), *feedback* (one-way communication), and *estimate* (decision). All techniques described in this chapter embody combinations of these three processes. Table 11.2 shows the techniques discussed in this chapter (plus brainstorming, which is covered in Chapter 7) with their classification relative to the three processes.

There are six factors (Nelms and Porter, 1985, pp. 57–58) to consider in choosing a technique or designing one especially for a given situation:

1. *Logistics*—Resources largely determine the method. If severely limited, a one-round estimate (that is, no feedback with additional input from participants) provided by a survey is the most appropriate. However, feedback

TABLE 11.2 Classification of Expert Opinion Techniques

Technique	Talk (T)	Feedback (F)	Estimate (E)	Process
Committees	X		X	TE
Brainstorming	X			T
Nominal Group Process	X	X	X	EFTE
Surveys			X	E
Delphi		X	X	EFE
Shang Inquiry		X	X	EFE
EFTE	X	X	X	EFTE
POSTURE	X		X	ETE

is desirable, so estimation plus feedback is usually the minimum desirable format.

2. *Feedback*—The format used for feedback does not seem important as long as extreme positions are not overemphasized. Minimizing delay between successive feedbacks seems desirable.

3. *Communication medium*—The medium of exchange appears to be a function of access to experts, resources, and time.

4. *Sample size*—While it usually is desirable to maximize sample size, studies indicate that groups of 12 or fewer converge on valid results even when individual judgments are only moderately valid.

5. *Stopping rule (multiple communications)*—The number of rounds used is a tradeoff of time and resources against the stability of group and individual responses. Consensus can be used as a criterion, but it occurs at the cost of increased group pressures.

6. *Interaction*—It appears that interaction is most usefully seen as a function of task complexity and of group social structure (for example, social complexity increases if there are serious disagreements that individual members care about deeply). The greater the complexities, the more interaction is needed.

The techniques discussed in the following section differ in their strengths and weaknesses, as well as in their time and resource demands. In many instances, a combination of techniques will be the most satisfactory.

11.4 TECHNIQUES FOR GATHERING EXPERT OPINION

Common methods for gathering expert input are considered in this section. Those used in the individual mode, such as a face-to-face (in person) or telephone interview, are detailed first. Those used in the immediate group mode—that is, face to face but with several experts at one time—are considered next, followed by a discussion of group methods usually applied at arms length. Many technology managers favor face-to-face methods because such techniques allow the manager to assess the candor, judgment, and creativity of the expert. However, it often is difficult to meet individually with numerous experts in person or collect all of them at a single place and time. Arm's-length techniques, such as surveys and the Delphi technique, can provide structure and depth that is often impossible in the give-and-take of a face-to-face meeting. Thus each approach has value in the right circumstances. (For an excellent review of various input techniques see Wolfe, 1987.)

11.4.1   Individual Input

Individual input can be obtained in person, by telephone, or by electronic mail. In-person and telephone interviews are quite common. These interviews can be structured, focused, or nonstructured.

A *structured interview* has an explicit schedule of fixed questions that usually are in closed form (the responses to open-ended questions are sometimes difficult to decipher and categorize). An example of a recent structured interview asked schoolchildren specific questions concerning the level of technology that they expect to experience at the beginning of the twenty-first century: "Do you think that you will be living and working in space in the year 2000?"

A *focused interview* is directed to respondents known to have pertinent knowledge. Prior to the interview, the interviewer must carefully study the topic. An interview guide should be prepared that specifies a set of topics to be discussed. The emphasis of this type of interview is to obtain subjective information regarding the situation under study. For example, the following question might be directed toward a computer chip manufacturer: "What advances do you see in the development of VLSI (very large scale integrated) chips in the next 10-year period?" This is a question that does not have one answer. It would be asked only after the interviewer has carefully considered the possibilities for future chip development and has studied the progression of integrated circuitry. The interviewer should have enough background knowledge to ask for clarification if necessary or to ask for expansion if the response is not sufficiently detailed or if the response goes off in a tangential direction.

The *nonstructured*, or nondirected, *interview* has no schedule and no prespecified set of questions. Respondents are encouraged to reveal their opinions and attitudes as they see fit. This type of interview is difficult to control, and the responses obtained are difficult to compare. A sample opening question for such an interview might be: "What advances can we expect to see in office automation in the next 10 years?" From this point, the interviewer must wing it, following up responses with additional leading questions. After talking with several respondents, the interviewer draws inferences from the collected responses.

When using either focused or nonstructured interviews to gather information, the interviewer must be clear on the objectives to be accomplished since interviewees may ask you to explain the purpose of a particular question. It also is important in all three types of interviews that only critical questions be asked; questions that are merely interesting should be saved for a cozy discussion by the fire. The interviewee's attention can be captured for a limited time, and any question that is not directly pertinent to the technology forecast could sacrifice more important information. Interviews should last no longer than 45 minutes, which limits the number of questions that can be asked.

It is highly advisable to conduct a *pilot interview* prior to the real thing. In the pilot, several respondents, or stand-ins, are asked the questions that will be raised in the actual interview. If they have problems answering these questions, or if they give responses in a different manner than that anticipated, revision of the interview is needed.

These revisions should remove ambiguity, maintain pertinence, and revise leading questions. Leading questions can bias responses severely; asking the same question different ways can provide strikingly different responses. "Office automation seems to have reached a plateau in 1990, do you agree or disagree?" is an example of a leading question. The respondent is inclined to agree, thinking: "Yes, by 1990 a lot did happen in office automation. Perhaps the interviewer knows a lot about this subject; I don't want to disagree with someone who is knowledgeable about this."

The sequencing of questions in a personal interview also affects the answers that are given. Questions can be funneled from the general to the specific, or, less frequently, in the opposite direction. Funneled questions lead from one to another: (1) "What do you think are going to be some of the most important technological developments in computer operating systems in the next 25 years?" (2) "Of all those developments mentioned, which do you think will be the most important?" (3) "On what basis did you select these as the most important?" In the inverted funnel sequence, narrower questions are followed by broader ones.

11.4.2 Committees, Seminars, and Conferences

These group techniques require experts to be in the same place at the same time. Proceeding from committee to seminar to conference, the formality of meeting format and the number of participants generally increase, while the opportunities for interaction decrease. Committees usually meet more frequently than seminars, and seminars meet more often than conferences. Generally, committees are charged with a specific task, whereas conferences and seminars exist to exchange information. Finally, committees most often have fixed membership; conference and seminar attendance is less rigid, although there usually is a core of frequent participants. Committees, seminars, and conferences are all potentially beneficial platforms for expert input; however, the committee is the most familiar, and it is the only one that will be discussed in detail.

The committee is ubiquitous in modern organizations. It is the most common method used when more than one expert's input is required. It is the easiest face-to-face meeting to convene (especially if experts are employees of the organization) and a comfortable format for participants. Martino (1983) notes two major advantages to the committee method: (1) the total knowledge of the group is at least equal to that of any one member and (2) the number of factors that it can consider at least equal those of any one member. The former is important because the meeting is intended to increase the knowledge base for a decision; the latter is significant because bad decisions often can be traced to neglect of a critical factor. While the committee also possesses at least as much incorrect information as any member, it is assumed that the group's pooled knowledge will cancel out this misinformation.

Although the committee meeting is common, it often is poorly conducted and frequently is not seen as productive relative to time invested. A successful meeting depends heavily on the skill and preparation of the chairperson and of the committee members. Although the meeting usually is conducted in an unstructured give-and-take format, the chairperson should prepare an agenda and ensure that members understand the topic and goals in advance. Members also should, but seldom do, prepare. Moreover, there are limitations inherent in the give-and-take process that can limit the success of the unstructured committee meeting.

In a face-to-face meeting no one is anonymous. Experts are identified with positions that they feel must be defended to preserve their credibility. Other members suppress input in deference to experts or to those of status, authority, or power. The pressure to conform often produces a false or premature consensus through bandwagon effects. Group conclusions sometimes can be influenced more by the repetition of an input than by its validity. Thus the premise that information will cancel misinformation often goes unrealized. Although some of these deficiencies can be countered by effective leadership, the complexity of the topic often defeats efforts to stay on track, and the meeting wanders from the issue.

There are benefits of the committee meeting that are overlooked when productivity is considered too narrowly. For example, if the pressures for consensus inherent in the meeting are handled effectively, a group identity and a commitment to the subsequent decision can be forged. Moreover, the meeting can educate committee members about dimensions of the decision and about developments and concerns in areas of expertise other than their own. Thus benefits from committee activities can sometimes surface in areas quite different from immediate meeting goals.

Returning to our example of the small engineering college, based on preliminary discussions with internal experts, the data systems manager forms a committee to explore the dimensions of the computer system decision. This committee is made up of representatives of the administration, students, and one faculty member from each academic department at SmEC. Each is expert in some dimension of the problem, and some also are knowledgeable about computer hardware, software, and applications. The data systems manager sends the first meeting agenda to each member and states that the committee's goal is to establish the characteristics required of the new computing system. Before the meeting, the manager discusses the agenda and goal with each member and suggests types of input that seem most useful. The data systems manager plans the first few meetings as opportunities for participants to educate each other about their needs and expectations. Each meeting is conducted so that everyone participates, and overly vocal or dominant individuals are neutralized by a statement such as "Thanks for that input, does anyone else have a comment?" Additional meetings (described in later sections) are planned to develop a substantive definition of the new computing system.

Brainstorming is a technique used to produce creative ideas without evaluating them (this technique is discussed in Chapter 7). Imagine that the data systems manager at SmEC might devote an early committee meeting to a brainstorming session to elicit ideas about the kinds of applications the new computer system should support. Later, the manager and/or the committee might evaluate the suggestions to determine those with promise. These might require additional expert input for full evaluation.

11.4.3 The Nominal Group Process

The Nominal Group Process (NGP) is a powerful group technique developed by Delbecq and Van de Ven (1971). It has been used effectively in the early stages of problem definition (Roper, 1986, 1988; Sachs and Clark, 1980); it also can be used to define the dimensions of a decision and to identify options and questions for conventional or Delphi surveys. Like brainstorming, the NGP is designed to overcome the unproductive aspects of unstructured, face-to-face meetings and to stimulate creative thinking by a group of experts. When carefully applied, undue pressures for consensus, bandwagon effects, and the negative effects of hostility between participants can be avoided. The process also is structured to build strong group identity and support for subsequent decisions.

The NGP is based on indications that small groups that *do not interact* are the most creative idea generators, while groups that *do interact* perform the best evaluation (Delbecq, 1974). Thus, unlike brainstorming, the NGP incorporates evaluation as well as idea generation. At the close of an NGP meeting, the manager not only will have a rich list of ideas, he or she will also have a preliminary rating of those the group feels are most important. The six steps of the NGP are presented in Exhibit 11.1. Also given there are brief statements of the objectives of each step. As in any group meeting, the manager should carefully explain the agenda and the meeting goals before starting.

■ **Exhibit 11.1 Steps in the Nominal Group Process (Based on Roper, 1988; and Delbecq, 1974).**

Step 1: Silent (Nominal) Idea Generation. Each participant is urged to list all of the factors that he or she believes are central. Participants work silently and individually from 10 to 25 minutes.

Objectives: "To give the participants time to think, to force them by tension and social competition to produce, to avoid polarization and premature evaluation and closure, and to avoid pressure for conformity and deference to status."

Step 2: Group Round-Robin Listing of Factors. Sequentially, each participant reads one factor from his or her list. This is done without explanation or advocacy and without question or comment by the group. Each factor is written on a flip chart and posted for all to see. This process continues until all factors have been listed; generally it takes 45 minutes to an hour. The round-robin sequence gives each participant control of the floor in turn and draws input from reticent individuals unlikely to contribute in the give-and-take of a normal meeting format. The process often triggers new factors not listed during Step 1.

> Objectives: "Equal sharing of ideas, depersonalization of ideas, production of a written record, focusing of the problem, visual and auditory concentration, polarization of conflicting viewpoints."

Step 3: Discussion and Clarification of Listed Factors. Each factor listed on the flip charts in Step 2 is discussed informally by the group. Factors are considered in sequence. The primary thrust is clarification. The group leader probably will find it best to discourage advocacy at this point, as the cooperative spirit of the group may still be fragile—especially if antagonism initially existed among participants. During the discussion, additional factors may be added to the list. These factors may be new or may combine factors already listed to reduce repetition and overlap. New factors may be formulated by individuals but are generally group products.

> Objectives: "Clarification, reduction of repetition, overlap, and ambiguity."

Step 4: Silent (Nominal) Individual Written Voting on Priorities. Each participant is asked to silently and independently select the most important factors from the list compiled through Step 3. This is not the final ranking of factors; rather it is intended to focus attention on the quest for the most important factors, thus reorienting the thrust of the first three steps during which breadth, not focus, was sought. The specific number of factors participants are directed to select is flexible. However, Shillito (1973) suggests having each participant vote on 20 percent of the number of factors on the list. Occasionally, this may prove an unwieldy number. If so, a smaller percentage can be selected.

> Objectives: "Focus on most important issues, rank ordering of items, consideration of alternatives."

Step 5: Discussion of Voting Results. Before the start of this step, the group leader tabulates the votes from Step 4 and displays the factors with the highest totals on the flip chart. The group is then asked to evaluate the result. Typically, they are not satisfied. This dissatisfaction sharpens the focus. A general discussion of why the preliminary result is unsatisfactory leads to a more general discussion of evaluation criteria and finally to the formulation of new factors that encompass the concerns of the group. New formulations often are produced by categorizing factors from the list under major areas of concern. By the time this step has been reached, a common group identity generally has been formed and a genuine concern to determine the most important factors is manifested. Antagonism is seldom displayed among participants. While this step often can take as long as the group leader is prepared to spend, a useful limit is 30 minutes.

> Objectives: "To compare perceptions of the approaching consensus and to clarify misinformation."

Step 6: Final, Silent (Nominal), Individual Written Voting. In this step, participants are asked to select the most important factors from the list as expanded through Step 5. Selection is done silently and individually. Often a rating scale that ranges from 0 (no importance) to 100 (absolutely necessary) is used, with each participant displaying the relative importance of the factors he or she selects. The number of factors that participants are directed to choose should follow the guidelines suggested in Step 4. It should be emphasized that the preliminary ranking of factors (Step 4) has served its purpose—that is, it has sharpened the focus of the group. Therefore, factors should be selected from the entire list without regard to the preliminary ranking. In fact, it is best to remove the flip chart reporting the preliminary results before voting.

Objectives: "Final re-ranking of ideas and closure." ∎

The NGP provides an effective platform for communication among experts from different fields. Comparisons (Bakus et al., 1982; Delbecq, Van de Ven, and Gustafson, 1975; Van de Ven, 1974) have shown that NGP meetings typically generate nearly twice as many ideas as unstructured, face-to-face meetings and somewhat more ideas than Delphi. The chairperson should run one or more trial NGP meetings before the real thing. He or she also may wish to review the detailed instructions given by Delbecq, Van de Ven, and Gustafson (1975).

Suppose the data systems manager at SmEC decides to use the NGP for a meeting of the committee of internal experts. Before the meeting the manager explains the process and that it will be used to examine academic objectives that should be sought through classroom/laboratory computer applications. The manager also explains that this information will form part of the input necessary to define the computing environment that the new system must foster. The goals of the meeting are to develop a broad list of objectives and an initial indication of which are most important. At the NGP meeting, the seven participants compile a list of 60 objectives. The final ranking of objectives is shown in Figure 11.1.

The NGP format can be modified to fit special circumstances. The most common is that the number of experts is greater than effectively can be accommodated in a single session. Delbecq (1974) suggests, for example, that the optimum number of participants is six. Others (see Roper, 1988) have found groups larger than ten unsatisfactory because the number of ideas generated is too large to be handled effectively in Steps 4 through 6 (see Exhibit 11.1). In this situation, Delbecq (1974) suggests dividing participants into subgroups that individually pursue Steps 1 through 3. These subgroups are then combined to report the results of their efforts. The combined group completes the remaining steps using a master list developed from individual subgroup results. However, this process is difficult to complete at a single session. Also, unless duplications are eliminated, the master list may be too ambiguous, overlapping, and repetitive to produce a satisfactory result. Therefore, Roper (1988) suggests planning two sessions and consolidating subgroup lists before beginning Steps 4 through 6 with the combined group.

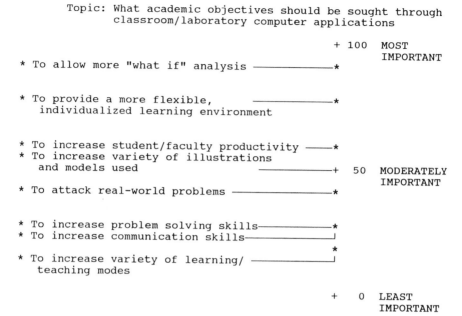

Figure 11.1   Final ranking of academic objectives from the Computer Committee at Sm EC. From materials developed during deliberations of the Commission for Computer Integration (1989).

When the primary purpose of the meeting is to develop breadth of perspective or to compile a rich list of ideas, Steps 1 through 3 can be very effective by themselves. In this format, the NGP resembles a structured brainstorming session. Limiting the process to these steps also can be used to gather input while delaying the perception or advent of closure. However, this weakens group identity and hence support for the final decision (Roper, 1988).

11.4.4 Surveys

Surveys are the most common method for soliciting input from groups of experts when face-to-face meetings are impractical. The method is popular because it is relatively quick, reasonably easy, and inexpensive; it can be used to collect input from a large number of experts in a variety of formats; and it avoids the negative dynamics of face-to-face meetings. Disadvantages (such as failure to produce consensus and problems with bad samples and bias) are shared to various degrees by all expert opinion techniques. Traditional surveys are one-time affairs and do not provide a formal feedback mechanism such as that built into the Delphi technique.

The survey embodies a number of basic assumptions. For example, group input is more likely to be correct than individual input. As in the face-to-face meeting, it is assumed that information will cancel misinformation. Since traditional polls do not incorporate feedback, however, canceling must be achieved by some scheme for averaging responses. This technique also assumes that questions have been posed clearly, concisely, unambiguously, and in vocabulary shared by the respondents. It is also assumed that surveys are unbiased and do not include unstated or misleading assumptions. Such qualities are not always easy to ensure. The reader may wish to pursue examples given by Emory (1976) for a better understanding of the problems. Generally, at least some questions should be open-ended. This allows respondents to "use their own words" and thus gives a clearer picture of meaning and rationale.

Surveys can be carried out in the personal mode (face-to-face), the impersonal mode (printed questionnaire sent by mail), or in a combination of these modes (Emory, 1976). They also can be executed by computer via an electronic network (such as Bitnet) if the means are available to most respondents. This electronic mode offers some opportunities for unstructured feedback even in the traditional survey format. It also provides a means to begin processing some of the responses quickly. The personal mode is likely to be most expensive and time-consuming, but offers opportunities for increased flexibility and first-hand evaluation of expert credibility.

Once information needs have been identified and experts chosen, the next step is to construct the survey. Before beginning, the manager should review a basic text in the field such as Emory (1976), Jolliffe (1986), or Raj (1972). The manager then should *begin at the end* by considering the expected results: What response format will be most useful? How will input be organized, processed, and analyzed? Will input analyzed in this way fulfill information needs? Next, a draft survey should be developed and tested on a sample group (perhaps colleagues). Finally, the survey must be revised to remove deficiencies that have been found.

To be effective, the survey must be carried out carefully, with due consideration given to sampling, bias, construction, and analysis. Before executing a survey, the manager must carefully consider the form and substance of the input to be gathered in the context of the decision it is to inform.

Martino (1983) gives hints for good survey questions. Compound statements, for instance, should be avoided because it is hard to decide what responses to them mean. Suppose a question is posed to determine how important it is to provide symbolic manipulation support for teaching calculus on SmEC's mainframe computer. A low rating may mean that the expert does not think symbolic manipulation should be used in calculus; on the other hand, it may mean that the expert thinks personal workstations are better solutions than the mainframe. Ambiguous questions should be avoided for similar reasons. Ambiguity often stems from terms that are not well defined or from terms that mean different things to experts from different fields.

Too much descriptive material in a survey can be as harmful as too little. Additional information appears to help with concepts unfamiliar to participants and confuse them when concepts are familiar. Individuals also seem to find it more difficult to deal with negatively posed questions and questions that use negative comparative adjectives (the term "less", for instance) than with those framed positively (Cassels and Johnstone, 1979). Finally, Lipinski and Loveridge (1982) note that many persons find it difficult to estimate probability. For example, "When will there be a 25 percent probability that something will occur?") is a difficult question to answer. Estimates of "nearly certain" (greater than a 90 percent probability), "an even bet" (50 percent probability), and "almost no chance" (less than 10 percent chance) appear to be easier to handle.

11.4.5   The Delphi Technique

The Delphi technique is a special form of survey. Delphi, however, is designed to ensure a participant's *anonymity, controlled feedback and iteration*, and a *statistical group response*. The process was developed in the early 1950s at the Rand Corporation for military applications. After declassification, it was recognized as a useful method for technology forecasting and planning by American business and industry. Delphi since has spread around the world. Unfortunately, its rapid diffusion often has led to careless use. This carelessness stems, in part, from uncritical acceptance of the technique without adequate attention to underlying assumptions and requirements (Rieger, 1986; Sackman, 1974).

A Delphi survey is composed of a series of questionnaires known as "rounds." Input from each round is gathered, analyzed, and fed back to participants in the next. Typical feedback includes the median and the interquartile range of group response and the rationale for agreement or disagreement with group input. The first round normally should be open-ended to ensure that important aspects of the topic have not been overlooked by the manager in preparing the survey. The process continues until response stabilizes—usually within four rounds. A sample application of the Delphi technique is presented in Exhibit 11.2.

Delphi participants generally act anonymously. This reduces tendencies for an individual to defend untenable positions in order to preserve credibility and for a vocal or powerful individual to dominate. Feedback is intended to inform participants of the current position of the group and arguments for and against different points of view inherent in that position. Martino (1983) notes that an important aspect of the feedback is to allow the group to concentrate on the goals of the *process* rather than on social goals "such as winning the argument or reaching agreement for its own sake" (p. 17). Statistical treatment of responses is intended to portray differences of opinion within the group; the arguments are intended to portray the rationale supporting those differences.

The growth of Delphi has not been without controversy. In the 1970s, many issues of contention were raised (such as the value of anonymity, superiority of pooled group judgments, the validity of qualitative versus quantitative research within the scientific paradigm). However, at this stage of Delphi development, questions of sound application appear more important to the technology manager. Rieger (1986) suggested that Delphi is most appropriate for normative forecasts and noted that areas that are value-laden and that demand participation by multiple publics are prime targets for Delphi.

Many users pay insufficient attention to basics such as population sampling, piloting procedures, questionnaire validity, and reliability. Population sampling and piloting procedures are common concerns of all survey techniques. Questions about validity and reliability, however, are more critical to Delphi because of its multiple-round character. These primarily deal with whether or not stability has been reached in the responses to succeeding rounds. Mouly (1963) suggested retesting to determine if responses change from the original survey. However, the time and work involved in Delphi surveys (for both manager and participants) make this impractical in most cases. A more practical approach is application of the chi-squared test for group and individual response stability suggested by Chaffin and Talley (1980) and Dajani, Sincoff, and Talley (1979).

Martino (1983) gives useful suggestions for Delphi surveying. First, the manager should obtain the agreement of experts to participate, which may require contracting and paying them. Careful selection of experts and attention to sampling are wasted if the survey has a low response rate. Second, the Delphi method, especially its iterative nature, should be explained carefully to all participants. Next, the manager should make the survey as easy to complete as possible. The number of questions also should be limited. Martino suggests a maximum of 25 in most cases. Sometimes mutually exclusive events or conditions may be included as a result of input from the unstructured round. The reason for these inclusions should be explained so that participants do not feel their consistency is being checked. Finally, managers must resist temptations to include their own input to avoid distorting survey results.

A lot of work is required to conduct a Delphi survey (Linstone and Turoff, 1975). To maintain an acceptable turnaround between rounds, computer analysis is desirable, even for modest-sized polls. Even then, each round of a mail survey will take about one month. (This time can be reduced considerably by using electronic mail or next-day delivery.) Internal surveys, however, can be completed in less than one-half this time.

■ **Exhibit 11.2   Application of Delphi to Determine Computer System Functionality (Based on Commission for Computer Integration, 1989).**

The data systems manager from our earlier examples has combined Nominal Group Process input with other input to define the ideal computing environment for SmEC. The next step is to determine the system characteristics necessary to deliver that environment. The manager would like to pose these characteristics independent of hardware and software considerations so as not to limit the range of possible vendor equipment or to cloud group perspective by what specific equipment will or will not do. Based on analysis of various inputs, the manager poses system characteristics in terms of *functionality*—that is, the capability of the system to deliver the services required to provide the ideal computing environment. The group decides functionality, F, can be described in terms of three components—ease of use, E; versatility, V; and capacity, C:

$$F = w_E E + w_V V + w_C C$$

where each w represents the relative importance (or weight) of each component.

The manager needs three types of information to flesh out the descriptor F:

1. The factors upon which the components depend
2. The appropriate weight for each component
3. The appropriate weights for each factor composing each component

Once this information is obtained, the manager can evaluate the existing system and determine specific areas in which changes are needed and the relative importance of those changes. F also will provide a metric with which to measure products offered by different vendors. Because of the complexity of the deliberation, the manager decides to use Delphi.

The manager begins by explaining Delphi, the type of input that is desired, and how it will be used. These descriptions are given in writing for future reference and are discussed. Each component of F is defined:

Ease of use—Measures the system's physical and operational characteristics that affect the ease with which it can be used to accomplish educational tasks.

Versatility—Measures the ability of the system to accommodate both a broad range of uses and changes in those uses—in other words, measures the system's robustness and flexibility.

Capacity—Measures system capability to perform the tasks necessary to support the academic goals of students and faculty.

Round 1 of the Delphi poll identifies 9 factors upon which participants feel E depends, 12 factors that determine V, and 7 that define C. Results for Rounds 2 through 4 are shown in Figure 11.2 for one of the factors defining E. (Figure 11.2 is shown in the same form used to provide feedback to the group.) The weights for each component determined by normalizing group responses are

$$F = 0.33E + 0.30V + 0.37C$$

Weights for each factor of each component are found in a similar manner. The values for E, V, and C would be unity for a perfect system. Such a system would produce an F of 1.

A second Delphi survey is conducted to evaluate the functionality of the system presently operated by SmEC. Tight control is possible since all participants work at SmEC and a 100 percent response is achieved for each round of each survey. Chi-squared tests for group stability (Dajani, Sincoff, and Talley, 1979) and individual stability (Chaffin and Talley, 1980) both are satisfied for the question shown in Figure 11.2. Differences in administrative, staff, faculty, and student responses are not probed.   ∎

The Delphi survey is a powerful method to obtain the subjective input of experts. However, it takes longer, requires more work, and is more expensive than conventional surveys. When conducting a Delphi survey, the manager must pay close attention to fundamental questions of technique and application as these are areas in which many surveys fail.

Round 2:

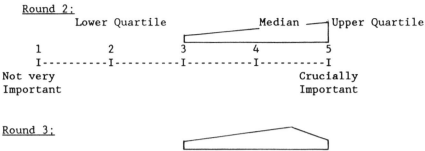

Round 3:

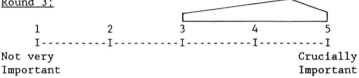

More Important: - The variety of operating systems now available that have a common user interface would greatly enhance ease of use and willingness to use the system for different applications.

Less Important: - All command structures are somewhat arbitrary. Uniformity and intuitiveness would be nice, but not as important as other factors.

Round 4:

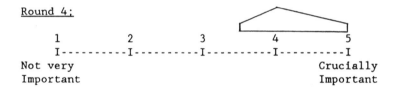

More Important: - The variety of operating systems now available that have a common user interface would greatly enhance ease of use and willingness to use the system for different applications.

- Intuitiveness of commands would result in greater acceptance and less fear.

Less Important: - All command structures are somewhat arbitrary.

Uniformity and intuitiveness would be nice, but not as important as other factors.

Figure 11.2   Delphi responses by round regarding uniformity and intuitiveness of command structure.

■ **Exhibit 11.3 The EFTE Procedure**

1. Participants are provided pertinent background information.
2. Questions about background information are resolved in a face-to-face meeting by the manager who discourages discussion.
3. Each participant completes a Delphi survey and returns it to the manager.
4. Group input is summarized and displayed in the usual format. In addition, responses are ranked from highest to lowest.
5. Group response is discussed freely among participants, but the anonymity of individual responses is preserved.
6. A second Delphi round is completed. Index cards are provided for anonymous questions and comments. Both survey and cards are returned to the manager.
7. Delphi results again are summarized and displayed. Anonymous comments and questions also are read and posted.
8. Participants once more discuss group input.
9. Delphi responses are examined for individual stability. If stability is not found, Steps 6 through 9 are repeated.
10. Final results are summarized and given to all participants. Additional statistical analysis may be performed.   ■

11.4.6 Delphi Variants

Many variations on the basic Delphi theme have been proposed over the years. The simplest involve detail (not necessarily unimportant) modifications to the process. For example, Martino (1983) notes variations such as (1) beginning with a list of events or concerns rather than an unstructured round designed to elicit them and (2) executing a technology forecasting Delphi within a predefined social, economic, and/or political context. Other variants limit the number of rounds, although this perhaps is best done by measuring response stability (see previous section). Even the medium has been changed by computerizing the entire process so that it is conducted in "real time." The most radical changes, however, involve breaching the anonymity condition by face-to-face discussion of intermediate or final results and/or feedback.

An interesting variant, *Shang Inquiry* (Ford, 1975), combines Delphi with a method used by the ancient Chinese to divine information from their gods (or, since few of us are students of Far Eastern social customs and all of us were kids, the game we called "yer gettin' hotter"). In this method, experts estimate the highest and lowest values of a variable in the opening round. The manager uses these to find the lower and upper bounds of the variable and the midpoint, R. In subsequent rounds, group modal response to the query "Is the correct value higher or lower than R?" is used to zero in on the correct value. For example, if the response is that the value should be higher, the midpoint for the next round is computed by averaging R with the upper bound; if the response is that the value should be lower, R is averaged with the lower bound. Ties are broken by a coin toss. Thus each round elicits a progressively more precise estimate.

Nelms and Porter (1985) suggest a technique they call Estimate-Feedback-Talk-Estimate or *EFTE* (see Table 11.2). This technique's most significant variation from classical Delphi is the free discussion of results and the relatively fast feedback embodied by the process. Ten steps are involved, excluding iteration (see Exhibit 11.3).

There are other modifications to Delphi, and new ones undoubtedly will continue to be proposed.

11.4.7 POSTURE

The Policy Specification Technique Using Realistic Environments (POSTURE) is based on concepts of policy capture (see Chapter 18) and social judgment theory. This technique is intended not only to determine the judgment of a group of experts, but also to model the judgment process itself. The rationale is

> If one understands the judgment process, there is a good chance of improving it. If one can improve the judgment process, one should be able to improve the resulting judgments and the decisions they imply. (Brockhaus, 1975, p. 127)

POSTURE is an attempt to formulate a process to accomplish these lofty goals. It combines some of the input methods described in earlier sections with the technique of multiple linear regression, described in Chapter 9, to capture and model the judgment process of individual experts.

To use the process, the manager must first identify, perhaps using an earlier method such as brainstorming, the factors used to formulate the judgment. Next, weights indicating relative importance must be assigned to each factor. This usually is done by formulating a diverse set of short scenarios that incorporate the range of possible combinations of factors. Each expert evaluates every scenario using a numerical scale to indicate its desirability. Then the manager calculates a multiple linear regression equation for each participant. These equations presumably make explicit how each individual weighted the various factors to reach a judgment of desirability. Next the manager clusters the regression equations to produce an expression for subgroups of participants. Finally, participants may be brought together to try to formulate a consensus judgment about the scenarios. If a consensus is achieved, an equation is developed to represent the group.

POSTURE results heavily depend on the quality of the scenarios, successful identification of all factors pertinent to the judgment, and the selection and quality of participants. Its advantages primarily accrue from (1) capturing the judgment process and (2) realizing the value of the process as a means for individuals to learn more precisely the factors on which different groups disagree. The former allows the manager to simulate responses to similar decisions. The latter emphasizes the importance of this and most other input methods as learning processes for participants. The technique is fairly resource-dependent and may appear unrealistic or contrived to participants and, perhaps, to upper-level management. It seems most appropriate to support situations in which a decision or similar decisions are made repeatedly (for instance, in the hiring process).

11.5 CONCLUSION

The basic concepts of obtaining expert opinion have been presented in this chapter. While it is impossible to cover all aspects of this complex and important task, concepts for selecting experts and methods have been presented as have the most common techniques. The manager should pay special attention to selecting experts who have the information required and to choosing a method that will provide the information needed for an informed decision. Care should be taken to review the six factors required to choose a technique (see Section 11.3) to ensure that the method chosen is compatible with information needs, resources, and time. The manager is cautioned that in the final analysis he or she, and not a group of experts, has the responsibility for a decision.

EXERCISES

11.1 You are the chief strategic planner for a major American aerospace company that seeks new commercial initiatives. You want to ascertain (1) if new trains (such as bullet trains and Mag-Lev trains) are likely to offer commercial promise in the coming decades and (2) whether your firm should enter this business. For each issue, discuss promising sources of expertise; whether individual or group expert opinion processes appear more compatible with your needs; and ways you could go about obtaining the cooperation of the appropriate experts in the appropriate format.

11.2 Prepare three survey questions that address the commercial promise of the new trains discussed in Exercise 11.1. Write whether these questions appear suitable for in-person interview, mail survey, and electronic mail administration. Set your written comments on suitability aside. Pilot test your three questions with three people (colleagues, friends) using the mail survey or electronic mail. Pilot the same three questions with three different people using an in-person interview. Consider the pilot test in terms of how well people understood your questions, surprising responses, and needs for revision. Compare these results with your written comments.

11.3 Form into groups of at least five persons and designate a manager. Your assignment is to conduct an NGP inquiry to identify and prioritize the factors necessary to make your campus (or company) more accessible to visually impaired persons. Follow the procedures outlined in Exhibit 11.1.

11.4 Conduct a three-round Delphi exercise on the following two questions. Identify an appropriate group of colleagues or fellow students to take the role of experts on "nanotechnology." Consider nanotechnology as the deliberate manipulation of materials at the atomic level to build purposive structures (such as nanomachines).

Question 1—In what year do you anticipate the first laboratory assembly of a functional, inorganic nanomachine from atoms using physical manipulation (not chemical processes)? ("Never" is a legitimate answer; think in terms of molecules being assembled to have mechanical properties such as wheels or levers.)

Question 2—What is the probability (to the nearest 5 percent) that such nanomachines will have been implemented for initial, limited applications (commercial or military) within 25 years? (Think in terms of the introduction to market of a nanomachine product that a user could buy off the shelf.)

Have each group member anonymously answer each question. Compile responses and provide statistical feedback (in the form shown in Figure 11.1) to everyone. For the second round, request that those who choose to respond outside the interquartile range provide a one-sentence rationale. Again, provide statistical feedback on the second-round responses along with comments (anonymously). Complete a third round; no comments needed. Report the final-round results statistically. Did the process move toward consensus? Had results stabilized adequately?

11.5 Describe how you could adapt Exercise 11.4 to perform a Shang Inquiry. Would this be a better process than ordinary Delphi for the nanotechnology issue? Explain. (Optional: Perform a Shang Inquiry on these same two questions.)

11.6 Describe how you could adapt Exercise 11.4 to perform an EFTE process. Would this be a preferable process to Delphi for this issue? Explain. (Optional: Perform the EFTE process on these same two questions.)

11.7 Describe a technology forecasting or assessment issue for which POSTURE would be a good candidate.

___12
SIMULATION

OVERVIEW

Simulations portray some aspect of a real-world system that can be used to study its behavior. This chapter introduces simulation and several simulation methods useful in forecasting and technology management. The first method discussed is cross-impact analysis, a relatively new technique for which some even newer perspectives are offered. Cross-impact in turn provides the basic information for the second technique, KSIM modeling. This method can be easily employed to provide information about complex real-world situations. System dynamics methods are also described as a philosophy for analyzing and understanding complex real-world systems. Gaming is introduced as a means for studying the behavior of decision makers in search of goals or in competition with each other.

12.1 INTRODUCTION

Simulation means different things to different people. To an airline pilot, for instance, it means physical emulation of the cockpit and analog or digital emulation of aircraft flight behavior. An operations researcher, however, generally thinks of simulation in terms of discrete-event computer models that emulate the system he or she is studying. Once such models have been verified and validated, they can be used to study the effect of changes in the real system. A military analyst may see simulations as war games with stochastic behavior and may use them to study strategy and tactics, the effects of new weapon systems, or other battlefield possibilities. War games are especially useful since the real system, fortunately, is rarely employed. The common notion among these different perspectives is that simulation simplifies a real-world system yet captures and portrays its major characteristics in ways that are useful for study and training.

While employed for the same purposes, different types of simulation are used in forecasting and managing technological systems. Several methods are considered in this chapter: cross-impact analysis, KSIM modeling, system dynamics, and gaming.

12.2 CROSS-IMPACT ANALYSIS

A basic limitation of many forecasting techniques is that they project events and/or trends independently (Stover and Gordon, 1978); thus they fail to account for the impact of events or trends on each other. For example, a successful nuclear fusion process could have a major effect on petroleum exploration. Likewise, the scarcity of petroleum resources holds great economic implications for the development of nuclear fusion. These two technologies do not exist in isolation. Each has a history; each is affected by developments in the other.

One approach to capturing interactions between events is to construct a model, that is, a formal representation of interactions among significant variables. There are several types that can be employed. A *mathematical model* uses equations to represent the system in which the events occur. Such models often require a major time investment to construct. Even with this investment, model coverage usually is limited (for example, mathematical models of inventory systems, of the economy, or of resource allocation systems). There are, however, special models that cut across disciplines and account for the effect of one event upon another. In the technology forecasting arena, one such model is *cross-impact analysis* (CI). Basic CI concepts are widely used and have applications in many areas, including natural resource depletion, strategy and tactics for warfare, institutional change, organizational goals, communication capability, and computer capabilities.

Since CI deals with the future, it involves uncertainty. Therefore, it is a stochastic rather than a deterministic model. Traditional CI is focused on the effects that interactions between events have on their probabilities of occurrence. Thus it deals with discrete events and incorporates no dynamic (time) dimension. While still discrete, the dynamic dimension can be added to CI using the concepts of Markov chains. This modification of traditional CI can be employed to study the chain reaction of events/trends on other events/trends over time. These CI approaches are described in the following subsections.

12.2.1 Traditional Cross-Impact

The concept of CI arose from a game called "Future" which Gordon and Helmer devised for Kaiser Aluminum in 1966 (Helmer, 1983). In the game, a future world was constructed in which some or all of 60 events might have taken place (technological breakthroughs, passage of legislative measures, natural occurrences, international treaties, etc.). Each event was assigned an initial probability of occurring. As play progressed; these probabilities changed. Part of the change was due to actions of the players, the remainder was determined by the occurrence or nonoccurrence of other events. Changes of the latter type gave rise to the concept of CI.

A specific example is useful to understand how traditional CI works. Suppose we are planning for a particular communication technology, say facsimile transmission (fax). Fax technology allows text and images to be transmitted over normal telephone lines. Hard copy input is provided by the sender; it is then transmitted, providing hard copy at the receiver's end. Transmission is fast, 15 seconds or so per page; thus it is much cheaper than conveying the same information verbally (even when that is possible) and much faster than using the mail service. We wish to know what the future of fax technology will be over the next five years. First, we must determine the ways in which the operating environment is likely to change over that time horizon. Then we must identify events that could have noticeable impacts on the use and/or on the planned uses of fax and the probabilities that these will occur.

Suppose the events are identified as $E_1, E_2, E_3, \ldots E_m$. These represent entirely *external determinants*—that is, natural or man-made events over which we have no control (such as a global economic depression or legislation imposing a large tax on each fax received). *Events completely under our control are not included.* These must be treated differently. If the number of events grows too large, it may be necessary to rank them and retain only those that are most important. This could be accomplished by Delphi polling (see Chapter 11) or by having interested parties assign points to each event on a scale of 0 to 100. For the example of fax transmission, suppose we have identified the four events shown in Table 12.1. For

TABLE 12.1 Occurrence Matrix for a FAX Example

The Probability of This Event Becomes:	If This Event Occurs:			
	Increased Taxes/Cost	Negative Legislation	Replacement Technology	Market Saturation
Increased taxes or costs on transmission (0.70)*	1.00 $P(1\|1)$	0.40 $P(1\|2)$	0.70 $P(1\|3)$	0.51 $P(1\|4)$
Negative legislation (0.40)*	0.20 $P(2\|1)$	1.00 $P(2\|2)$	0.38 $P(2\|3)$	0.31 $P(2\|4)$
Development of a replacement technology (0.60)*	0.90 $P(3\|1)$	0.72 $P(3\|2)$	1.00 $P(3\|3)$	0.33 $P(3\|4)$
Market saturation (0.45)*	0.33 $P(4\|1)$	0.35 $P(4\|2)$	0.05 $P(4\|3)$	1.00 $P(4\|4)$

* Initial marginal (ceteris paribus) probability

convenience, we have arranged our work space in an occurrence matrix with the events E_1 through E_m ordered both across the top and down the lefthand side of the array. The next step is to estimate the probability that each will occur. These estimates are called the *marginal probabilities*. (They also are sometimes referred to as ceteris paribus—all else equal—probabilities to indicate that they are estimated without considering any of the other events.) These probabilities are subjective and might be estimated through some Delphi-like procedure. For our example, Table 12.1 shows that we have estimated the probability that there will be increased taxes or costs on fax transmissions to be 0.70.

We have completed two components of the CI matrix: the events critical to the forecast have been identified and their initial (marginal) probabilities of occurrence have been estimated. The cells of the matrix will be used to record the *conditional probabilities* (that is, the probability that event i occurs given that event j occurs). These probabilities are the heart of CI: they portray the impact that the occurrence of any event has on the probability that any other event will occur.

The conditional probabilities must be estimated next. First note, however, that the matrix diagonal entries all will be 1.00, for it is certain that event i will occur given that it has occurred. The first step is to compute the statistically acceptable range of conditional probability for each cell (pair of interactions) *above the diagonal*. These ranges will provide guidelines if we have no other basis from which to estimate the conditionals. This can be done using the marginal probabilities established previously for each event. To explain how to compute this statistical range, we must first introduce some statistical notation.

$P(i)$	=	probability that event i will occur (the marginal probability of i)
$P(i\|j)$	=	probability that event i will occur given that event j has occurred (the conditional probability of i given j)
$P(\bar{i})$	=	probability that event i does not occur
$P(i\|\bar{j})$	=	conditional probability that event i will occur given that event j does not occur
$P(i \cap j)$	=	probability that both events i and j will occur (the intersection of events i and j)
$P(i \cup j)$	=	probability that event i or j or both will occur (the union of events i and j)

By using the laws of conditional probability and the probability of compound events, Sage (1977) showed that limits exist to the range of statistically acceptable conditional probabilities. If the occurrence of event j enhances (increases) the probability that i will occur, then

$$P(i) \leq P(i|j) \leq [P(i)/P(j)] \tag{12.1}$$

On the other hand, if the occurrence of j inhibits (decreases) the probability that i will occur, then

$$1 + \{[P(i) - 1]/P(j)\} \leq P(i|j) \leq P(i) \qquad (12.2)$$

Note that only the initial marginal probabilities $P(i)$ and $P(j)$ are necessary to compute these ranges, and they already have been estimated.

Next we must estimate a conditional probability for each of the cells above the diagonal and compare the estimates to the ranges computed from Equation 12.1 or 12.2. *Estimates that violate the computed ranges should be retained if a solid rationale for them can be given.* For example, in Table 12.1, the conditional probability $P(1|2)$ has been estimated as 0.40, which is within the statistically acceptable range, 0.25 to 0.70, computed from Equation 12.2. However, if we had estimated that it should be 0.15 and had evidence to support our estimate, we would enter 0.15 instead. Alternately, we could elect to assign one of the extreme values of the range to such a probability. Thus, lacking strong evidence to support our estimate of 0.15, we might choose $P(1|2)$ to be 0.25 instead.

Now that conditional probabilities above the diagonal have been estimated (the $P(i|j)$s), we can turn to those below the diagonal (the $P(j|i)$s). Here, we can use Bayes' rule to help. If the $P(i|j)$ was in the range established by Equation 12.1 or 12.2, Bayes' rule says that the corresponding probability below the diagonal should be

$$P(j|i) = [P(i|j)/P(i)]P(j) \qquad (12.3)$$

If $P(i|j)$ was not in the range or if we do not agree with the value produced by equation 12.3, we should subjectively estimate the value of $P(j|i)$. In other words, if the values computed using Bayes' rule are reasonable, keep them. Otherwise, estimate values believed to be more appropriate. For example, in Table 12.1, the conditional probability $P(3|4)$ was estimated as 0.33, within the range of 0.11 to 0.60 computed from Equation 12.2. Therefore, Bayes's rule can be applied to give a value of $P(4|3) = [P(3|4)/P(3)]P(4) = 0.25$. Table 12.1 indicates, however, that we apparently had a strong rationale to support a lower estimate, 0.05.

Just as the occurrence of an event can affect the probability that another will occur, its nonoccurrence can have an impact as well. In our fax example, for instance, if increased taxes or costs of transmission fail to materialize, then the impetus for and probability of replacement technologies will decrease. Thus we need to construct a nonoccurrence matrix (see Table 12.2).

Our last step is to estimate the entries for the nonoccurrence matrix, using the same philosophy as we did for the occurrence matrix. First we will compute the entries statistically from the following equation:

$$P(i|\bar{j}) = [P(i) - P(j)P(i|j)]/[1 - P(j)] \qquad (12.4)$$

Lacking evidence to the contrary, these values will be entered. However,

TABLE 12.2 Nonoccurrence Matrix for a FAX Example

The Probability of This Event Becomes:	If This Event Does Not Occur:			
	Increased Taxes/Cost	Negative Legislation	Replacement Technology	Market Saturation
Increased taxes or costs on transmission (0.30)*	0.00 $P(1\|\overline{1})$	0.85 $P(1\|\overline{2})$	0.60 $P(1\|\overline{3})$	0.90 $P(1\|\overline{4})$
Negative legislation (0.60)*	0.87 $P(2\|\overline{1})$	0.00 $P(2\|\overline{2})$	0.35 $P(2\|\overline{3})$	0.40 $P(2\|\overline{4})$
Development of a replacement technology (0.50)*	0.48 $P(3\|\overline{1})$	0.52 $P(3\|\overline{2})$	0.00 $P(3\|\overline{3})$	0.78 $P(3\|\overline{4})$
Market saturation (0.55)*	0.73 $P(4\|\overline{1})$	0.56 $P(4\|\overline{2})$	0.75 $P(4\|\overline{3})$	0.00 $P(4\|\overline{4})$

* Initial marginal (ceteris paribus) probability of nonoccurrence: $P(\overline{i}) = 1 - P(i)$

if evidence supports a different estimate, that estimate will be entered instead. Returning to the example

$$P(2|\overline{1}) = [P(2) - P(1)P(2|1)]/[1 - P(1)] = 0.87$$

If we have no reason to estimate some other probability, then 0.87 should be entered into the nonoccurrence matrix.

Note that the diagonal entries in the nonoccurrence matrix will all be 0.00 since the probability of an event given that it has not occurred is 0. Negative probabilities predicted by Equation 12.4 should be set at 0, while predicted probabilities greater than 1 (certainty) should be set to 1.

The next stage in CI analysis is to simulate the effects of these conditional relationships. We must determine whether the initial estimates of event marginal probabilities are mutually consistent given these perceptions of how events impact each other.

If all the entries in the two matrices agree with results computed from Equations 12.1 through 12.4, then the initial marginal and conditional probabilities are mutually consistent. However, if one or more of the conditional probabilities differ from computed results, we will have to "play" the CI matrices to determine a consistent set of marginal probabilities. A computer-based Monte Carlo simulation can be used to do this:

1. An event is selected randomly (say Event 2 in Table 12.1).
2. A random number between 0 and 1 is generated and compared to the marginal probability of the event to determine if it occurs. Suppose the random number is 0.26, since 0.26 ≤ 0.40, Event 2 is assumed to occur. If the random

number were greater than 0.40, it would be assumed that Event 2 did not occur.

3. The marginal probability of each remaining event is replaced by its conditional probability given that the event in Step 2 occurs or does not occur. That is, in our example $P(i)$ is replaced by $P(i|2)$ if Event 2 occurs, or by $P(i|\overline{2})$ if it does not ($i \neq 2$). Thus, since Event 2 occurred in Step 2, the replacement values will be $P(1) = 0.40, P(3) = 0.72, P(4) = 0.35$.

4. A second event is selected randomly from those remaining (Events 1, 3, and 4), and Steps 1 through 3 are repeated. In this play, the probability used in Step 2 is the value produced in Step 3 of the previous play. Thus, if Event 2 occurred in the first play and Event 4 is selected in the second, the probability of Event 4 used in Step 2 of the second play is $P(4|2) = 0.35$.

5. The process described in Steps 1 through 4 is repeated until all four events have been selected. All marginal probabilities are then returned to their initial values and the game is "replayed," typically 1,000 or more times.

6. Each time the game is "played" the events that occur are noted. The total number of occurrences divided by the number of games is taken as the final (marginal) probability for each event. The initial marginal probabilities are then replaced by the final marginal probabilities, which account for event interaction.

The conditional probability of an event occurring given two or more other events is required in Step 3 after the first two events have been determined. For instance, the examples woven into these six steps suggested that Event 2 occurred in the first round of play. If in the second round Event 4 is found to occur, then we need conditional probabilities such as $P(1|2 \cap 4)$ to proceed. The occurrence and nonoccurrence matrices only specify pairwise interactions—that is, joint probabilities of one event conditioned on one other event. Joint conditionals are called *second-order conditionals*. There also are third-order and higher-order conditionals, for example, $P(i|j \cap k)$ and $P(i|j \cap k \cap l)$. These probabilities are difficult to determine. Instead, they usually are approximated by averaging second-order probabilities. For instance,

$$P(1|2 \cap 3) = [P(1|2) + P(1|3)]/2$$

A similar averaging procedure is used for higher-order, nonoccurrence probabilities, and probabilities such as

$$P(1|2 \cap \overline{3}) = [P(1|2) + P(1|\overline{3})]/2$$

These approximations are acceptable when the conditional probabilities being averaged are close in value.

Software in the Technology Forecasting (TF) TOOLKIT, which supplements this book, can be used to play this CI game. Of course, the game is probabilistic and so results vary somewhat from game to game. However, after 1,000 plays this program produces final marginal probabilities of $P(1)' = 0.60, P(2)' = 0.50, P(3)' = 0.51$, and $P(4)' = 0.51$ for our fax example, where the prime symbol has been added to distinguish the final marginal probabilities from the initial values. Note that there have been changes in the marginal probabilities of all events. After 10,000 plays, the resulting marginal probabilities are 0.61, 0.50, 0.51 and 0.50 for Events 1 through 4, respectively. If Equations 12.1 through 12.4 were used to generate all the conditional probabilities in the two matrices, we likely would have seen little difference between initial and final marginal probabilities. This is because using those equations tacitly assumes that the conditional probabilities are consistent with the initial marginal probability estimates.

The CI game is attractive since it can be generated from relatively little data. It can be used to examine the interaction of events and to ensure, insofar as possible, that the probabilities to be used in, say a scenario, account for those interactions. The technology manager also could use the CI matrices to plan strategies to enhance or inhibit the probability that key events occur. Moreover, constructing the CI matrices can provide a useful format within which to frame discussions of interactions between events. Equally important, merely constructing the matrices forces the manager or forecaster to consider interactions between events. However, it is important to note that, even though the probabilities produced by CI are consistent with our perceptions of how they interact, they still may be incorrect.

Halverson, Swain, and Porter (1989) point out that the effect of an event on the probabilities of other events is not necessarily determined by the size of its marginal probability. This stems from the somewhat trivial observation that an event with a low probability of occurrence has a high probability of nonoccurrence. Since the nonoccurrence of an event can have a major effect on other events, the effect of an unlikely event thus may be quite high.

Halverson, Swain, and Porter (1989) also observe that the conditional probability estimates may be more important than the marginal probability estimates. The initial marginal probability estimates likely will be changed by the conditional estimates during the game. Therefore, we should concentrate on accurately estimating the impact of events on each other (that is, estimating the conditional probability) and expend less effort on estimating the marginal probabilities.

Despite its advantages, classic CI has a number of deficiencies. These primarily stem from the fact that it is a discrete model with no time dimension; thus it cannot deal with cases in which the time dimension is important. For instance, in the fax example, suppose that one event cannot occur unless another has occurred and that the second event cannot occur earlier than three years from today. CI, as presented, is powerless to portray this situation. Nor can it portray trends since they, by definition, are time dependent. Modifications of classic CI designed to address these limitations are described in the following subsections.

12.2.2 Time-Dependent CI

As noted previously, the traditional CI approach includes no dynamic (time) dimension. The probabilities of occurrence and nonoccurrence are assumed to be valid for all time. However, it is possible, even likely, that the probability of most events will change as time passes. For instance, in the fax example, Event 3 (the development of a replacement technology) likely will become more probable as time passes. Further, while Event 1 (increased taxes or costs of transmission) could occur at any time, its effect on Event 3 may be cumulative over time—that is, as the costs grow, other means of communication may become more and more desirable. Rather than state the CI game in terms of $P(i : j)$, we could use time-dependent probabilities such as $P(E_i^{t+1} : E_j^t)$. This notation indicates both time-dependence and ordering by specifying the probability of Event i at time $t + 1$, given that Event j has already occurred by time t.

The fax example model can be reformulated as a Markov process, which will allow us to answer questions about the probabilities of events over time, the sequences of events over time, and the expected times of occurrences of events. This also will allow us to model how the effects on other events vary with time.

A Markov process is a system of states governed by a matrix of *transition probabilities*.[1] These are the probabilities that the system will go from any state $N_t = i$ to some other state $N_{t+1} = j$ during the next step in time. It is assumed that these transition probabilities only depend on the current state—that is, the Markov process has no memory. For n events there are 2^n states which define a $2^n \times 2^n$ Markov transition matrix. Each of the numbered states represents a distinct scenario defining the occurrence or nonoccurrence of the n events. Suppose that there are three events. We will adopt a shorthand notation to indicate any scenario, call it N. If the scenario, N, is $(1\,\bar{2}\,\bar{3})$, then Event 1 occurred, but Events 2 and 3 did not. We will write this as (100). The order of the three digits represents Events 1, 2, and 3; a 1 indicates occurrence, a 0 indicates nonoccurrence.

Because there are three events, $(2^3 = 8)$, the 8×8 transition matrix shown by Table 12.3 will be needed to order this work. In that matrix, the probability of moving from one state at time t to another state at time $t + 1$ (that is, the transition probabilities) are displayed in the cells. For instance, the probability of going from state 010 (state number $N = 3$ in the matrix) at time t, to state 110 ($N = 7$) at time $t + 1$ is given by $p(3{:}7)$, which is recorded in the cell formed by row 3 and column 7. In general notation, the probability of transition from state i to state j in a one-unit passage of time is $p(i : j)$, recorded in cell (i, j).

Since each row in the array contains the probabilities of going from state i to each of the eight possible states in the next step, the sum of the probabilities in each row must be unity (that is, one or the other of the states must occur at time $t + 1$). Note that diagonal entries are the probabilities that the state will not change. If these probabilities are large (close to 1.0) the system will change slowly. The probabilities below the diagonal may be set to 0 if events cannot be undone (cannot be reversed).

A Markov matrix requires a great deal of data. A matrix for n events will require that we estimate 2^{2n} transition probabilities. Even if entries below the diagonal can be set to 0, we will need $(2^n - 1)2^n/2$ probability estimates. For example, a six-event model would require a minimum of $(2^6 - 1)2^6/2$ or 2,016 entries. Such

---

[1]This discussion draws heavily on Swain et al. (1989).

TABLE 12.3   Markov Transition Matrix for Three Events

Transition From This State at Time t	To This State at Time $t + 1$							
	(000) $N = 1$	(001) $N = 2$	(010) $N = 3$	(011) $N = 4$	(100) $N = 5$	(101) $N = 6$	(110) $N = 7$	(111) $N = 8$
(000) $N = 1$	$P(1{:}1)$	$P(1{:}2)$	$P(1{:}3)$	$P(1{:}4)$	$P(1{:}5)$	$P(1{:}6)$	$P(1{:}7)$	$P(1{:}8)$
(001) $N = 2$	$P(2{:}1)$	$P(2{:}2)$	$P(2{:}3)$	$P(2{:}4)$	$P(2{:}5)$	$P(2{:}6)$	$P(2{:}7)$	$P(2{:}8)$
(010) $N = 3$	$P(3{:}1)$	$P(3{:}2)$	$P(3{:}3)$	$P(3{:}4)$	$P(3{:}5)$	$P(3{:}6)$	$P(3{:}7)$	$P(3{:}8)$
(011) $N = 4$	$P(4{:}1)$	$P(4{:}2)$	$P(4{:}3)$	$P(4{:}4)$	$P(4{:}5)$	$P(4{:}6)$	$P(4{:}7)$	$P(4{:}8)$
(100) $N = 5$	$P(5{:}1)$	$P(5{:}2)$	$P(5{:}3)$	$P(5{:}4)$	$P(5{:}5)$	$P(5{:}6)$	$P(5{:}7)$	$P(5{:}8)$
(101) $N = 6$	$P(6{:}1)$	$P(6{:}2)$	$P(6{:}3)$	$P(6{:}4)$	$P(6{:}5)$	$P(6{:}6)$	$P(6{:}7)$	$P(6{:}8)$
(110) $N = 7$	$P(7{:}1)$	$P(7{:}2)$	$P(7{:}3)$	$P(7{:}4)$	$P(7{:}5)$	$P(7{:}6)$	$P(7{:}7)$	$P(7{:}8)$
(111) $N = 8$	$P(8{:}1)$	$P(8{:}2)$	$P(8{:}3)$	$P(8{:}4)$	$P(8{:}5)$	$P(8{:}6)$	$P(8{:}7)$	$P(8{:}8)$

data requirements greatly exceed those of the traditional CI game and greatly diminish the number of practical applications for this approach. However, Swain et al. (1989) present a scheme for estimating these probabilities with data requirements more nearly equal to those of the traditional CI approach. For example, consider a Markov matrix for a two-event model with the following entries:

$$P^{(t,1)} = \begin{bmatrix} 0.550 & 0.300 & 0.098 & 0.052 \\ 0.000 & 0.850 & 0.000 & 0.150 \\ 0.000 & 0.000 & 0.650 & 0.350 \\ 0.000 & 0.000 & 0.000 & 1.000 \end{bmatrix}$$

The same scheme used to order the larger matrix in Table 12.3 has been used here so that the first row and column represent the state 00 (or $N = 1$), the second row and column represent state 01 ($N = 2$) and so forth.

For convenience we will use $P^{(t,1)}$ to represent the entire matrix. The first superscript indicates that the matrix gives the transition probabilities for changes starting at time t; the second superscript indicates that it applies for one time step (from t to $t + 1$). In this example, entries below the diagonal have been set to 0 and the transition probabilities in each row sum to 1. Note also that $P(4 : 4) = 1.0$. If the system reaches this state, it will not change in subsequent steps. This is called the *trapping state*. Depending on the situation being modeled, there may be one trapping state, more than one, or none at all.

One problem noted with the classic CI is that state probabilities are assigned a single (marginal) probability that does not change with time, which is an unrealistic formulation. However, the transition matrices allow us to represent such variations. Suppose the probabilities of the four states in our sample matrix are estimated to be 0.7, 0.3, 0.4, and 0.1 at time t. For convenience, we will represent this as the row matrix

$$S^t = [0.7, 0.3, 0.4, 0.1]$$

The transition matrix allows us to compute the state probabilities at $t + 1$ as

$$S^{t+1} = S^t P^{(t,1)} \tag{12.5}$$

or

$$S^{t+1} = [0.7, 0.3, 0.4, 0.1] \begin{bmatrix} 0.550 & 0.300 & 0.098 & 0.052 \\ 0.000 & 0.850 & 0.000 & 0.150 \\ 0.000 & 0.000 & 0.650 & 0.350 \\ 0.000 & 0.000 & 0.000 & 1.000 \end{bmatrix}$$

so that

$$S^{t+1} = [0.385, 0.465, 0.329, 0.321]$$

By using the transition matrices and the initial probabilities, we can compute the changes in state probabilities sequentially over the time period of interest.

To time market entry and/or resource allocations, we often are interested in knowing when a state might be reached. The time it takes to first reach a particular state is called the *first passage time*; the associated probability is called the *first passage probability*. We also may be interested in knowing how the cumulative probability of occurrence of a particular state may change over time. Both can be investigated using the time-dependent CI approach as described in the following paragraphs.

The probability that the system will make a transition from any state i at time t to any state j sometime during n steps is given by

$$P^{(t,n)} = P^{(t,1)}P^{(t+1,1)}P^{(t+2,1)} \ldots P^{(t+n-1,1)} \tag{12.6}$$

For simplicity, we will examine the special case in which the transition matrices do not change from step to step (that is, the entries in the matrix do not depend on t).

$$P^{(t,n)} = [P^{(t,1)}]^n$$

The n-step transition matrix is obtained by multiplying $P^{(t,1)}$ by itself n times. For example, to compute $P^{(t,2)}$, the matrix $P^{(t,1)}$ would be multiplied by itself to produce

$$P^{(t,2)} = \begin{bmatrix} 0.302 & 0.420 & 0.118 & 0.160 \\ 0.000 & 0.722 & 0.000 & 0.278 \\ 0.000 & 0.000 & 0.422 & 0.578 \\ 0.000 & 0.000 & 0.000 & 1.000 \end{bmatrix}$$

Thus the probability of remaining in state 1 for the period from t to $t+2$ is 0.302, and the probability of going from State 1 to State 2 sometime in that period is 0.420. Note that the probability of staying in state 4 (the trapping state) is 1.0. Therefore, it is impossible to retreat to a lower numbered step from this state.

Following the process outlined above, $P^{(t,n)}$ for this example is given by

$$P^{(t,n)} = \begin{bmatrix} [0.55^n] & [0.85^n - 0.55^n] & [0.98(0.65^n - 0.55^n)] & [1 - 0.85^n - 0.98(0.65^n - 0.55^n)] \\ 0 & [0.85^n] & 0 & [1 - 0.85^n] \\ 0 & 0 & [0.65^n] & [1 - 0.65^n] \\ 0 & 0 & 0 & 1 \end{bmatrix}$$

We can use this matrix to study various transitions of interest. For instance, we may wish to know how long it might take to go from a base state ($N = 1$) where neither event has occurred (state 00) to a state ($N = 4$) where both events have

TABLE 12.4 $P(1{:}4)$ **and** $f(1{:}4)$ **for Selected** n

n	$P(1{:}4)$	$f(1{:}4)$
0	0.000	-
1	0.052	0.052
2	0.160	0.108
3	0.280	0.120
4	0.393	0.113
5	0.492	0.099
...	...	...
19	0.954	0.008
20	0.961	0.007

Source: Swain et al., 1989. Reprinted by permission

occurred (state 11). The probability that we have arrived at state 4 at some time during the first n steps is represented by

$$p(1{:}4) = 1 - 0.85^n - 0.98(0.65^n - 0.55^n)$$

Solving this expression at various values of n produces the results shown in Table 12.4. Also displayed in that table are the incremental changes in probability from step to step, $f(1{:}4)$, which indicate the probability of reaching state 4 *in that step*. The table shows, for instance, that the probability of having reached state 4 *by* the fifth step is nearly 0.500, while the probability of reaching state 4 *in* the third step is 0.120. The table indicates that the most likely first passage time is three periods since $f(1{:}4)$ has the highest first passage probability, 0.120. The mean first passage time (that is, the mean number of steps to reach state 4 from state 1) would be $1(0.052) + 2(0.108) + \cdots = 7.28$ periods.

The time-dependent CI approach adds considerable power to cross-impact studies. Using it, we can more realistically portray the variation of probabilities and impacts with time and handle the sequencing of event probabilities representative of real-world situations. The price we pay for this extension comes from the increased data requirements and manipulation complexity. The classic and time-dependent CI approaches are compared for a specific case in Exhibit 12.1.

■ Exhibit 12.1   Classic and Time-Dependent CI Compared—The AIMTECH Case

AIMTECH is the U.S. Air Force project acronym for a forecast of the development of artificial intelligence (AI) technology. Although the AIMTECH study was much broader, we will focus on a subset intended to augment pilot decision making in combat situations. An AI system capable of planning against actions by an adversary requires development of two other technological capabilities. Thus the subset of (technological) events that concern us here are

ANA—An AI system capable of reasoning by analogy

COM—An AI system embodying common sense

ADV—An AI system capable of planning action against an adversary

The bottom line of the AIMTECH study was to determine which AI targets were most attractive for R&D development.

The AIMTECH forecast (Reitman et al., 1985) was used by Swain et al. (1989) as an example of the Markov-based CI approach. We rely heavily on the Swain et al. (1989) study to compare and contrast analyses by the classic and time-dependent CI approaches. Data for this example are only illustrative and, in some cases, result from assumptions beyond the study's scope.

In classic CI, we would develop initial marginal probabilities for each event and conditional probabilities to display the effects of their occurrence or nonoccurrence on each other. Such a set of matrices is shown in Table 12.5.

These matrices represent very limited CI. For example, note that COM's occurrence does not change the initial marginal probability of ANA and vice versa. Thus ANA and COM are independent of each other. Likewise, the nonoccurrence of ADV does not affect either ANA or COM. Further, the occurrence of ADV implies the occurrence of both ANA and COM, and ADV cannot occur unless both ANA and COM do. Following the procedure outlined in Section 12.2.1, the CI matrices in Table 12.5 can be played using the Monte Carlo technique. After 10,000 plays the final marginal probabilities were found to be 0.59, 0.60, and 0.77 for ANA, COM, and ADV, respectively.

The problem with the classic CI solution is that since the model is entirely independent of time so are the results. Thus we cannot take into account that certain events must precede others. We have no way of knowing when any might occur; therefore, we have no way to determine if increased R&D funding, for instance, would cause an ANA or COM to be realized significantly earlier. Nor has the classic model allowed us to portray impacts between events that grow or decay over time. This is unrealistic. Such effects are apparent when we view the probabilities estimated for the three variables over time by the AIMTECH study (see Figure 12.1). For example, as Figure 12.1 shows, ANA cannot occur before 1986 or after 1994 and COM must occur in the window from 1993 to 1997. This time dimension is lost in classic CI. These are precisely the kind of dynamics for which time-dependent CI is intended.

TABLE 12.5   AIMTECH Cross-Impact Matrices

The Probability of Event Becomes:	If This Event Occurs			If This Event Does Not Occur		
	ANA	COM	ADV	ANA	COM	ADV
ANA (0.85)*	1.0	0.85	1.0	0.0	0.85	0.85
COM (0.65)*	0.65	1.0	1.0	0.65	0.0	0.65
ADV (0.37)*	0.44	0.47	1.0	0.0	0.0	0.0

Source: Swain et al., 1989 Reprinted by permission.

* Initial marginal probability.

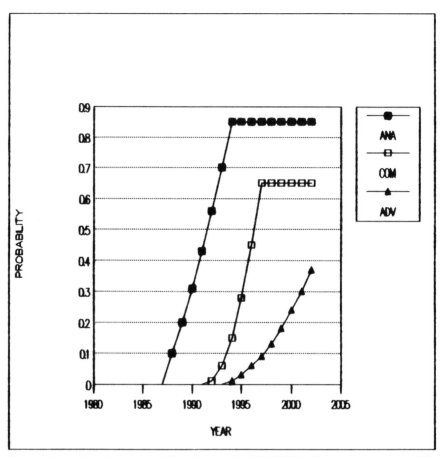

Figure 12.1 Cumulative probabilities forecast for AIMTECH. (*Source*: Swain et al., 1989. Reprinted by permission.)

Swain et al. (1989) have estimated the transition matrix data from information given in the AIMTECH study. Two of the transition matrices are given in Table 12.6. In studying that table, recall that Event 3, ADV, can only occur after Events 1 and 2, and no events un-occur. Because of the ADV constraint, only five of the possible eight theoretical states are modeled: the remaining three cannot occur. Ordering and numbering of the matrices in Table 12.6 follow the scheme used in Table 12.3. It is important to note from the table that, unlike the example given in the last section, the transition matrices are different for different times. Therefore, we must use Equation 12.6 to compute the n-step transition matrix.

Swain et al. (1989) estimate the state probabilities in 1993 to be $S^{1993} = (0.26, 0.04, 0.68, 0.02, 0)$—that is, the probability of being in state 1 in 1993 is 0.26, and so forth. These probabilities, of course, are derived from those shown in Figure 12.1. For example, the probability of ANA having been reached by 1993 is 0.70 in that figure. In the example then, the probabilities for state 5 (0.68)

TABLE 12.6   Markov Transition Matricies, $p^{(1993,1)}$ and $p^{(1996,1)}$ for AIMTECH

$p^{(1993,1)}$	Transition from This State at Time t	To This State at Time $t + 1$				
		(000)* $N = 1$	(010) $N = 3$	(100) $N = 5$	(110) $N = 7$	(111) $N = 8$
	(000)* $N = 1$	0.2692	0.2154	0.5154	0	0
	(010) $N = 3$	0	0.6000	0	0.4000	0
	(100) $N = 5$	0	0	0.9500	0.0500	0
	(110) $N = 7$	0	0	0	0.5000	0.5000
	(111) $N = 8$	0	0	0	0	1
$P^{(1996,1)}$						
	(000)* $N = 1$	0.5000	0.5000	0	0	0
	(010) $N = 3$	0	1	0	0	0
	(100) $N = 5$	0	0	0.6415	0.3585	0
	(110) $N = 7$	0	0	0	0.8846	0.1154
	(111) $N = 8$	0	0	0	0	1

Source: Swain et al., 1989. Reprinted by permission.

* Order of events = (ANA, COM, ADV)

and state 7 (0.02) must total 0.70. Likewise, since the probability of achieving COM by 1993 is 0.06, the sum of the probabilities for states 2 and 7 also must be 0.06. As outlined in the previous section, the initial state probabilities and the transition matrices can be used to "bootstrap" our way through the time horizon, calculating state probabilities as we go. For instance, using Equation 12.5,

$$S^{1994} = S^{1993} P^{(1993,1)} = (0.07, 0.08, 0.78, 0.06, 0.01),$$

while S^{2002} can be shown to be (0.01,0.14,0.34,0.14,0.37). Thus the probability of reaching any state is dynamic, as shown by Figure 12.2.

In contrast to the classic approach, the time-dependent CI provides information about the process throughout the time span. A number of analyses can be performed. For example, using Equation 12.6 the nine-step transition matrix $P^{(1993,9)}$ can be obtained. We can examine the last column of this matrix to find the probability of attaining ADV by 2002, given the various initial states in 1993. These probabilities are 0.2041, 0.4184, 0.3377, 0.9221, and 1.0 for beginning states 1, 3, 5, 7, and 8, respectively. Thus, if we start in state 1 (000) in 1993, there is a probability of about 0.20 of attaining ADV *by* 2002; if we start in state 3 (010), there is a probability of about 0.42. This is contrasted with the probability of achieving ADV *in* 2002 of 0.37, which is averaged across all starting states. As outlined in the previous section, this analysis could be extended to probe the most likely or the mean first passage times to ADV.

In this section, the Markov-based, time-dependent CI analysis has been contrasted with classic CI. The classic approach provides the marginal probabilities for ANA, COM, and ADV in 2002. The Markov approach provides information

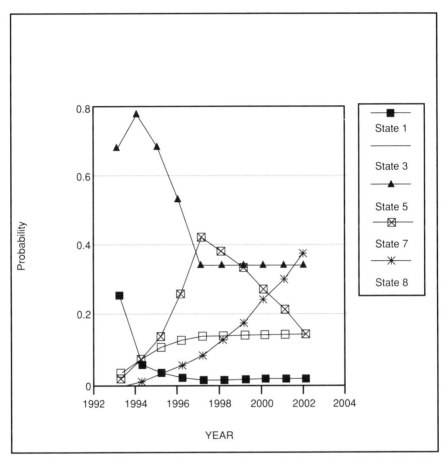

Figure 12.2 Probabilities of attaining states. (*Source:* Swain et al., 1989. Reprinted by permission.)

throughout the process and a means of dealing with impacts that vary with time—a major improvement. However, expansion of CI to this dynamic dimension could only be made at the expense of developing nine transition matrices spanning the period from 1993 to 2001—each one of which requires at least as much input as that provided for the classic CI process. Clearly, the increased resources implied are not appropriate to every topic. ■

12.2.3 Other Time-Dependent CI Schemes—XimpacT

Enzer and Leschinsky (1986) have extended classic CI to approximate time steps, although the causality implied is not explicitly modeled. This approach is incorporated in the XimpacT CI software package.[2] Inputs to XimpacT include a set of

---

[2]Distributed by XimpacT, contact Selwyn Enzer, IBEAR Program, University of Southern California, Los Angeles, CA 90007.

trends (including their nominal forecasts), a set of events (including their probabilities), and cross-impact factors. Outputs are termed *alternative futures* (or scenarios) to acknowledge the uncertainty or probabilistic nature of the problem. Their purpose is to encourage communication about alternatives among policy analysts and decision makers. In addition, following the scenarios year by year can help decision makers to understand the underlying dynamics.

Output information about scenarios is available both in tabular and graphical form. Tabular output includes summary reports about the scenarios and statistical reports about trends. The summary reports indicate which events occur and their year of occurrence as well as year-by-year values of the simulated trend. Statistical reports for each trend indicate the maximum, minimum, average, and standard deviation over all scenarios for each year. Graphical output includes the same types of data with some flexibility for the user in choosing data to be plotted.

12.2.4 Closure

While the discussion of CI presented in the preceding sections is extensive, the manager should by no means mistake it for the last word. Rather, it provides the background that he or she will need to understand ongoing CI development as well as one or two extensions of the classic approach. The manager is advised to refer to Sage (1990) and Martino (in press) for further information.

Sage (1990) points out four problems in CI analysis. First, there is no guarantee that those filling the matrix of CI relationships will not violate Bayes' rule. Second, the relationship between $P(i:j)$ and $P(i)$ should at least be symmetrical, but even this may not happen. While revised approaches using likelihood ratios instead of the simple enhancing/inhibiting/independent schemes can provide symmetry, they do not ensure that the probabilities so generated will be consistent with Bayesian probability theory. Third, if the occurrence of an event affects the probability of another event, then the nonoccurrence of that event should also affect the probability of the second event. However, the simplest approaches to CI assume that nonoccurrence has no impact. Finally, Sage notes that the time element is important, but CI often does not include a mechanism for handling the timing of events.

The following eight-step procedure for CI analysis is recommended:

1. Use collective inquiry methods to generate a set of appropriate events, elements and relations for a particular problem.

2. Use a structural modelling method to evolve a structure for the events under consideration.

3. Develop a scale so that one may translate expert opinion and knowledge concerning the likelihood of occurrence of events into probabilities. This scale should be discussed, understood and approved by the group supplying the probability estimates.

4. Obtain an initial set of probabilities of the type in which estimators have most confidence. Every effort should be made to obtain as many of the direct tree-related conditional probabilities as possible. For direct tree-related conditional probabilities

not estimated, obtain sufficient marginal and lower-order conditional probabilities, and impact assessments, to bound the direct tree-related conditional probabilities. These upper and lower bounds are computed.

5. Display the set of direct tree-related conditional probabilities or bounds computed on these probabilities if they are not directly estimated. Obtain final probabilities with which to determine future probabilities.

6. Determine future event probabilities.

7. Incorporate the results of the cross-impact analysis into a system simulation model of realities in which events are assumed to occur or not occur as predicted from the results of the cross-impact analysis. Inputs or policies connected with the events must be determined in order to accomplish this simulation. Iterate these simulation runs for other likely futures.

8. Incorporate the results of the cross-impact analysis and system simulation exercises as the systems analysis phase of systems engineering activities for the particular problem under consideration. (Sage, 1990, p. 262–263)

Sage then emphasizes the complexities of adjusting the analysis when new information is presented. The basic approach is Bayesian in that the prior probabilities are adjusted for posterior observations or perceptions stated as observations. However, human beings are not particularly good at adjusting hierarchical inference (that is, logical interconnections of elements in a net or tree). Sage shows how quantitative procedures based on a few assumptions and using the computational power of computers might improve the handling of new information for complex systems developed in the CI analysis. Developing an approach based upon hypothesis evaluation, he suggests that inference can be done on the basis of imprecise knowledge that can be either hypothetical or evidential.

Joseph Martino has developed an approach to handle both the probability and timing of events in the CI format (Martino, pending). He uses the timing of events to reduce the number of cross-impacts that must be considered. For example, if Event 1 must precede Event 2, the occurrence or nonoccurrence of the second event cannot affect the probability of the first. Martino also notes that implicit assumptions the manager makes about the occurrence of earlier events affect the estimates he or she makes about later ones. Thus some impacts are implicit in the estimates themselves and should not be accounted for a second time by CI. For instance, suppose we estimate a 70% probability that the New York Yankees will win the 1995 World Series. This estimate clearly assumes that they will win the American League pennant that year. If they do, it has no impact on winning the series because that has been assumed. We need only estimate the impact of not winning the pennant on the probability of winning the series. Combining timing and structural factors such as those outlined here can significantly decrease the amount of information that must be developed and improve the validity of a CI analysis.

The CI approach described by Martino recognizes both events that must occur (or not) at a specific time and those that can occur with varying probabilities over a span of years. It also accommodates sets of events that are yoked together in such a way that one, but not more than one, event in the set may occur. Finally, the approach can be used to identify events that are especially pivotal or especially sensitive to the occurrence of other events.

12.3 KSIM

KSIM is a deterministic simulation model developed by Julius Kane (1972). KSIM extends the concepts of CI to produce a dynamic simulation that is easy to use yet sufficiently powerful to provide meaningful analysis of many real-world problems. The model retains the concept of the impacts of events on each other characteristic of CI. However, this concept is married to a differential equation that portrays an S-shaped (logistic) growth or decline of the variables being modeled. This equation provides the continuous, dynamic (time-dependent) characteristics of KSIM. The logistic variation is a loose analogy to biological system growth (the support and rationale for this analogy are presented in Chapters 9 and 10). Since impact magnitudes are estimated subjectively, KSIM in effect utilizes both "hard" (objective) and "soft" (subjective) input. Thus it is an appropriate implementation of Kane's premise that experience, opinions, and judgments control decision-making.

The variables modeled by KSIM, X_i, are first identified, defined, and quantified. The maximum value of each variable is determined so that each can be normalized on a scale of 0 to 1. The initial value of each is also estimated. The simulation marches forward from these initial values a step at a time using the differential equation

$$\frac{dX_i}{dt} = \sum_{j=1}^{N} (\alpha_{ij} + b_{ij} \frac{dX_j}{dt}) X_i \ln X_i \tag{12.7}$$

where

$$X_i = \text{the variable described}$$
$$N = \text{the total number of variables considered}$$
$$X_j = \text{the impacting variables}$$
$$\alpha_{ij} = \text{the long-term impact of } X_j \text{ on } X_i$$
$$b_{ij} = \text{the short-term impact of } X_j \text{ on } X_i$$

The solution to this logistic CI equation is

$$X_i(t + \Delta t) = X_i(t)^{P_i(t)} \tag{12.8}$$

where

$$X_i(t + \Delta t) = \text{value of variable at end of the time period}$$
$$X_i(t) = \text{value of } X_i \text{ at the start of time period}$$
$$\Delta t = \text{the time period}$$

and

$$P_i(t) = \frac{1 + \Delta t(\text{sum of inhibiting impacts on } X_i)}{1 + \Delta t(\text{sum of enhancing impacts on } X_i)}$$

or, mathematically,

$$P_i(t) = \frac{1 + 0.5t \sum_{j=1}^{N} [|I_{ij}(t)| - I_{ij}(t)]X_j(t)}{1 + 0.5t \sum_{j=1}^{N} [|I_{ij}(t)| + I_{ij}(t)]X_j(t)} \tag{12.9}$$

and

$$I_{ij} = \alpha_{ij} + \frac{b_{ij}}{X_j(t)}[dX_j(t)/dt]$$

While the equations appear formidable, operationally the concept is relatively straightforward. One must estimate the impacts of the level of each event (that is, level = value of X_j) on all other variables. This is the α_{ij}, which is determined in much the same manner as impacts in CI. Then, the impacts of the rates of change of each event (dX_j/dt) and the slope of the *trend* in X_j on the other events (b_{ij}) are estimated in the same fashion. Once these impact magnitudes have been determined, a relatively simple computer program (such as that included in the TOOLKIT) can be used to solve the equations and perform the forecast.

The characteristics of KSIM are pretty much what would be expected of a logistic curve. For example, when the sum of the inhibiting impacts is greater than that of the enhancing impacts, the power $P_i(t)$ in Equation 12.8 will be larger than one. And, since $0 \le X_i(t) \le 1.0$, $X_i(t + \Delta t)$ will be smaller than $X_i(t)$. Further, all else being equal, the larger the variable causing the impact, the greater the magnitude of that impact will be. Note also that a given value of $P_i(t)$ will have less effect on the magnitude of X_i if X_i is near either 0 or 1. This produces the S-shaped variation we expect of growth or logistic curves.

KSIM is one of the few dynamic models that can be constructed and used with relatively limited time and resources. The general procedure that a group of technology forecasters or managers would use is as follows:

1. Discuss the problem and agree on the scope and boundaries of the simulation (such as level of aggregation, spatial boundaries, and time frame).
2. Identify, define, and label the important variables and determine their initial values, ranges, and maximums. Normalize each variable on a 0 to 1 range.
3. Structure the long-term and short-term impact magnitudes and array them in matrix form as in CI. Impacts that increase the size of a variable (enhance it) are positive, those that inhibit it are negative. Numerical values for mag-

nitudes are proportional to the size of the impact. For example, if X_1 is not impacted by X_2, then α_{12} and/or b_{12} will be 0. This work sometimes is cut in half by considering only short-term (b_{ij}) or long-term (α_{ij}) impacts.

4. Run the model and refine the impacts, variable definitions and/or values until the outcome is satisfactory. Usually a base case is run and the output is compared to a similar situation or to theoretical behavior. The process is repeated as often as necessary to produce acceptable results.

The model can now be run to investigate the effects of changing initial values or basic assumptions or of introducing new assumptions. In this way, alternative futures can be examined and forecasts and trade-offs can be determined. KSIM also provides for a very useful extension, allowing external events or policy decisions to be added to the model as variables in the CI matrix. This is done by formulating the impacts of, say policy options, as additional columns but not rows in the CI matrices. Thus a decision to invoke a policy option impacts the variables, although the option is not itself impacted by the variables. Using this approach, the decision maker can systematically investigate the effects of policy decisions on the behavior of the system.

KSIM should be viewed as a process as well as a product. The benefit of KSIM accrues from building the model, as much as from operating it and analyzing the results. Building the model provides the format in which a team can structure the discussion of a complex issue. In that format, experience, opinion, and judgment can be incorporated along with hard data. Further, a completed model allows alternatives to be quickly formulated and their consequences to be assessed. Thus KSIM can provide an environment within which the manager can study and learn about complex situations.

The process makes a number of assumptions that imply limitations as well. KSIM assumes that a satisfactory model can be devised and that the variables and their interactions can be accurately defined. It also assumes that realistic bounds can be placed on the variables; that a growth curve adequately represents the change patterns being studied; and that opinions, experience, and other subjective information can be formulated mathematically. Equally important, KSIM assumes that the pairwise relationships portrayed by the matrices adequately represent true causal interaction, a much more complex situation. Finally, KSIM models a deterministic world; however, the technology manager, the forecaster, and the rest of us live in a probabilistic one. For an example of how to apply the KSIM model, see Exhibit 12.2.

■ Exhibit 12.2   KSIM Applied to Model Fax Transmission

To clarify concepts, consider the example of transmission used earlier. We will examine four variables:

1. Number of fax machines (irrespective of sophistication), N

2. Median cost of fax machines purchased, C
3. Number of pages of fax transmission, T
4. Cost/transmission (regardless of length), $

We also assume a single policy option—taxing fax transmissions. Suppose that impact magnitudes were estimated on a scale from 0 (no impact) to ± 3 (major impact) and that we have defined a major impact for each variable as one that causes a 10 percent change in the level of the variable. The short-term and long-term impact matrices that are estimated appear in Table 12.7. Note that these impacts are merely presented as representative for the purposes of this example; they are not careful estimates.

Note that the policy impacts are represented as long-term impacts only and that they are incorporated by adding a column to the long-term impact matrix. The results of running the base case (no policy intervention) are shown in Figure 12.3. Note that the model predicts that the number (N) of fax units in operation quickly approaches the maximum value. Other variables grow rather quickly as well. However, the median cost (C) of a fax unit grows slowly and then declines. Before this model is used for forecasting and decision making, it would be necessary to verify, insofar as possible, and modify the variable initial levels and impacts. This process might begin by setting year 1 as some time in the past and checking to see if model predictions track historical variable behavior. It is easy to see that even this simple check might be difficult because of problems associated with gathering the necessary historical data. These problems might cause us to redefine variables to more readily fit the available data.

Assume the model has been fine-tuned and verified to our satisfaction and now we are interested in finding the changes in variable behavior that might be caused by a policy intervention. In the example, that intervention is a tax imposed on the normal cost of fax transmission. The impacts portrayed in Table 12.7 for this policy option were constructed assuming that a tax of approximately 5 percent is imposed. We can model various degrees of policy intervention by choosing different values on the range 0 to 1 for the initial level of the policy variable. Note

TABLE 12.7 Long- and Short-term KSIM Impacts on FAX Transmission Variables

On This Variable	Long-term Impact of This Variable					Short-term Impact of the Rate of Change of This Variable			
	N	C	T	$	Policy	N	C	T	$
$N(0.2)$*	$+3$	-1	$+3$	-1	-1	$+2$	$+1$	$+2$	-1
$C(0.35)$*	-2	0	-1	0	0	0	0	0	0
$T(0.15)$*	$+2$	0	$+3$	-2	-1	$+2$	$+1$	$+2$	-2
(0.30)*	0	$+1$	-1	0	$+1$	0	0	-1	0

* Initial values; variables scaled from 0 to 1.

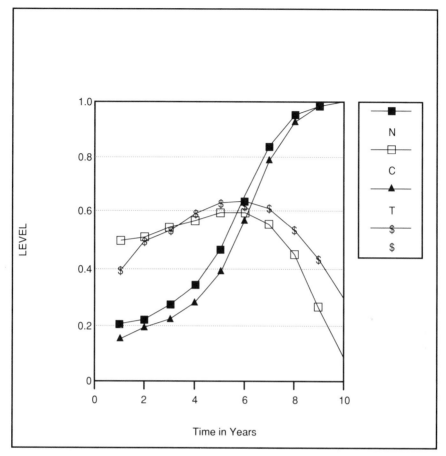

Figure 12.3 KSIM model of fax growth—base case.

from the matrices that policies are represented by columns but not rows in the matrix. Thus policies impact, but are not impacted by, other variables. For this reason, the value of the policy variable does not change with time. We might choose a policy implementation of 0.2, for instance, to indicate that we will tax only those transmissions that involve documents of 10 or more pages. The results of this policy intervention are shown in Figure 12.4.

Figure 12.4 shows that the policy intervention has significantly changed the picture from the base case. Neither the number of fax units *(N)* nor the pages of transmission *(T)* rise as fast as before. The median unit cost *(C)* rises to a higher value before falling, but the peak is delayed about two years. Even with the tax, the cost per transmission ($) changes little over the first four years. However, it peaks at a much higher value about a year later than without policy intervention. It is important to note that the policy impacts must be verified and modified just like the model itself. Only when the behavior seems reasonable to the manager or forecaster can a degree of confidence be placed in the results. This point is

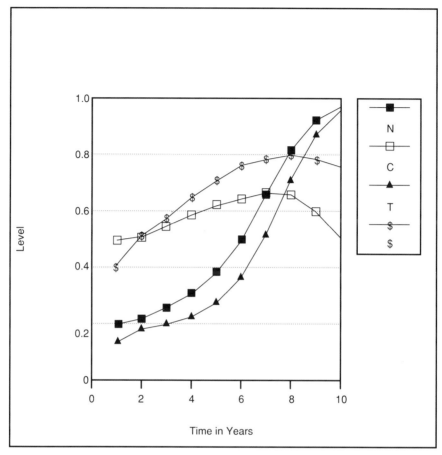

Figure 12.4   KSIM model of fax growth—policy intervention.

easily overlooked, for the model may assume a spurious credibility because of the sophistication of the computer equipment used to produce it. Beware!

The procedure could be extended to several policy options, or the policy option procedure could be used to model external events.   ∎

12.4   FORECASTING WITH SYSTEM DYNAMICS[3]

System dynamics (SD) is a quantitative simulation approach used to forecast and modify the dynamic behavior of important human systems. Both the philosophy and methods of SD are important. SD embodies a philosophy of physical and human

[3] This section was contributed by Professor Willard R. Fey, Georgia Institute of Technology.

causality that is focused on systems that are complex, nonlinear, and aggregated and that involve information-feedback control processes with fuzzy goals, decisions, information gathering and transfers, production functions, and time delays. The variables that characterize the operation of such systems have time histories composed of complex combinations of trends, oscillations, and random variations.

SD methodology is used to: identify the dominant feedback loops; represent loop relationships with nonlinear difference equations; analyze loop dominance and forecast future performance; synthesize structure changes to improve patterns of behavior; and implement changes effectively in real systems. Features of the methodology flow from the nature of causality in human systems described by the philosophy.

SD most often is used to improve the performance patterns of real systems. Occasionally, it is used to design new systems or to forecast future system behavior of an existing system without modifying it. In either case, a model of system causal feedback structure must be built. Then this model must be analyzed to establish its validity. Only when sufficient confidence in the model has been established can it be used to modify or forecast system operation. This is done by simulating model performance over a period of time sufficient to reproduce a significant portion of system history and to project future behavior. Ordinarily, SD analysts do not attempt to forecast particular variable values at specific times. Instead, they try to forecast a pattern of variation over a period of time. In forecasting technology development, therefore, SD would not be used to forecast the appearance of particular technologies, breakthroughs, or production processes. Instead, the process within which the technology develops and the impacts on human systems would be represented.

This is accomplished by writing equations to model the technology creation process and the social and economic environments of interest. This model then is simulated to re-create past time histories for the variables that are affected by technology (such as productivity, material use efficiencies, costs of production/mining/growing, diversity of consumer products, speeds of transportation, etc.) to provide verification. Then these histories are projected to forecast future patterns. In this way the goals, resources, limits, and opportunities in the entire socioeconomic environment that affect technology development are taken into account.

12.4.1 The System Dynamics Philosophy of Human Systems

Human systems (families, companies, institutions, nations) are organized into goal-seeking, information-feedback loops as shown in Figure 12.5. Components of these systems incorporate the activities of perception, decision making, and action; accumulation of things, ideas, and emotions; and the concepts of goals, expectations, and beliefs. Loops do not function separately. They are connected to form complex, internally interacting feedback systems. Thus changes in one part of a system eventually are felt everywhere in it.

System performance cannot be produced by, nor deduced from, the parts separately. Some loops, however, are more influential than others, although dominance often shifts from one loop to another and all contribute and are affected. Most

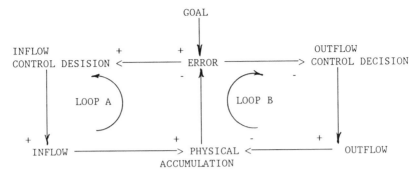

Figure 12.5   Feedback loops in a simple control system.

accumulations (information accumulations excepted) are deterministic. However, the deterministic behavior of system operations and concepts is modified by random forces, inconsistencies, and a partial lack of time continuity. In most real situations, control is imposed at the higher levels of aggregation for which goals are established. Therefore, SD models generally track aggregate flows rather than individuals or things. Most decisions in SD models are repetitive, and a fixed time between repetitions (or a continuous flow) is assumed even though the interval actually may vary. Thus computer-based SD simulations (such as DYNAMO and STELLA) use fixed time intervals between calculations instead of a variable interval based on random event occurrences, as do stochastic simulations.

Loop characteristics (such as number and types of accumulations, algebraic sign, time delay, strength of response, nonlinearities, etc.) determine the type and magnitude of the time pattern that an isolated loop produces. There also are aspects of the loop linkages that determine the complex combination of patterns that compose system time histories. Human systems are complex, nonlinear, and quasi-periodic with nonstandard objective functions. Therefore, neither rigorous deterministic nor stochastic mathematical analyses can be used to obtain closed form solutions (time responses), optimum parameter values, or optimum structure modifications. Thus simulation is necessary.

Since system behavior patterns are produced by interaction of all system components, they can be changed by altering the feedback structure. Structure change occurs often in real systems (for example, firing the coach, trading a player, or selling an unprofitable product line). However, there is seldom a loop analysis such as SD provides to justify it. Whether it can be analyzed or not, the structural change activity forms a metaloop, as shown in Figure 12.6, that may complicate the process considerably. The metaloop contains the system feedback loop structure that creates the performance patterns. These patterns are perceived and evaluated, and, when they are unsatisfactory, structure changes are developed and implemented. Figure 12.6 shows that SD becomes part of the metaloop when it is used for analysis and change in a real human system. Thus SD is a part of the system that is being studied; it is not an external, independent, objective, and passive activity,

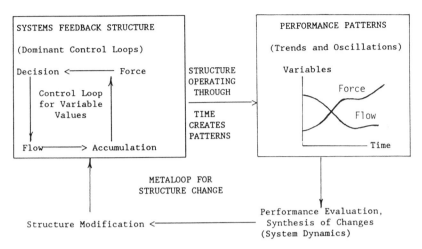

Figure 12.6 The metaloop that controls system structure change.

as scientific studies usually are assumed to be. Since it is an integral part of the system, the SD activity might be included in its own analysis.

12.4.2 System Dynamics Methodology

SD is a science-aided art usually practiced by a consultant or advisor to a human feedback system (company, city, nation, etc.). The consultant executes a series of prescribed activities in a logical sequence to solve a particular dynamic problem (for example, to create growth in a stagnant company or to reduce inventory oscillations). The usual SD problem involves changing the behavior pattern of an existing system by altering its feedback structure. However, occasionally the structure of a new system is synthesized or the future time patterns of an existing system are forecast without attempting to improve them. Whatever is done, the majority of the following SD activities will be required.

The first step is to identify the important variables about which the client is concerned. Then data are gathered to determine the historical time patterns for those variables. Next the operation of the system is observed and measured to determine how people, materials, products, money, and ideas are stored and flow through the system; how information is obtained and used; and how decisions are made and implemented. The goal is to understand the performance (time patterns) and operation (feedback structure). This understanding provides the basis for writing a hypothesis (the dynamic hypothesis) that identifies the feedback loops and linkages the analyst believes are most important. Since there is noise in the system and less important parts of the structure are omitted to simplify the analysis, the hypothesis is neither complete nor perfect. Therefore, the success of the analysis depends on the analyst's ability to distinguish between dominant and unimportant loops and between patterns caused by random forces and those caused by the operations of reasonably deterministic loops.

The loops identified by the dynamic hypothesis are modeled with first order, nonlinear difference equations. Usually these cannot be solved analytically, so system behavior must be simulated using one of the SD computer programs (such as DYNAMO or STELLA). Simulations are performed for a variety of conditions (different exogenous inputs, initial values of accumulation variables, parameter values and tables, decision functions, and/or noise sequences). These variations are specified by an experimental design developed by the analyst to determine if the model adequately emulates the real system. The analyst also seeks to determine if model performance patterns are created by the loops identified in the dynamic hypothesis for the reasons he or she hypothesized. Corrections are made to the model and hypothesis until they are "accurate enough" to be acceptable and until the causes of system behavior are understood.

Next modifications in feedback structure to improve system performance are synthesized. This is a creative process not an algorithmic one, unless the system is linear and very simple. Synthesis is difficult, and there are many errors that can be made, so it must be completed carefully. Structural changes (such as elimination of loops or creating new loops) usually are more effective than adjusting parameters. It is possible for different SD analysts to propose different changes for the same system or to overlook simple, highly effective ones. Nevertheless, reasonably effective changes do emerge from most SD studies.

At this point a "transition" analysis is performed to forecast system behavior from the time that recommended changes are made until improved system behavior is expected. This is necessary because system performance does not instantly improve when the structure is changed. Rather, in feedback systems, there is a transition period during which behavior differs from both the old and new patterns. This analysis involves simulation of the old and new system models together with the introduction of the changes to create one entire life cycle. This is an analysis of the metaloop shown in Figure 12.6. Once the transition analysis is complete, the recommended structure changes are introduced into the real system. Care must be taken to avoid undesired changes in other parts of the system and to see that the recommendations are implemented properly. Continuing observation and evaluation are necessary to correct implementation errors and to be sure that the expected transition and new system behavior patterns develop as anticipated.

When SD is used for forecasting only, the synthesis, transition analysis, implementation, and evaluation parts of the methodology are not required. However, all other activities described in this section still must be carried out.

12.4.3 Technology Forecasting with System Dynamics

Technology is particularly difficult to forecast. Previously unknown physical laws, materials, processes, and applications will be discovered and new technologies will affect society in many unprecedented ways both now and in the future. Changes in technology stem from research—a creative and expansive process. Basic research focuses on new discoveries about the laws of nature in physical or biological relationships. Advances occur in places where nature hides them, not necessarily where researchers seek them. Applied research concentrates on solutions to specific

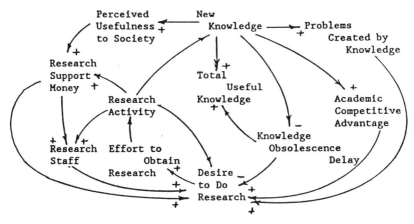

Figure 12.7 Fey's causal loop diagram for research in U.S. higher education.

problems. Sometimes these problems are solved, sometimes they are not. Sometimes applied research reveals new problems or unexpectedly solves unrelated ones.

Until World War II research was largely an individual or small group effort. It was randomly focused on subjects of interest to relatively few people who were usually unsuccessful. Only a few researchers, called inventors, were consistently successful. Research in this era was largely applied. However, the research process changed drastically during World War II. Under the pressures of wartime, governments and universities found that they could create knowledge in a systematic way in preselected areas through properly managed efforts by groups of highly trained people. It was through such processes that radar and the atomic bomb, for example, were developed.

As a result of such discoveries, the U.S. government created the National Science Foundation after the war to fund organized research. Corporations and universities came to believe that research could be enormously effective and instituted active research programs to obtain the benefits. The chaotic, individualistic quest for knowledge that led to the Industrial Revolution, the automobile, and the airplane suddenly became organized and institutionalized. Technological development suddenly was transformed from a random process controlled by the genius of a few people to a systematic feedback process controlled by the objectives of human systems and the resources that they could bring to bear. Thus, it became a target for SD analysis. The first SD modeling of research was done by Abraham Katz (1958). It was followed by the work of Edward Roberts (1964) at MIT, which was related to product development within a firm.

In the early 1970s Willard Fey studied the U.S. higher education system. His model included the feedback structure of university research and its impact on knowledge expansion as a major factor in the dynamic performance of the aggregated university system. A causal feedback loop diagram for some of the research forces in the university system is shown in Figure 12.7. The research sector shown in the figure contains nine feedback loops, all of which are positive. Positive loops tend to reinforce existing performance trends. Thus it is not difficult to understand

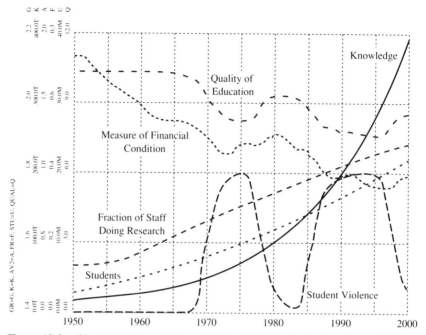

Figure 12.8 Simulated 50-year history (1950 to 2000) for Fey's education model. (*Source:* Fey, W.R. and Knight, J.E., The Dynamics of Educational Institutions, *Proceedings, 1971 Summer Computer Simulation Conference,* p. 1118, 1971. Reprinted by permission. © Simulation Councils, Inc.)

why the institutional setting has contributed so much to knowledge growth. Simulated time histories for several educational system variables are shown in Figure 12.8. The first 20 years of that simulation (1950 to 1970) approximate the historical patterns quite well. The last 30 years (1970 to 2000) constituted a forecast at the time the model was developed.

Morecroft (1986) examined the fledgling high-technology growth market for automated storage and retrieval systems (AS/RS) and developed a model to examine how marketing effort, capacity expansion, and industry reputation interact to produce eight- to nine-year growth cycles in the demand for these systems. In his model, three subsystems portray AS/RS market growth: (1) customer interest and industry reputation; (2) studies done by firms preparing to bid on systems; and (3) new competition. These are incorporated in an overall feedback model (see Figure 12.9) that can be used to design and interpret simulation experiments in market dynamics.

These three models illustrate the conventional SD principles associated with forecasting systems with important technological feedback loops. These principles include:

1. Technology cannot be studied by itself. It occurs in a context of social, political, financial, personnel, and educational factors that must be considered when the dominant control loops are identified.

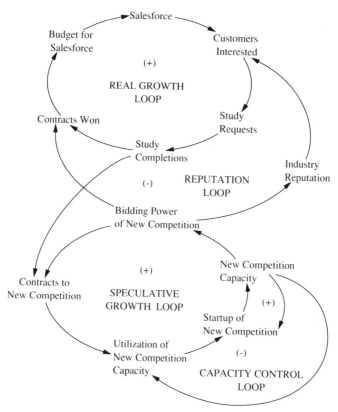

Figure 12.9 Morecroft's feedback loop structure of the AS/RS growth market. (*Source:* Morecroft, "Dynamics of a Fledgling High Technology Growth Market," *System Dynamics Review*, vol. 2, pp. 36–61. 1986. Reprinted by permission of the System Dynamics Society.)

2. Technological development is not a random process. While there is noise in the process and research outcomes are not perfectly predictable, the long-run socioeconomic performance patterns are created by feedback control processes.

3. Dominant control tends to be exercised at aggregate levels, so SD models for systems with technological aspects usually are highly aggregated. Individual research projects and individual researchers normally are not represented.

4. Forecasting is done with a causal model of the dynamic feedback control process within which technological development takes place. There is no extrapolation of past data time histories or sampling from probability distributions except when noise is used in the feedback system.

5. The objective is to forecast patterns of change over a period of time (even a century or more), not specific research outcomes or values of variables at particular times.

6. The standard SD methodology is used to determine past performance and system structure. A dynamic hypothesis is written to describe the domi-

nant feedback control loops and to indicate how their operation creates the performance patterns. A corresponding quantitative model is developed. The model is analyzed (simulated under a variety of conditions) to establish its validity and verify the hypothesis.

7. When the model is accepted as "valid enough" (it cannot be perfect) and the loop structure is thoroughly understood, the forecast is made by simulating the model over a period of time that includes past history and a future period.

8. Forecasts are usually not made for technological variables themselves, but rather for important system variables. In education, for instance, the growth of knowledge is not the pattern of major concern. The quality of education, financial condition, and student body size would be the subjects of study.

9. Since technology is so pervasive in our civilization, there is almost no significant system that can be studied using SD without the consideration of some technological variables.

This last point has an important bearing on the most widely known SD study ever conducted. In the late 1960s, a group directed by Dennis and Donella Meadows conducted an SD study of the world system. They attempted to represent the dynamics of the collision between expanding population and industrial output, on the one hand, and "fixed" limits of nonrenewable resources, pollution assimilation capacity, and arable land, on the other. The simulation forecasted continued rapid growth in population, industrial output, and most other variables until approximately 2050. Then one by one the variables peaked and collapsed to levels from 20 percent to 60 percent of their peak values within 30 years (see Meadows et al., 1972). (Bremer—1989—describes six other world-simulation models, although none are based on SD.)

Economists correctly criticized the Meadows world model for omitting financial variables such as prices, interest rates, money supply, and debt levels. However, omitting technological development was an even more serious error. Many of the constant values and constant transformation functions in the model (such as industrial capital/output ratio, service capital/output ratio, inherent land fertility, food processing loss, etc.) depend on technology. To change these constant factors to variables would require including the people and resources devoted to creating new knowledge as well as the ways in which they are used in the model. Clearly, such an addition is needed. It was the development of medical and industrial technology over the last few centuries that extended the limits that restrained population expansion and the quality of life. It is not surprising that the old limits and some new ones appeared with a vengeance in a model that froze technological development. There may be insurmountable limits of some kind but they cannot be assessed credibly unless realistic technological development factors are included in the analysis.

12.5 GAMING SIMULATION

It is possible to use simulation for technology forecasting and management through game playing. Gaming employs an entirely different approach than the methods described in other sections of this chapter. It involves constructing a realistic set of rules and then observing the behavior of players who either compete or cooperate to achieve a goal within the context provided by the rules. Forecasts or management decisions are then conditioned on the behavior of the players and/or on the outcomes they are able to achieve.

The game Monopoly® provides such an example. In Monopoly the goal is to achieve economic dominance. The rules are loosely patterned on real-world business and economics, and uncertainty is introduced by the throw of the dice and randomly shuffled and drawn instruction cards. Thus Monopoly loosely can be styled as a simulation game. We certainly can learn something of the motivation of our friends (and ourselves) by observing the behavior of the players.

Many simulation games have been constructed to investigate organizational change, business development, war tactics and strategy, policy making, communications, learning, environmental planning, health care, diplomacy, gambling, and so forth. An International Simulation and Gaming Association publishes the proceedings of their annual conferences (see Crookall et al., 1988; Klabbers et al., 1989), and two journals publish developments in the field—*Simulation & Games* and *Simulation/Games for Learning*.

Gaming is a powerful method for treating issues that are complex and ambiguous. In many instances, it has proven an effective alternative or adjunct to strategic planning methods used by large institutions. The United Nations, Upjohn Pharmaceutical, Chase Manhattan Bank, Xerox, and Conrail, for instance, have used gaming to solve problems in a complex environment. Gaming is especially appropriate for situations characterized by

1. Numerous interacting variables
2. Variables and their interactions that are difficult or impossible to quantify
3. The absence of a conceptual model for decision making
4. Decision making in a sociopolitical context in which the players may be idiosyncratic and unpredictable

While it is not possible here to consider all the caveats associated with good gaming, there are a number of general rules that can be given. These are discussed in four general areas—designing, playing, evaluating, and correcting the game.

In designing a game, the manager must first set the objectives for the simulation and the goals toward which players will work jointly or competitively. Next the rules governing player actions and interactions and other aspects of the game must be devised. For the simulation to be effective, these should emulate the most important aspects of the situation being gamed. Measures of player success also must be provided. These are usually translated into indicators displaying the degree to which the goal has been achieved. Indicators might be token money (as in Monopoly), judgments made by other players, mathematical payoff functions, and so forth. Further, a display system must be devised to show the progress of the game. This might be a board, as in Monopoly, or a video display, as used by Nintendo® and other electronic games. In the business environment, a computer screen often may be the most appropriate display. Finally, the game must include mechanisms for introducing the randomness that results from exogenous (outside) events.

It is also important that players represent as many of the different groups involved in the situation being simulated as is necessary to achieve reality. Each player will bring a somewhat different culture and perspective, which is important to the context of the game. The game's rules and goals should be carefully documented, each player should have a copy and should understand the goals and rules before beginning. A nonplaying observer should oversee the game and make decisions in cases not covered by the rules.

Once the game is completed, the results must be evaluated and compared with the objectives of the game. The indicators can be used to help determine how well the objectives of the simulation have been met. Players should be encouraged to provide feedback on the realism of the game, the clarity of the rules, and so on. Also, since one of the objectives usually is to learn something about the situation,

---

■ Exhibit 12.3 A Manufacturing Technology Game: ADVANTIG

It is useful to consider one of the many games in common use to clarify the contexts in which gaming can be useful. Exhibit 12.3 provides an example.

Advanced Manufacturing Technology, or ADVANTIG[4] is a simulation that has been used by manufacturing companies trying to transform their operations using advanced manufacturing technology such as computer aided design/computer aided manufacturing (CAD/CAM), robotics, automated material handling, and material requirements planning (MRP). The goal of the game, as the name implies, is to win the automation game by gaining a manufacturing advantage over competitors.

To play ADVANTIG, a team from the company tries to transform a fictitious firm, F.L.I. Casting, Inc., from a low-technology to a high-technology organization. Key management plays the roles of F.L.I CEO, production manager, head of engineering, head of finance, and so on. There typically are from 8 to 21 participants who should have some knowledge of CAD, computer numerical controlled machines, robotics, assembly techniques, and MRP. Decisions about strategy, technology, and implementation are made by the team. Their objective is to select advance manufacturing technology and to implement it profitably while meeting present customer obligations and capturing new business.

Four aspects of F.L.I. operations are simulated: production, management, engineering systems, and dealing with advanced manufacturing technology vendors. Management sets F.L.I. strategy, bids on contracts, and maintains financial records such as contract forms, balance sheets, and material requisitions. Engineering buys and implements new technology with the help of the vendors. The game typically lasts from four to five hours and is followed by a two-to-four-hour evaluation (debriefing). ■

---

[4]ADVANTIG is available from the Industrial Technology Institute, P.O. Box 1485, Ann Arbor, MI, 48106.

players should be encouraged to interact with each other and analyze their actions and interactions.

The evaluation likely will show that the objectives, rules, and/or conduct of the game need to be revised for it to more nearly represent the situation being simulated. Once revisions are completed, the revised game should be conducted again with new players and revised again based on a new evaluation. Thus the game is iteratively made to more nearly simulate the situation it portrays. Once the manager is satisfied with the simulation, it may be possible to replay the game with a set of historical circumstances. Although player response and exogenous events may occur differently, it should be possible to calibrate and fine-tune rules and objectives somewhat to increase confidence in the game.

12.6 CONCLUSION

Simulation is a powerful tool for the forecaster and technology manager to apply in the proper circumstances. However, he or she must realize that simulation generally requires a significant expenditure of time and resources. It is important when weighing the strengths and weaknesses of various approaches to view the building of simulation models as a significant aid to understanding complex situations. Therefore, simulation must be considered for its value as a process as well as its usefulness as a predictive and/or analytical method.

Simulation has different meanings and can be accomplished by applying many different techniques. All simulation, however, seeks to simplify complex situations while retaining those aspects that are important for understanding system behavior. Thus all simulation models contain critical core assumptions that must be carefully tested. Moreover, all simulation models require modification and fine-tuning when first completed to produce system emulation that is satisfactory to the user—that is, validate the model. One means of validation is to see if the model can "predict" historical behavior patterns. It is important to note, however, that no model of a complex situation such as technology growth can be completely validated. Success in predicting historical behavior is no sure guide to successful prediction of the future. Thus the technology manager or forecaster must retain a healthy skepticism when employing simulation as a tool.

EXERCISES

12.1   Play the CI game in the fax example (see Table 12.1) for 100, 200, 500, 1,000, 2,000, 5,000, and 10,000 runs using the computer program in the TOOLKIT. Plot the values of the cross-impacted probabilities. Draw inferences from this exercise.

12.2   Suppose we want to forecast the future of an electric car that reaches a speed of 60 miles per hour, travels 120 miles between recharges, and requires 12 hours for a full recharge. Prepare occurrence and nonoccurrence matrices similar to those in Tables 12.1 and 12.2.

12.3 Manually play the AIMTECH matrices shown in Table 12.5 20 times. Compare the cross-impacted probabilities to the original values.

12.4 If you have access to XimpacT, execute the program for a 10-year horizon using PERSONAL.CIM as input. Display all the tabular and graphical outputs.

12.5 Consider the Markov transition matrix information given in the AIMTECH Exhibit 12.1.

(A) What is the probability of remaining in State 1 for another step?

(B) What is the probability of going from State 1 to State 3 on the next step?

(C) What is the probability of going from State 3 to State 2 on the next step?

12.6 Using the same information as in Exercise 5,

(A) What is the probability of going from State 1 to State 3 *on* the third step?

(B) What is the probability of going from State 1 to State 3 *by* the third step?

12.7 For the AIMTECH example, compute S^{1997} if $S^{1996} = (0.02, 0.13, 0.53, 0.26, 0.06)$.

12.8 Consider Exhibit 12.2. Using the program in the TOOLKIT, vary the impact of the median cost, C, of a fax unit on the number of fax transmissions, T, by increments of 0.5 and compare your results to those in Figure 12.3. Investigate the impact of different initial variable levels on the result.

12.9 Develop a policy implementation of your own and investigate the effects that varying its degree of implementation (initial value) has on the base case shown in Figure 12.3. Repeat this using an external event instead of a policy.

12.10 Develop a KSIM model for some older technology (such as television). Choose the variables and initial values. Ask your classmates to help you estimate the cross-impacts. Use the KSIM program in the TOOLKIT to run the model and compare the forecast behavior with the historical trends of development. Modify the model until you have what you consider to be an acceptable match.

12.11 Obtain a copy of Morecroft (1986) and analyze the article.

12.12 Morecroft (1986) contains a copy of the DYNAMO program used to draw the inferences given in the article. If you have access to DYNAMO, first enter the code and obtain output similar to that presented. Modify the inputs and examine the output to gain greater insight into the model.

___13
SCENARIOS

OVERVIEW

Scenarios are outlines of some aspect of the world—in our case, the future world. This chapter discusses the nature of scenarios, presents the various types that can be useful to the technology manager, and suggests how the manager might use them. Scenario construction is also considered in some detail. Utility and validity, which are useful criteria by which to evaluate scenarios, are presented.

13.1 INTRODUCTION—WHAT ARE SCENARIOS?

Webster's New World Dictionary defines a scenario as: "an outline or synopsis of the plot of a drama, opera, etc., indicating scenes, characters, etc.; or an outline of a motion picture indicating the action in the order of its development, the scenes, the cast of characters and their appearances, etc."

Rather than outlines of dramas, operas, or movies, for our purposes, scenarios are outlines or synopses of some aspect of the future. Although they may be presented in narrative form, they often include numerical data and graphics. Scenarios even may incorporate multimedia presentations of still and motion pictures, audio, and so forth. There is no reason why scenarios should not be as appealing to their audience as *Star Wars* was to its.

Scenarios incorporate and emphasize those aspects of the world that are important to the forecast. These are the *dimensions* or important factors of the scenario. Scenarios may integrate several specific forecasts concerning individual dimensions to create a multidimensional forecast.

Some of the most interesting and well-known future scenarios are the better works of science fiction. By imaginatively presenting stories of the future, authors encourage their readers to open their minds and understand conditions or environments that may appear alien or "far out." While Jules Verne's descriptions of long undersea voyages and travel to the moon must have struck nineteenth-century readers as speculative, to say the least, analogous technologies now are commonplace. Aldous Huxley's vision of a genetically programmed society in *Brave New World* has not yet been realized. With each passing year, however, that sort of society becomes increasingly possible as biomedical technologies at the molecular and molar levels develop rapidly.

The systematic use of scenarios in forecasting, planning, and strategic analysis began after World War II. Becker (1988) divides the period from 1950 to 1986 into three parts. From 1950 to 1965 scenarios were used in the United States to set the context within which action took place. In France they were taken as portrayals of strategies and their outcomes. During this period the predictive power of forecasts was thought to be high. However, from 1965 to 1975 users began to realize that scenario-based predictions were not as accurate as formerly believed. From 1975 to 1986 forecasters began to integrate strategies and contexts into single scenarios. Thus more realistic views of the use and accuracy of scenarios for forecasting have eventually prevailed. During this later period, impact assessments (such as technology assessments and environmental impact assessments) were especially common scenario forms. Evaluation of old forecasts began during this period as a means of improving the quality of forecasts (see, for example, Ascher, 1978).

Scenarios usually incorporate uncertainty. This is expressed by describing a range of possibilities, usually by a few individual scenarios that, taken together, constitute the range of futures considered likely. Alternately, uncertainty can be expressed as the probability of a single scenario or, at a micro level, as a range of projections for the dimensions within a single scenario. It is important for forecasters to avoid the bane of scenarios—fantasy. While science fiction may incorporate fantasy, the forecaster must maintain close contact with reality. The discussion of critiquing scenarios presented in Section 13.5 should help maintain a firm basis in reality.

13.2   TYPES OF SCENARIOS

Scenarios can be distinguished based on their temporal orientation—that is, as future histories, snapshots, or combinations of both. *Future histories* map a trajectory from the present state to some future time, focusing on a particular part of the world and tracking its development during some period. "The Energy Age" (Mathisen, 1977), for instance, presents a hundred years of future energy history from 1977 to 2077. A classic work on conflict escalation emphasizes stages or "rungs" in developing escalation scenarios (Kahn, 1965). Future histories stress the dynamics of a system, possibly at the expense of its structural or cross-sectional characteristics. Asimov's *Foundation* series is a fictional example of a future history. *Snapshots*, on the other hand, present a cross-sectional view at a single point in time, and structural, rather than dynamic, characteristics are emphasized. Orwell's *1984* and Huxley's *Brave New World* are fictional snapshots. Considered from a planning perspective, snapshots portray goals or end states. Future histories portray plans or paths to end states. These views can be integrated in a single scenario consisting of a trajectory that leads to an end state.

Another distinction between scenarios is whether they are descriptive (that is, extrapolative) or normative. Descriptive scenarios address the issue of what *will* come to pass. Normative scenarios describe what *should* come to pass. These scenarios respond to planning and policy-making concerns by representing goals and plans. Descriptive scenarios simply lay out the range of possible futures determined in the forecast without attempting to evaluate them as desirable or undesirable.

Scenarios very often are constructed in sets. When a single dimension is dominant, the most common grouping is a set of three. The *baseline,* or most likely scenario, results from extrapolating current conditions and trends. This scenario tracks business as usual. The set also includes two extreme scenarios, referred to as *optimistic* and *pessimistic*. These evaluative designations refer to changes in the central dimension from the manager's perspective — positive changes in the optimistic and negative changes in the pessimistic scenario. Of course, when there are many important dimensions, the "space" in which the scenarios are embedded becomes multidimensional, and the selection of appropriate scenarios is more complicated since all relevant dimensions must be considered.

13.3   HOW ARE SCENARIOS USED?

Scenarios can provide intellectual stimulation for imagining the range of possibilities that the future may hold. This is exemplified by science fiction literature and by "visionary futures." This stimulation is valuable. It can provide forecasters, forecast users, and the public with a range of images that will open their minds and expand their imaginations to encompass a future that may be surprising and unexpected. Exhibit 13.1 shows a potential use of trends and scenarios to help form public policy.

As discussed, scenarios may present descriptive forecasts or they may offer normatively based plans or goal states. However, they have other uses in the planning and policy-making processes as well. Becker (1983) points out three distinct uses for scenario forecasts. First, scenarios can be used to assess whether various policies or actions will assist or inhibit the realization of conditions described in them. This approach can be used to construct future histories that link the implementation of policy options to desired future states. Second, scenarios can be used to assess how well alternate policies and strategies would perform under different conditions. This involves creating snapshots of future policy impacts and future histories leading to them. Finally, scenarios can provide a common background for individuals involved in planning within an organization, an exercise in environment creation using scenarios as the vehicle for creation and presentation. In the case of normatively based plans and goal states, snapshots can be used to identify the goal states and future histories can be used to describe the workings of the plans designed to achieve them.

Scenarios have been used widely in business and government applications. Businesses can use them to pull together the data and assumptions underlying their strategic plans. This can help to communicate the plan to employees, to financing sources, to stockholders, and, in some cases, to the public. Governments can obviously use them to explain plans for the future and to build public support for their realization. Exhibit 13.1 describes legislation that will increase the formal use of scenarios in the work of the U.S. Congress.

■ Exhibit 13.1 Alternative Scenarios in the U.S. Government

Senator Albert Gore has proposed a bill, the Critical Trends Assessment Act, that would give the federal government a highly visible source of major alternative futures, or scenarios (Gore, 1990). The bill recognizes the complexity and interdependence of the modern world, exhibited by issues such as global environmental change, homelessness, and international economic competition. It notes the richness of information sources available on many critical trends, and it targets the need to integrate existing sources of information to posit alternative, long-range (20-year) futures.

The bill would provide for ongoing assessment through a new Office of Critical Trends Analysis, which would report on alternative futures every four years. In addition, the Joint Economic Committee of Congress would produce a similar, independent report every two years.

This initiative extends a pattern of increasing use of scenarios by various levels of government and by large corporations. They have proven to be a valuable tool for understanding the uncertainties and the options of the years ahead (Bezold, 1990). ■

The scenario form has two generic uses that can be exploited within the forecasting, policy, and planning processes—*integration* and *communication*. First, the scenario offers the potential to integrate information from diverse sources and of different character into a single forecast. The results of trend extrapolation and expert opinion studies, for instance, can be woven together in narrative form. Further, quantitative data and qualitative information can appear in a scenario side by side with values. Thus the limitations of structure that trend analysis, expert opinion studies, and models exhibit individually can be overcome by incorporating their forecasts in a scenario. Broad features of a situation can be expressed clearly and in an integrated and understandable way, which brings us to the second generic use of scenarios—communication. The technical report does not convey information effectively to nontechnical audiences. In format, length, content, and style the technical report is not usually a good vehicle for mass communication. The scenario, which expresses a technical analysis in literary form, is a far better approach to such an audience. It can engage their imagination without sacrificing sound technical analysis.

In cases for which there are no data, credible experts are lacking, and the assumptions required to develop good models are not available, scenarios usually are the only way to create a forecast. In such cases, a good deal of imagination is required to build on whatever information is available. The pitfall is that these scenarios may be more fantasy than forecast since their basis in fact, while the best available, may not be sufficiently robust. Yet even such an effort should narrow uncertainty about the future and focus the attention of decision makers and other involved parties more precisely than if scenarios were not attempted.

13.4 CONSTRUCTING SCENARIOS

The procedure for constructing scenarios presented here should not be viewed as a rigid prescription, but rather as a checklist or outline that systematically guides forecasters without rigidly binding them. Scenarios are an opportunity for creativity both in thought and expression, but they are likely to be more useful if they have structure. Exhibit 13.2 outlines a typical set of steps for successful scenario construction.

■ **Exhibit 13.2   Checklist for Scenario Construction**

1. Identify topical dimensions
2. Identify intended users' interests and the appropriate style of information presentation
3. Specify time frame
4. Specify general societal contextual assumptions and specific (technology) assumptions
5. Set out the key dimensions
6. Decide on the number of scenarios and their emphases
7. Build and present the scenarios   ■

13.4.1 The Topical Dimensions

The first consideration in constructing a scenario is to identify the dimensions that it will incorporate. These are typically determined by the purposes for which the scenario is being constructed. However, there are situations in which the forecaster has some flexibility in identifying the dimensions to be addressed. This flexibility should be used (1) to lay out dimensions that are relevant for the purposes of the forecast and (2) to assess the forecaster's ability to build a sound forecast from the information that is available or that can be developed in the course of scenario construction.

13.4.2 The User

Scenarios are constructed for a specific user or group of users. These might be the forecasters themselves, a government agency, a corporation, a public interest group, an individual, or the public. The focus of the scenario must relate to the needs and interests of the user. The format in which it is organized and presented should reflect not only the user's interests, but also the type of information and style of presentation that are most appropriate. For example, scenarios intended for a popular audience should not be presented in a technical report format; a literary or multimedia presentation would be more appropriate. By the same token, scenarios intended for a professional audience, such as economists, should incorporate all the economic sophistication and technical style required to satisfy their critical demands. Although the forecaster should be guided by the needs and interests of the user, professionalism sometimes demands that factors the forecaster believes are important should be considered even if the user may believe them to be irrelevant or wish to ignore them.

13.4.3 The Time Frame

The time frame places limits on the character of a scenario. For example, a very short frame, say a year, usually implies that current conditions and trends will dominate. Unless conditions are extremely unstable, this leads to immediate extrapolation of current conditions and to scenarios that tend to bunch around the baseline. On the other hand, a very long time frame, such as a century, almost ensures that accurate information and forecasts will be lacking even for relatively well-defined parameters, such as population. Thus the task of the scenario writer becomes speculative; success depends on his or her ability to integrate imagination and judgment with a very narrow information base. The intermediate case, perhaps 5 to 20 years, is more common for the technology manager. It allows a variety of forecasting techniques, such as trend analysis, expert opinion, and modeling, to be integrated into the scenario. Current conditions and trends may, of course, change even over the time frame of an intermediate forecast. Therefore, the uncertainty inherent in them is higher than that in very short term forecasts.

13.4.4 The Assumptions

The assumptions on which scenarios are based are the most important factors in determining the quality of the forecast (Ascher, 1978). Time spent examining and refining assumptions at the beginning is time well spent, as it can help eliminate errors that may force the entire forecast to be redone. The time frame is a major determinant of the stability of assumptions. The longer the time frame, the less likely it will be that current conditions and trends will continue. Therefore, as the time frame increases, progressively greater care is required to formulate the underlying assumptions.

Assumptions usually fall into two categories. The first, *general societal assumptions*, usually relate to broad conditions of political decision making and stability, the state of economies, societal values, global environmental conditions, and demographic trends. They also may involve the rates of development of various families of technologies and the rate at which they are implemented on a broad scale. The second category consists of *specific assumptions* relating to the forecast at hand. For example, in an energy forecast for the United States, these might involve governmental energy policies, the conditions in certain industrial sectors, societal attitudes toward energy use and conservation, and the status of critical technologies (including the possibility of technical breakthroughs).

Assumptions of both types will vary with the time frame of the forecast and with the perspectives of interest groups involved in the forecast as actors, sponsors, or users. Different interest groups generally have different biases. For example, there is an obvious difference in perspective on the relationship of smoking to cancer for the American Cancer Society and the tobacco industry. The forecaster will find it impossible to eliminate every bias—nothing would be left. However, it is worthwhile to make the biases contained in the assumptions as explicit as possible.

13.4.5 The Dimensions

The users and the purposes of the scenarios determine the dimensions of their structure. Scenarios can be one-dimensional or multidimensional. The more dimensions, the more complicated the scenarios and the more complex their construction and presentation. Although completeness is important, unneeded complexity can dull presentation clarity. Therefore, the forecaster must decide how each dimension will be expressed and what level of detail will be required for each.

The structure of each scenario must define interrelationships between dimensions as well as relationships between dimensions and the boundary conditions of the scenario. These can be formatted as a kind of template or pattern and used for all scenarios in the set. This template also must incorporate the underlying assumptions. Dimensions can be represented quantitatively or qualitatively, and the level of aggregation at which each is expressed must be determined. For example, a set of scenarios dealing with future U.S. energy use might incorporate three dimensions: yearly energy use, principal sources of energy, and major end uses of energy. The first dimension could be expressed quantitatively in quads of energy; the second and third dimensions could be expressed qualitatively. Sources could include coal, petroleum, renewable sources, nuclear fission, and nuclear fusion. End uses might include commercial, industrial, transportation, and residential uses.

Representations of the dimensions may involve forecasts. Thus the scenarios may integrate a number of forecasts dealing with different topics and incorporating different methodologies. In the preceding energy scenarios, energy use could be forecast by trend extrapolation and demand modeling (see Chapter 10) with some consideration of the potential impact of major events with significant probability of occurrence (see Chapter 12). Energy sources could be described by combining trend extrapolation based on time series data that describe the amount of energy produced by each (see Chapter 10). Expert opinion about their relative future importance, perhaps gathered with a Delphi poll (see Chapter 11), could be added. The sources of data and forecasts for each dimension must be considered in light of available time and budget. For a quick, low-budget effort, new forecasts are precluded; however, when significant time and money are available, sophisticated models may be built as part of the forecast.

13.4.6   The Number and Emphasis of Scenarios

The number and emphasis of scenarios can be determined by a procedure that uses morphological analysis (Chapter 7). First, consider each dimension of the scenario. For those to be represented qualitatively, determine the major emphases that the dimension may assume. For dimensions that are quantitative in character, select appropriate ranges of values including the baseline and extremes as well as various intermediate ranges if they are appropriate. The emphases and/or appropriate value ranges of each dimension establish the morphology of the scenario space. This morphology defines the maximum number of possible scenarios. The problem now is to identify the scenarios that will be significant in meeting the goals of the forecast.

For forecasting a single quantitative dimension, one to three scenarios almost always are adequate. The scenarios for a single qualitative dimension depend on the major emphases of that dimension. Multi-dimensional scenarios require more complex representations in which the sets of values for the various dimensions are considered as wholes. The most useful combinations of these are selected as the final set of scenarios. The major concern is to represent a complex situation without using too many scenarios.

Selecting scenarios from the morphology requires linking potential scenarios to the goals of the forecast. Initially it is best to choose high-probability scenarios whose outcomes will make a substantial impact. Often this set is satisfactory. Sometimes, scenarios that are too similar must be eliminated. At times, arbitrary judgments will be necessary to arrive at a final set.

For example, in the energy scenarios discussed previously, it may be convenient to largely ignore the mix of end uses by assuming that it will be more or less stable in all scenarios and similar to the current mix. Absolute values for energy use could be selected as high, medium, and low with suitable quantifications for each. Potentially dominant sources of energy then could be selected. In this instance, we might choose petroleum, coal, solar and renewable, and nuclear fission as candidates for dominance and reject nuclear fusion as unlikely. Thus we would have a maximum of 12 scenarios—three values × four sources. Next we need to decide how to eliminate unlikely or redundant cases. We might select petroleum/medium as the baseline scenario. In addition, we probably would want at least one high and one low energy use scenario as extremes. Likewise, each energy source should be highlighted. Thus we would have a minimum of four scenarios to work from. Now we rely on the background information and forecasts available to make the final selection. For example, in addition to petroleum/medium (baseline), the scenario set might include nuclear fission/high, solar/renewable/low, and coal/medium. The procedure for selecting scenarios is similar in other cases.

13.4.7 Building and Presenting Scenarios

The first requirement is to construct the template that each scenario will follow so that all will be comparable. Next the template must be filled in for each scenario using available forecasts and background information. The resulting scenarios then must be checked for utility and validity (see Section 13.5). When the scenarios meet the criteria established for them, they can be presented to the users.

At this point, the style and media used for presentation become important considerations. Whether as a technical report, a set of short stories, or a movie, the intellectual content of the scenarios should be packaged to effectively reach the intended audience.

13.4.8 Examples of Scenarios

A couple of examples should help clarify the process of building scenarios. Lough and White (1988) discuss the issue of decommissioning commercial nuclear power plants and consider two options, DECON and SAFSTOR. DECON involves immediate decontamination and dismantling. In this option, the plant and site are completely cleaned up, all radioactive materials are removed to a radioactive waste disposal facility, and the plant and site are released from all licensing restrictions. The SAFSTOR option involves safe storage and mothballing the facility, followed by deferred dismantling. Here, the plant and site undergo minimal clean up and then remain under physical security and licensing restrictions for a period of, typically, 20 to 50 years. The overall impact on reactors of either of these alternatives is similar. SAFSTOR generates lower occupational doses of radioactivity and lower levels of radioactive waste, however, which compensate for the costs and uncertainties involved in controlling the site for a long period.

Lough and White developed three decommissioning scenarios. The most important dimension is the institution that dominates decommissioning policies. The policies and outlook of the institution determine the criteria for selecting among the decommissioning alternatives. These criteria constitute the second dimension of the scenarios. The preferred alternative is the third dimension, taken as optional.

The federal government, in particular the Nuclear Regulatory Commission (NRC), dominates decommissioning policies in the first scenario. In accordance with the NRC staff preferences regarding timeliness and planning criteria, the two options that dominate the technical choice set are DECON and the SAFSTOR option with dismantling deferred up to the maximum of 50 years.

In the second scenario, state authority dominates nuclear activities. Because state regulators place high value on timeliness and on the planning criteria associated with decommmissioning, DECON is the only acceptable technical choice. Also, the radioactivity levels of the site are expected to be returned to preconstruction values.

The utility dominates decommissioning policies in the third scenario. All technical options are considered by the utility with the criteria for choice being low cost.

Another example is provided by a set of contextual scenarios that deal with the health care of the elderly in the Netherlands from 1984 to 2000 (Becker et al., 1986). Medical, biological, and technological developments are assumed to be identical for all three scenarios of the set. Further, economic conditions serve as a backdrop and do not greatly influence any scenario. Social, political, and demographic conditions are the dimensions that vary from scenario to scenario. The scenarios are abstracted in Exhibit 13.3.

■ **Exhibit 13.3   Three Alternative Scenarios of Health Care for the Elderly through 2000 in the Netherlands (Based on Becker et al., 1986).**

Scenario 1 (Baseline): Autonomous developments resulting in about the same pressure on health services for the elderly as currently felt:

1. Continuation of the emancipation of women resulting in a decrease in informal help from women to the elderly
2. Emancipation of the elderly as a result of higher levels of education
3. Increasing political influence of the elderly
4. Increase in the proportion of elderly belonging to ethnic minorities
5. Continuation of the debate with regard to euthanasia and an increase in passive euthanasia
6. Decrease in the number of children (important with regard to children helping their parents when they are old)
7. No major changes in norms and values with regard to health and health services

Scenario 2: Autonomous developments resulting in more pressure on health services for the elderly:

1. Growing professionalization of care for the elderly
2. An attitude of the population toward health services that results in a relatively high consumption of these services
3. Less solidarity between generations (less care by children, etc.)
4. Less solidarity within the elderly generation (less informal help from seniors to seniors)
5. Norms and values in society put the elderly more and more into an isolated position (disengagement, mandatory retirement)
6. The elderly are less financially dependent than in 1984, so they can pay for professional help
7. No new types of living together of the elderly

Scenario 3: Autonomous developments resulting in changes in pressure on health services for the elderly.

1. A more critical attitude toward health services diminishing the quest for medical consumption

2. Growth of a sense of responsibility of individuals for their own health, resulting in changes in lifestyle involving less high-risk behavior

3. Less pressure on institutional care and cure; the elderly live on their own for a longer time

4. Less solidarity between generations

5. More solidarity within the generation of the elderly; help from senior to senior increases

6. Norms and values in society incorporate the elderly more fully into the dynamics of social life

7. The elderly are less dependent than in 1984, not because of financial position, but because the social climate has changed

8. New types of living together by the elderly grow in importance ■

Critical incidents could be added to these elderly healthcare scenarios to produce substantial perturbations to the contexts described in them. Scenarios, with impacts of critical incidents considered, need to be linked to common background assumptions in medicine, technology, and economics to provide the basis for developing options to guide health care policy for the elderly in the Netherlands.

13.5 CRITIQUING SCENARIOS

There are two instances when it is necessary to critique a set of scenarios: (1) forecasters critique scenarios before presenting them to a user group or (2) technology managers critique scenarios presented to them. The task in either instance is to determine whether the scenarios form an adequate basis for making decisions about the future—decisions that may require resources to be committed today. As usual, such decisions involve risk.

Forecasts cannot be evaluated on the basis of their outcomes, since these are in the future. The only viable option is to assume that the inputs—in this case, the scenarios—are surrogates for the output. Thus the manager must establish criteria on the input side for evaluating the scenarios and then apply them. The result of this evaluation determines the credence to be placed in the scenarios.

Following Porter and Rossini (1977), the major criteria for evaluation are utility and validity. *Utility* addresses the usefulness of the forecasts to the user group. It is determined by the goodness of fit between what the forecaster has produced and what the user needs. *Validity* is determined by whether the forecast predicts what it claims to predict. Useful subcriteria for validity include:

1. The degree to which techniques used in the scenarios reflect the state of the art

2. The quality of information used in the scenarios, as compared to the best information available

3. The internal consistency of the scenarios

4. The plausibility of the scenarios (Do the future histories used make sense? Do they lead to snapshots that are credible? Can the snapshots be reached by plausible future histories?)

Using these criteria does not guarantee the validity of any scenario. However, if applied carefully, they should allow a manager to assess the level of confidence that can be placed in the forecast. When applied by the forecasters, they will provide insight into revisions, if any, that should be made.

13.6 CONCLUSION

Scenarios are outline sketches of some aspects of the future that are useful in forecasting. They may be future histories—that is, dynamic time paths from the present to some point in the future. They also may be snapshots—structural cross-sections of a future state. Combinations of these are possible as well.

Scenarios are useful as stand-alone (independent) forecasts if data for time series are lacking, if expertise is weak or nonexistent, and if there are no solid bases for model building. Even if other forecasts are possible, scenarios may be used to integrate disparate forecasts and all sorts of qualitative and quantitative information. They also can be extremely effective vehicles for communicating forecasts to non-technical audiences because they can use a wide variety of literary and multimedia techniques.

Scenarios provide a basic technique for the forecaster, planner, technology manager, or policy maker because of their flexibility in dealing with a wide range of forecasting contexts. Scenarios always must be checked for utility and validity. They should be based on the best available information and techniques and prepared so that they will be consistent and plausible to the user.

EXERCISES

13.1 Changes in the political and economic order in Eastern Europe and the Soviet Union imply short- and long-term impacts on farmers and grain dealers. Construct a morphology for a set of scenarios requested by a major commodities exporter.

13.2 High-definition television technology can be a major factor in determining the direction of home electronics producers and retailers operating in the United States. Construct a template and a set of three scenarios to be used by a major home electronics manufacturer operating in the United States in deciding what resources should be invested in the R&D of this technology today. Consider regulating and marketing as well as technological issues.

13.3   In many midwestern states the vast majority of electrical power generation is fired by coal (for example, in Indiana over 90 percent). Local coal resources are primarily of medium- and high-sulfur content. Environmental legislation, especially the 1990 Clean Air Act, will have a major impact on power generation and coal mining in these states. Prepare a set of scenarios for a technology manager charged with making decisions about the expansion of a major surface mine in the Midwest.

____14
ECONOMIC FORECASTING AND ANALYSIS

OVERVIEW

Technology and economic activity are inextricably intertwined. Technological change is perhaps the most significant source of economic growth, and economic factors are crucial to the success of new technology. This chapter describes the interrelationship between economics and technology and shows how economic forecasting can aid technology forecasting and management. Tools of economic analysis and their use by managers are presented. The chapter begins by describing how technological progress has enhanced economic well-being. Demand, its determinants, and forecasts are considered next and illustrated by example. Finally, input/output analysis is introduced as a powerful tool to aid in decision making.

14.1 INTRODUCTION

World resources are scarce relative to human needs and desires; therefore, choices are necessary. Economics is really a science of choices—how they are made and how they ought to be made. Because economics must deal with the complexities of human behavior, accurate prediction is difficult. The tools of the economist and the technology manager are not as precise as those used by the physical scientist. Nonetheless, economic tools can be used to aid in decision making. Economic inquiry is crucial to understanding how technologies become integrated into the fabric of economic activity.

To the economist, demand is the willingness to buy plus the ability to pay. Technological innovation only occurs when consumers, businesses, and governments demand it. On the one hand, a technology launched when profits are high and unemployment is low has a greater chance of success than one offered when consumers are insecure and businesses are struggling to survive. On the other hand, new technology has led societies out of recessions and depressions when production has stimulated investment and recovery. Technology managers must understand these relationships and how to integrate economic analysis with management decisions.

14.2 TECHNOLOGY AND THE ECONOMY

Over 30 years ago, Nobel laureate economist Robert Solow (1957) concluded that most of the increase in the standard of living in the United States was due to technological progress. Specifically, he concluded that 87.5 percent of increased output per capita from 1909 to 1949 was due to technical change; the remainder was due to capital investment. Subsequent studies have refined this analysis. However, most economists would agree with Edwin Mansfield (1989, p. 700) that "the rate of technological change is perhaps the most important single determinant of a nation's rate of economic growth." Clearly, technology has created new products to enhance the quality of life. It also has dramatically improved the productivity of labor and capital resources.

People can produce more in less time if they know how to use resources more effectively. In the early twentieth century, for instance, Henry Ford could raise the wages of automobile assemblers to $5 a day by production innovations. Later, American factory workers could reduce their work week to 40 hours while continuing to improve their incomes. More recently, Japan, South Korea, Taiwan, and other Asian countries have gone from devastated nations to world economic powers thanks to the modern technology used in their production processes.

New technology can overcome inherent limitations imposed by the *law of diminishing marginal returns*. This law says that if the input of a resource is increased in a production process for which supplies of some other resources are fixed, at some point the additional output that results from one more unit of the input will decline. For instance, machinery and equipment are scarce (fixed resources) in China, but there are a lot of workers. Thus simply adding workers to the production process will not add much to output. On the other hand, Japan has a large inventory of capital equipment in relation to the number of workers. Therefore, adding more of the same equipment could stretch the limits of the labor force and not increase output very much. With technological improvement, both capital and labor resources can be used more efficiently. Thus increased inputs together with the improved technology can mean greater productivity and higher worker income.

It is important to stress that a new idea must be incorporated into a new product or process to produce a benefit. This process of innovation does not automatically follow the discovery of new knowledge. In fact, *Business Week* (1989, p. 15) noted that according to Rustum Roy "The more inventive ideas the U.S. dreams up, the further it will fall behind. Each one will be just another opportunity for a foreign rival to out-innovate a U.S. company in producing it." Evidently if you invent a better mousetrap the world does not necessarily beat a path to your door— it may take your idea and beat you to the marketplace.

A conclusion that inventions are counterproductive is too strong. However, it illustrates that merely increasing research and development (R&D) expenditures to produce more inventions is too simplistic. In fact, Drucker (1985) stresses that new knowledge is not the most attractive source of innovation. The chances of success are lower and lead times longer for technology based on new knowledge than for that based on other sources of innovation. He suggests that the characteristics of the market and the production process are better indicators of opportunity than the newness of the knowledge base. Drucker sees the entrepreneur as one who systematically analyzes opportunities and develops strategies in response to inevitable change.

Understanding the role of technology in the economy is crucial to the success of firms in that economy. The technology manager needs an entrepreneurial perspective that welcomes change as a source of opportunity. No markets are static, and their dynamic nature can topple dominant firms and raise new, innovative enterprises. Further, the manager must understand the need for technology to be appropriate.

The phrase "appropriate technology" was coined by economist E. F. Shumacher (1973). His purpose was to adapt the thinking of Ghandi to the subject of economics. India and many other underdeveloped countries had suffered because their leaders had imposed Western technology. Shumacher called for smaller-scale production processes that take into account a people's characteristics and culture. Including this human dimension broadens the view taken by traditional economics; however, the lessons are not inconsistent. Conventional economic theory would agree that the large, unskilled, underemployed labor force of an underdeveloped area is not effectively utilized by large-scale, sophisticated, labor-saving production approaches. Furthermore, the effects of products and manufacturing processes on the people who run them are known to affect productivity and quality as well as workers' feelings about themselves and their lives. Using a multidimensional perspective in selecting and developing technology has been part of the reason that the Japanese have succeeded so well in the world market. Therefore, this perspective is clearly very relevant to technology managers.

14.3 MARKETS AND INNOVATION

Traditional economics asserts that market conditions determine the success of a technology. Resource prices, for instance, are guides to the current scarcity of the inputs to production. If labor is becoming scarce (and therefore more expensive), innovations that make labor more productive will succeed. Robotics, for example, may be widely adopted because labor costs are high in developed economies such as those of Japan, Western Europe, and the United States. On the other hand, if interest rates (the resource price of money) are high, then innovations that improve or extend the life of existing equipment are more attractive.

Successful technology responds to market conditions. The development of high-rise building technology, for instance, was a response to high rents in congested central business districts. Likewise, many energy conservation technologies (such as light-weight, high-strength materials in automobiles) were adopted because of large petroleum price increases in the 1970s. As raw material and disposal costs rise, innovations to facilitate recycling will become increasingly attractive.

Changes on the demand side of markets also invite innovation. For example, the exercise craze of the 1980s created an attractive environment for the Sony Walkman®. And high-definition television may become attractive in the 1990s if families prefer entertainment at home and are willing to pay for enhanced picture quality. Such changes in tastes may be hard to quantify, but other determinants of demand are more quantifiable and predictable. Rising incomes, for instance, increase the demand for luxury goods and gadgets. Likewise, the demographic reality of increasing numbers of the elderly will make innovations in prosthetic devices (such as effective artificial hips) attractive. In a similar fashion, falling computer hardware prices make developments in software application more lucrative.

Knowing market conditions obviously is not enough. A manager must be able to visualize how a new technology will fit into future conditions. This is difficult. In fact, Drucker (1985) advises against innovating for the future because an economically effective technology needs quick application to produce returns on investment. However, the manager who does not innovate for the future forfeits the opportunity to leapfrog competitors and be the first to take advantage of emerging opportunities.

Forecasting techniques (see Chapter 10) provide ways to extrapolate current trends in important market variables. Hilbrink (1989) suggested decomposing technologies into their component parts and examining trends in each component to give indications of potential for growth and economic impact. For example, data processing systems could be divided into components such as the central processor unit, transmission systems, memory, storage, and so forth. Hilbrink notes that large data processing system architecture is sensitive to the balance between memory and data transmission costs. His analysis suggests that transmission costs will fall faster than memory costs. If so, the design of system architecture may take advantage of this future development by emphasizing telecommunications connections rather than on-site processing. Since changes are occurring very rapidly in many technologies in addition to computing, approaches such as the decomposition method suggested by Hilbrink offer advantages over just innovating for now.

Another approach to analyzing future conditions was suggested by McIntyre (1988), who noted that innovators should examine the infrastructure needed to support a new technology, in addition to examining the usual growth curves for it. In the short run, there may be no market for a new product because the infrastructure is not available for its effective use. Television, for example, was not very attractive to consumers before broadcasting facilities had been widely established. More recently, while home computers have not met the ambitious sales projected for them, McIntyre says that they, like the automobile, could become a necessity if some reasonable applications are widely adopted.

Actually, the importance of infrastructure to new technology was recognized much earlier by Schumpeter (1934). Among the many important contributions of this economist and innovation scholar was an insight into the way markets for new products develop. In his study of the U.S. aluminum industry, Peck (1961) drew on Schumpeter's insight to show how the market for a new product develops more rapidly if its uses are demonstrated. Alcoa used this demonstration effect to help expand its market in the early twentieth century. The firm developed new applications for aluminum and priced them attractively to get manufacturers to use the metal. The increased demand made Alcoa's monopoly on primary ingot production much more profitable.

McIntyre (1988) proposed analogous strategies for today's high-tech firms. He stressed the need to stimulate market adaptation and described some specific approaches to do it. One approach is to ensure that product design includes the potential for growth or enhancement. For example, IBM designed expansion slots into its early PCs so that other vendors could manufacture plug-in circuit cards to provide additional features. Similarly, firms (such as computer manufacturers) can help create infrastructure by stimulating the development of other firms that make complementary products (such as software). At times, innovators may wish to form strategic alliances to establish compatible standards, a form of infrastructure.

The growth of a new technology also critically depends on competing and complementary technologies. Therefore, it is important to carefully analyze the markets for these associated technologies and to understand the markets on which the new technology itself depends. While the complete development of a market analysis is beyond the scope of this text, there are some basic tools of economic analysis with which the technology manager should be familiar. Economic forecasts, for example, are important to the timing of product introduction, and input/output analysis is very helpful in understanding the interdependencies of the various markets. The following sections provide an introduction to these concepts.

14.4   FORECASTING THE ECONOMY

New technology can stimulate the economy and help bring about higher levels of investment and economic growth. However, an individual innovator must respond to existing and projected economic conditions because, alone, he or she cannot have much impact on them. Therefore, it is important to be aware of economic conditions and how they will affect the development of a new technology.

Over the past century, industrial economies have generally grown; however, surrounding this basic growth trend there have been business cycles of varying severity. In an expansion phase, output grows, employment is high, and businesses see rising profits. Consumers have good incomes and are likely to be more receptive to new products. Business profitability makes it possible to invest in new capacity that incorporates new technology. On the other hand, a recession can lead to business losses and rising unemployment. Consumers tend to limit buying to necessities and to make do with old durable products. Businesses are faced with excess production capacity, and, as a result of reduced consumer demand, they feel less pressure to install new technology. It is dangerous to generalize because other factors may influence the success of a new technology. However, it is clear that recessions are probably not good times to launch new technologies; economic expansions, on the other hand, may help establish new technologies.

There are many theories about business cycles and their control (see, for example, Dornbush and Fischer, 1987, especially Chapter 10; Volland, 1987). Although the technology manager does not need the depth of analysis required of the corporate economist or financial manager, he or she should have a basic understanding of what affects the level of economic activity and how to get reasonable forecasts. This is important information as the vitality of the economy, the level of inflation, or changes in interest rates could have significant effects on a new technology. Although economic forecasting, like weather or technology forecasting, is not an exact science, it can help.

The basis of most forecasting of national economies is the model given by the following:

$$\text{Output} = \text{Consumption} + \text{Investment} + \text{Government} + \text{Net exports} \atop \text{spending}$$

or

$$Y = C + I + G + N \tag{14.1}$$

Consumption (C) includes everything from *services* (such as haircuts) to *nondurable goods* (such as food) to *consumer durables* (such as cars and appliances). Investment (I) is output that goes into structures, (such as homes, factories, office buildings), equipment (such as computers, robots, rolling mills), and inventories of materials or products that are held for later sale. Government spending (G) includes federal, state, and local purchases. These may be financed by taxes or by borrowing. The former directly affect consumers and investors. The latter affect credit markets and interest rates directly and the levels of C and I indirectly. Net exports (N) are sales of production to foreign buyers, less purchases of goods and services from abroad. Higher exports mean higher production and incomes in the domestic economy, while imports replace domestic production with foreign-made goods and services. Since the level of demand effectively determines the level of output (Y), the terms "output" and "demand" often are used interchangeably.

Most forecasts directly or indirectly estimate future levels of the four components (C, I, G, and N) of the aggregate demand for production output. Some forecasters (such as Data Resources, Inc., Chase Econometrics, Federal Reserve Board/MIT, and The Brookings Institution) use huge, computer-based, econometric models to make these estimates. These models employ many equations developed statistically in a manner analogous to the regression techniques discussed in Chapter 9. However, such models have not demonstrated significant superiority over more qualitative approaches.

For this reason, judgment or opportunistic forecasting is still very common. In such forecasts, the analyst uses the best information available to estimate each component and sums the estimates to predict demand, as in Equation 14.1. However, the individual components are not independent of one another. For example, personal consumption depends on disposable income, which, in turn, is affected by the level of government spending. The more the government spends, the higher incomes will be. In other words, consumption depends on government spending. Thus estimates made independently of one another can lead to internal inconsistencies and an erroneous forecast. To avoid this, the forecaster must iterate estimates to establish the internal consistency of the forecast.

Other forecasts use indirect approaches. Table 14.1 lists the leading, coincident, and lagging indicators provided by the U.S. Department of Commerce in its *Business Conditions Digest*. As the names imply, these data provide some indirect insight into what is happening to the components of economic activity over time.

TABLE 14.1   Economic Indicators

Leading Index Components

Average workweek, production workers, manufacturing
Layoff rate, manufacturing
Net orders for consumer goods and materials (1972 dollars)
Vendor performance, percentage of companies receiving slower deliveries
Net business formation (index: 1967 = 100)
Contracts and orders for plant and equipment (1972 dollars)
New building permits, private housing units (index: 1967 = 100)
Net change in inventories on hand and on order (1972 dollars)
Change in sensitive crude materials prices
Change in liquid assets
Stock prices, 500 common stocks (index: 1941–1943 = 100)

Roughly Coincident Index Components

Employees on nonagricultural payrolls
Personal income less transfer payments (1972 dollars)
Industrial production, total (index: 1967 = 100)
Manufacturing and trade sales (1972 dollars)

Lagging Index Components

Average duration of unemployment
Manufacturing and trade inventories (1972 dollars)
Labor cost per unit of output, manufacturing (index: 1967 = 100)
Average prime rate charged by banks
Commercial and industrial loans outstanding, reported weekly by
  large commercial banks
Ratio of consumer installment credit to personal income

Source: Compiled from indicators provided by the U.S. Department of Commerce, *Business Conditions Digest.*

For example, if the average workweek is lower and the layoff rate is higher, one would expect this to lead to less demand for, and hence production of, consumer goods at some later time. Changes in money supply also can precede economic change. For instance, monetarist economists, such as Nobel laureate Milton Friedman, place great importance on the role of money in economic activity. Therefore, they maintain that changes in money supply can be powerful predictors of later levels of prices and economic activity.

Technology managers are unlikely to find it in their best interest to generate their own economic forecast. However, relying on forecasts by a single expert is unwise because none have amassed consistent records of superior accuracy. Perhaps the best approach is to invest small amounts of time and money to examine the consensus outlooks that are readily available. *Business Week, Fortune, Forbes, Industry Week,* and other business magazines present forecasts, as do the *Wall Street Journal, New York Times,* and other major newspapers. Generally, each quotes a range of economists, and collectively they provide a broad view of the economic outlook. Businesses also can buy a consensus forecasting service such as *Blue Chip Economic Indicators.* This can be a cost-effective way to get a balanced picture of predictions by recognized people.

14.5 INPUT-OUTPUT ANALYSIS

The preceding section described tools for understanding the general economic climate for technology. Recall that the discussion of forecasting (see Chapter 4) noted the interaction between the various parts of the economy. There are tools that explicitly take account of these interactions and provide information about the way in which changes in one sector can affect others. One tool, input-output analysis, is useful both to identify the market potential for products and to show the impacts of new technologies on the economic system.

Input-output analysis was created and continues to be developed by Wassily Leontief (1985). In this analysis, the economy is modeled as a static structure that represents the flows of production from one industry to another and to the final demand components described in Equation 14.1. Input-output analysis breaks down the interacting transactions (flows) necessary to make a product to show which components of the economy must supply what types of production. The transactions must be measured in some unit, with monetary terms the most convenient. While it may be possible, for example, to show how tons of steel and glass and kilowatt hours of electricity can be added to make one car, it is much easier to add so many dollars worth of steel, glass, and electricity to the costs of other components to arrive at the cost of producing a car.

Consider an economy that is composed of only two industries, X_1 and X_2. Table 14.2 shows what happens to the output of these two industries. (Table 14.2 is similar to the interindustry transactions tables published by the U.S. Department of Commerce in the *Survey of Current Business.* While our example is limited to two industries for simplicity, the same principles apply to the government's data.)

TABLE 14.2   Interindustry Transactions

			Transactions in $				
	X_1	X_2	C	I	G	N	Total Output
X_1	100	200	400	100	100	100	1,000
X_2	500	600	200	300	200	200	2,000
Wages	300	700					
Other	100	500					
Total	1,000	2,000					

The first row, for example, shows how much of the total output of X_1 ($1,000) is sold from one firm to another within industry X_1 ($100), to industry X_2 ($200), and to the four final demand sectors indicated in Equation 14.1. The second row gives a similar breakdown for industry X_2 ($2,000 total output). Although the analysis could be made in terms of physical units (such as tons of steel), measurement in monetary units will be very useful in subsequent steps of the analysis.

The "Wages" and "Other" entries in the table are incomes that represent the distribution of the value added by each industry. *Value added* is determined by sales minus the cost of the intermediate goods purchased from other industries. Note that the total of each industry's column equals the total of that industry's row.

The direct requirements of one industry can be calculated for the output of another from the information in the interindustry transaction table. Table 14.3 presents these requirements for our hypothetical two-industry economy as computed from the information in Table 14.2. Entries show the share of each dollar of output of an industry (column) represented by input to the other (row). For example, a dollar of output from X_1 requires $0.10 (100/1,000) of its own output as an intermediate input plus $0.50 of industry X_2 output plus $0.30 of labor and $0.10 of other income, such as rents and profits. (The direct requirements table for the United States also is published in the *Survey of Current Business*.)

While the direct requirements table is useful, it does not tell the whole story. The limitations of the direct requirements table can be seen if one asks: "How

TABLE 14.3   Direct Requirements Table

	Output of Industry in $	
Input To	X_1	X_2
X_1	0.10	0.10
X_2	0.50	0.30
Wages	0.30	0.35
Other	0.10	0.25
Total	1.00	1.00

much would industry X_1 have to produce to sell an additional $100 of output to consumers?" Since $0.10 of each dollar of X_1 output goes for intermediate goods produced within the industry (see Table 14.3), its total output must be at least $110 to produce an additional $100 of output for consumers. However, there is more to consider. Table 14.3 shows that each dollar of output by X_1 uses an input of $0.50 from industry X_2 and that X_2 requires $0.10 \times 0.50 = $0.05 of output from industry X_1 to produce it. Thus the total output of X_1 must be at least $115 — but that requires additional input from X_2 — and so forth. A convergent pattern of increasing output levels has clearly been established. This simple example helps explain why the direct requirements table is a building block for the more useful total requirements table.

This problem can be better understood by examining it in terms of two equations in two unknowns. First, assume that the final demands for the outputs of industries X_1 and X_2 are as portrayed by the interindustry transactions table (Table 14.2), *except* that the level of consumer purchases of X_1 output is $100 higher (that is, $C = $500 instead of $400). This results in the following equations:

$$X_1 = (100/1,000)X_1 + (200/2,000)X_2 + 500 + 100 + 100 + 100$$

or

$$X_1 = 0.10X_1 + 0.10X_2 + 800 \qquad (14.2)$$

Likewise,

$$X_2 = 0.50X_1 + 0.30X_2 + 900 \qquad (14.3)$$

These equations are constructed using direct requirements data from Table 14.2 and the increased level of consumer purchases. They define actual industry output levels for the $100 increase in consumer demand for output of X_1.

Solving Equations 14.2 and 14.3 simultaneously gives values for X_1 and X_2 of $1,120.69 and $2,086.21, respectively. Comparing these values with those in Table 14.2, the result of a $100 higher level of consumer sales by X_1 would be to increase the output required of X_1 by $120.69 and of X_2 by $86.21. Thus every dollar of increased final demand for output from X_1 would increase the output for that industry by $1.2069 and would increase the output of X_2 by $0.8621. A similar exercise, assuming that final demand for X_2 increases, reveals that the direct and indirect effects on outputs total $0.1724 of additional output for X_1 and $1.5517 of additional output for X_2.

The solutions to these equations can be used to construct a total requirements table (see Table 14.4). This table shows the direct and indirect effects of a one-dollar increase in the final demand for each industry's output. The total effects of a one-dollar increase in final demand for the output of an industry can be found by reading down the column for that industry in Table 14.4.

TABLE 14.4 Total Requirements Table: Direct and Indirect Effects per Dollar of Final Demand

An Increase in Demand for Output of This Industry Is Produced	By an Increase of $1.00 of Demand for Output of This Industry	
	X_1	X_2
X_1	1.2069	0.1724
X_2	0.8612	1.5517

Matrix algebra can be used to apply these principles to the more realistic case in which the number of industries is quite large. Equations 14.2 and 14.3 can be written as

$$X = AX + Y \tag{14.4}$$

where X is a 2×1 matrix, the elements of which are X_1 and X_2 in the example; A is a 2×2 matrix, the elements of which are the entries of the first four cells in the direct requirements table (Table 14.3) expressed as fractions of industry total output; and Y is a 2×1 vector made up of the sums of the final demands for each industry's output from the components C, I, G, and N. For the two-industry example,

$$\begin{bmatrix} X_1 \\ X_2 \end{bmatrix} = \begin{bmatrix} 0.10 & 0.10 \\ 0.50 & 0.30 \end{bmatrix} \begin{bmatrix} X_1 \\ X_2 \end{bmatrix} + \begin{bmatrix} 800 \\ 900 \end{bmatrix}$$

Rearranging, the final demand, Y, is related to total output, X, by subtracting out intermediate goods (AX) so that

$$Y = (I - A)X \tag{14.5}$$

where I is the identity matrix. Solving for X uses the standard approach of inverting the $(I - A)$ matrix to get

$$X = (I - A)^{-1}Y \tag{14.6}$$

The elements of the $(I - A)^{-1}$ matrix are the entries in the total requirements table. They show the direct and indirect effects of an increase in demand for the industry heading each column. For our simple example,

$$(I - A)^{-1} = \begin{bmatrix} 1.2069 & 0.1724 \\ 0.8621 & 1.5517 \end{bmatrix}$$

Since computers are well suited to this kind of matrix manipulation, it is possible for the U.S. Department of Commerce to consider not just two industries but nearly

100 industry groups. Their computations produce the total requirements table in the February 1979 issue of the *Survey of Current Business* in the manner shown here. For example, that table reveals that a one-dollar increase in final demand for paper board containers and boxes leads to direct and indirect output increases of $1.05608 for the industry itself, a $0.51828 increase in paper and allied products output and a total increase of $0.06008 in the output of the chemicals and selected chemical products industry. Effects on all the other industry groups could be examined as well.

The implications of this total requirements table for those doing market planning should be clear. Each element in the row for the industry reveals the relationship of that industry to the final demands for output from each industry listed in the table. For example, suppose a new box-making technology is expected to double the output of boxes in the next decade. Since 6 percent ($0.06008/$1) of box-making costs are due to the chemicals industry, this change will be important for at least some firms that produce chemicals. Total requirements tables have been used to analyze the effects of economic changes ranging from decreases in defense spending to expanded investment in environmental protection technology. Isard (1972), for example, showed how input-output analysis could be used to analyze the environmental effects of economic development. This analysis involved an extension of the model to explicitly include ecological inputs in a manner analogous to industrial and factor inputs.

There are some severe limitations on the use of input-output tables by technology managers. Perhaps most important is that the table assumes that the proportions of inputs to outputs remain constant and that relative producer prices do not change. These problems are somewhat alleviated by frequent updates of the tabulated data, but as the technology manager is well aware, change can be rapid. This is not to say that the information provided is not helpful, but rather that the conclusions from input-output analysis should be treated as approximate. Leontief and others are actively working to show how technological change can be incorporated in the analysis and better understood (see Leontief, 1985, for discussion of the analysis of technology).

14.6 CONCLUSION

Technology and the economy are closely interrelated. In fact, the rate of technological change is considered by many to be the primary variable in economic growth. However, from the technology manager's perspective, the influence of the economy on the success of a new technology is of more central concern. It is important for the manager to understand how economic conditions affect the chances for success of the technology he or she manages, and it is crucial that decisions take into account the present and expected future states of the economy.

Economists study how scarce resources are allocated to satisfy human needs. It is unlikely that many technology managers can or should generate their own economic forecasts. The track record of economic forecasts suggests that managers should follow consensus forecasts for the short term made by relatively simple methods and frequently updated. These are widely available from several sources. Managers must understand economic factors sufficiently well to determine what the forecasts portend for their technology. Long-term economic forecasting is not well established. "Long wave theories," mentioned in Section 2.1 and 10.4.2, deserve consideration.

Managers must visualize how new technology will fit into present and future conditions, and they must understand how changes in the economy can offer opportunities. Therefore, the technology manager should develop strategies to take advantage of inevitable changes in the economy.

Finally, managers must develop their own philosophy about the time horizon for which they will develop new technologies—or understand and apply the philosophy of the corporation for which they work. Although innovating for the future implies greater risk than developing for the present, it also offers significantly greater rewards.

EXERCISES

14.1 Technology can increase productivity. List the technological developments of the past 30 years that make students more productive.

14.2 Events of the 1970s dramatically increased the price of energy. How have higher energy prices affected the technological characteristics of three products in use in the 1990s?

14.3. Pick a familiar technology. Conceptually break it into component parts. How are the prospects for technological progress different for these parts? How might you take advantage of these differences as a manager in an industry that produces this technology? In an industry that uses it?

14.4 Trace the important infrastructure and related technologies that developed to make the automobile so central to modern life in industrialized countries. How would a firm use this information when considering the development of a vehicle for use in third-world nations?

14.5 Use the sources suggested in Section 14.4 to develop a forecast of the U.S. economy for the next year. Identify several new technologies whose introduction and growth will be especially inhibited or helped by these economic conditions.

14.6 Find the most recent total requirements input-output table for the United States (examine the most recent issues of the *Survey of Current Business*, published by the U.S. Department of Commerce). Specify information from that table to use in market planning in an industry of interest.

14.7 Construct a total requirements table:

(A) If the interindustry transaction table is as follows:

Transactions (in $)

	X_1	X_2	C	I	G	N	Total Output
X_1	150	150	300	160	80	160	1,000
X_2	550	350	150	250	300	200	1,800
Wages	200	700					
Other	100	600					
Total	1,000	1,800					

(B) If I = $300 for X_1 output and everything else remains the same, what are the new values of X_1 and X_2?

(C) If C = $270 for X_2 output and everything else remains the same, what are the new values of X_1 and X_2?

(D) If G = $250 for X_2 output and everything else remains the same, what are the new values of X_1 and X_2?

PART III
ASSESSMENT TO MANAGE TECHNOLOGICAL CHANGE

____15

IMPACT ASSESSMENT: GENERAL ISSUES AND THE IDENTIFICATION OF THE IMPACTS OF TECHNOLOGY

OVERVIEW

This chapter begins Part III—consideration of the effects of technological change. It offers an overview of impact assessment, including a 10-step approach to perform it, and it describes problem definition and study bounding. Scanning and tracing techniques for impact identification are introduced as are issues in selecting significant impacts, both in likelihood and severity, for detailed analysis.

15.1 INTRODUCTION

The effects of changing technology are of concern at two levels. First, a valid technology forecast must attend to the impacts of the technology's introduction and adoption as well, because reactions to the technology will impinge on the development of the technology. Second, the technology manager needs to consider product acceptance by customers, indirect reactions by others, and potential regulation.

Two extreme cases serve as examples of the importance of impact assessment. At one extreme, the personal computer (PC) was broadly accepted in the United States. This acceptance stimulated the development of PC tools and toys, which fed back positively on PC sales. The rate of PC sales growth far exceeded the early forecasts. At the other extreme, Whirlpool introduced the trash compactor without taking time to assess its likely impacts. Concerns arose in the community about disposal of compacted waste; the media picked up on this as an environmental cause; and communities began to ban the use of compactors because of perceived danger to their municipal incinerators. The product was pulled off the market while studies were done. These eventually led to ways to solve the negative impacts, and trash compactors reappeared in the market. However, the costs of this interruption were excessive, and trash compactors never regained their momentum in the marketplace. Early impact assessment could have remedied the problems before they occurred.

15.1.1 General Considerations

Forecasting technological developments and assessing the impacts of technologies are closely related. There is a feedback relationship between them. Technology forecasting is a component of impact assessment that, as will be discussed in more detail later, typically precedes impact identification and analysis. However, impact assessments may identify technologies with desirable impacts, thereby leading into technology forecasts that ascertain more carefully the likely directions and rates of technological change.

There are many varieties of impact assessment, each with its own community of practitioners: technology assessment is central to technology management; environmental impact assessment is often required for major projects in the United States and other countries; and social impact assessment and risk assessment are also important.

In impact assessment the most difficult consequences to study and understand are the higher-order impacts. These can best be described as impacts of impacts, or links in a causal chain. Many of these effects are indirect and unintended (just as striking the cue ball on a pool table can lead to balls striking other balls whose consequences, such as the incidental pocketing of an 8-ball, are indirect and unintended). There are several uses of the impact assessment of technological change (see Arnstein and Christakis, 1975; Markley and Hurley, 1983). These include:

- Providing support for a technological development
- Deferring or stopping the implementation of a technology
- Stimulating research or development to remedy adverse effects of a technology
- Providing a reliable base of information for use by any parties concerned with the development of the technology

Impact assessments are done by various parties. The Office of Technology Assessment (a small staff agency of the U.S. Congress) has published over 400 studies of various technologies, their implications, and possible policy actions. These offer an excellent, publicly available source of information. Environmental impact statements (EISs) are prepared to meet the requirements of the National Environmental Policy Act of 1969 for assessment of the potential impacts of "actions significantly affecting the quality of the human environment" (section 102C). Initial reports are prepared by companies or consultants and are followed by governmental agency review. Many companies assess technologies in conjunction with market analyses, product reviews, and/or strategic planning.

15.1.2 Some Distinctions

Over the years, different meanings of technology assessment have emerged. One, just discussed, refers to *assessing impacts*—studying the effects of new and developing technologies. Another meaning, common in military and some industrial circles, pertains to the evaluation of alternative technologies—that is, *comparing technologies*. For instance, someone planning commercial space missions might want to assess alternative delivery technologies: Should we use the space shuttle or conventional, single-use rockets? Similarly, someone designing the space shuttle would have to select various component technologies: What sort of information displays work best (color monitors, black and white monitors, or dials)? Likewise, someone designing a color monitor would be called upon to choose among various subassemblies, components produced by different manufacturers, and so on.

Comparing technologies differs from assessing impacts in two important ways:

1. Comparisons often consider current technologies, whereas assessing impacts focuses on future possibilities.
2. Comparisons usually focus narrowly to assess how well technologies fulfill a specific mission, whereas assessing impacts addresses a broader set of effects.

One of the most difficult distinctions to convey about impact assessment is the distinction between impacts *of* a technology and impacts *on* a technology. The thrust of impact assessment is to analyze the likely effects of a technology—the impacts *of*. For instance, planners of commercial space applications might ponder impacts on the atmosphere, on net employment opportunities, on international relations, and so forth.

Conversely, technology forecasting concerns the impacts *on* a technology from other forces. Someone attempting to forecast the likelihood of commercial space applications would want to forecast its social acceptance. Issues that might impact the development of commercial space applications include concerns about environmental damage, political opposition because of poor return on taxpayer dollars, and legal uncertainties over what institutions hold jurisdiction in space.

In actuality, impacts of and impacts on represent two aspects of a complex whole. A well-done assessment of commercial space applications might find a significant probability of atmospheric impact. This information, in turn, could be used by concerned parties to oppose space development. Likewise, estimates of the jobs likely to result from commercial space development might be high and could trigger support for governmental subsidies for such developments.

While impacts on and impacts of are apt to be interdependent, it is essential to keep the distinction in mind in performing technology forecasting (TF) or technology assessment (TA). For instance, social forecasting within a TF/TA focuses on forces potentially exerting impact *on* the technology in question. As such, the forecast would focus on a future driven by any pertinent changes, including, but not limited to, the technology in question. In contrast, social impact assessment within that same TF/TA would concentrate on the social changes caused by the technology in question—impacts of. This requires a sound social forecast on which to overlay changes due to the technology in question. Similar "impacts of" versus "impacts on" distinctions must be maintained in dealing with all the contextual factors interacting with the technology—for instance, legal influences on whether the technology will be developed versus legal changes likely to result if the technology is developed; or organizational forces affecting development of the technology versus organizational changes likely from implementation of the technology. This distinction should be kept in mind particularly in Chapter 16, which deals with both impacts of and impacts on many areas.

15.2 GENERAL ISSUES IN IMPACT ASSESSMENT

The important steps in carrying out an impact assessment (IA) are introduced in this section. The key quality considerations and options in the scale of effort of an impact assessment study are also detailed. The critical steps of problem formulation and study bounding are addressed as well.

15.2.1 Assessment Steps

A 10-step approach to IA is depicted in Exhibit 15.1.

■ **Exhibit 15.1   10 Steps of IA**

Problem Definition

Technology Description　　Societal Context Description

Technology Forecast　　　Societal Context Forecast

Impact Identification

Impact Analysis

Impact Evaluation

Policy Analysis

Communication of Results　　　■

Several observations are in order concerning implementation of these steps. First, they should not be considered a linear progression. Given the complexities involved in significant IAs, often there are steps that must be redone based on knowledge gained in subsequent steps. For instance, impact evaluation may suggest mitigation efforts to alter the technology and thus change the technology description and forecast, which, in turn, would require that IA be redone for the altered technology.

Second, emphases on the 10 steps vary greatly among assessments. Indeed, in some cases, it may be appropriate to omit or truncate some steps. In addition, there are many other study strategies proposed for the various forms of IA. For instance,

Joseph Coates (1976) recommended another set of steps, of which the following could augment our 10 steps in some cases:

- Specify system alternatives
- Identify decision apparatus and appropriate action options
- Identify parties at interest
- Identify exogenous variables or events

The analyst should choose those steps that contribute information needed for management decisions.

15.2.2 Issues in Assessment Quality

There are three generic sources of material useful in forecasting and assessment: theory, data, and methods. Chapter 2 discussed the severe weaknesses of the common theories of social change that render them unsuitable for understanding sociotechnical change. DiSanto and Frideres (1986) consider this issue on a disciplinary level, noting that there are a plethora of theories in each of the disciplines commonly involved in impact assessment. Serious problems lie in finding an appropriate theory and in applying general propositions to concrete cases.

Hoos (1978) challenges data and method as well:

> The study of the future, once the bailiwick of seers and visionaries, has developed to a state far beyond that of the art of interpreting smoke patterns from a sacrificial altar or of fathoming strange sounds out of a Delphian cave .... Even though the magic is gone, the mumbo-jumbo remains, with data the driver, analysis the watchword, quantification the rule, and model-building the prime occupation. (p. 54)

From an environmental impact assessment perspective that is useful for all forms of IA, Ross (1987) translates these generic concerns into specific criteria for sound IA. A sound assessment *must*

- Focus on matters that make a difference to management, indicating the technical questions that guided the assessment
- Identify methods used and explain them so that others can duplicate the research
- Reflect the data, which must be of sufficient quality to support conclusions drawn; gaps in available data must be identified
- Provide explicit impact predictions, indicating the extent of uncertainty, and show the basis upon which these have been drawn

Additionally, Ross suggests that a sound assessment *should*

- Consider cumulative impacts of all pertinent, changing technologies
- Specify impact mitigation measures and their rationale

- Discuss time horizons and use dynamic models
- Quantify threshold criteria

These criteria can be modified for application to a wide range of impact assessments at various scales of effort.

The difficulties inherent in interdisciplinary IA should not be underestimated (Rossini et al., 1978; Burdge and Opryszek, 1981). Jargon, pecking order tensions, and differing data requirements and differences in acceptable bases of knowledge make collaboration difficult. Selective perception seems to be reflected in Idso's (1986) review of reports by the U.S. National Research Council and Environmental Protection Agency: "More CO_2 should be good for the world. It should be welcomed as the blessing which decades of sound agronomic research have shown it to be" (p. 98).

Some scientists reject forecasting out of hand. In one IA (sponsored by the National Science Foundation and reviewed by two of the authors), agricultural scientists provided past trend data showing strong and steady plant yield improvements over time, yet these same scientists absolutely refused to project that trend to predict future yields.

15.2.3   Impact Assessment at Different Scales of Effort

Rossini et al. (1976) distinguished a family of assessment studies according to the level of effort invested (see Exhibit 15.2).

A number of interesting approaches to *brief assessments* have been used. They offer no pat formula to shortcut the assessment process, but rather a rich set of options. These may prove especially adaptable for industrial IAs. Markley and Hurley (1983) present an approach to a brief assessment of a large-scale issue—increase in

---

■ **Exhibit 15.2   Different Scales of IA**

- *Macroassessment (comprehensive, full-scale)*: Full range of implications and policies considered in depth (expect an order of magnitude of 5 to 10 person-years of work).

- *Miniassessment*: A narrow but in-depth, or broad but shallow, focus (perhaps with an order of magnitude less effort invested).

- *Microassessment*: A thought experiment, or brainstorming exercise, to identify the key issues or establish the broad dimensions of a problem (perhaps an order of magnitude smaller than a miniassessment, say, 1 person-month of effort).

- *Monitoring*: Ongoing gathering of selected information on a topic; suitable for TF or as a follow-up to IA (indeterminate, but modest effort required).

- *Evaluation*: Assement of the performance of prior or ongoing projects and programs; evaluations can identify needed changes and can provide vital feedback on the validity of previous IA efforts (perhaps, 1 person-month of work). ■

---

atmospheric CO_2. Deadlines required the assessment in about two months, and the project team knew little about CO_2 effects. Their approach is outlined in Exhibit 15.3.

Hunsaker and colleagues (1983) offer a useful suggestion suitable for brief IAs—seek related studies that can help with technology description and contextual setting and that can help screen for important impact categories.

■ **Exhibit 15.3   Strategy for a Brief Assessment of Atmospheric** CO_2 **(Based on Markley and Hurley, 1983).**

1. Initial inquiry using a "snowball survey" wherein a knowledgeable person is contacted about the issues and asked to identify other experts; these, in turn, are contacted (typically by phone) and asked about the same issues.

2. Revision of study objectives and strategy based on initial inquiry; despite the short time available, the assessors went through a second iteration.

3. Bracketing the ranges of plausible variation using a high-impact and a low-impact case; this proved an economical way of dealing with the uncertainties involved and fit the sponsor's need for identification of potential show-stoppers—that is, impacts so severe that they would interfere with projected uses of fossil fuels.

4. IA based on the following activities:

   • Identifying physical effects and their relationships based largely on the literature

   • Flowcharting these effects and their interactions leading to environmental, economic, and sociopolitical impacts

   • Estimating magnitudes of effects based on the literature as high and low error bands

   • Considering impacts for the high-impact and for the low-impact cases

5. A useful tool for the atmospheric CO_2 impact estimation proved to be an impact matrix that was analyzed for each cell linking one of five important physical effects (such as altered climate patterns and oceanic effects) to one of six impact categories tailored to this topic:

   • Location/viability shifts

   • Supply/cost factors

   • Demographic migration

   • Health and well-being

   • Governance, regulation, and planning

   • Systemic, interregional, and international relations

This IA necessarily considered impacts and policy implications at a high level of generality or aggregation, except for those of particular interest to the client.   ■

In another approach, Mason (1986) suggests ways that a community can assess high-tech firms' suitability for economic development. His approach takes advantage of multiple uses of available information:

1. Identification of parties at interest, social description, and social forecasting (including community goals) should serve other economic development assessments and could prove valuable as marketing aids for economic development agencies.
2. Technology description and forecast can begin from a firm's business plan; furthermore, a firm's market research may provide useful impact data.
3. Impact evaluation may be done by a qualitative strengths/needs matrix. Community strengths in labor, technology, education, culture, environment, transportation, finance, and utilities are rated from -5 to $+5$. The importance of each of these to a given firm's needs is scaled from 0 to 5. Multiplying strength ratings by needs score and adding these for each of the strength categories gives a basis for comparing how well different firms fit the community. A similar matrix can be formulated for each firm, crossing community strengths with firm characteristics (such as employees, facilities, and customers) to bring out the multidimensional implications of a specific firm locating in the community. These matrices can help assess how well the community meets a firm's needs; they can also identify the higher-order impacts of a firm locating in the community (for instance, its employees would provide a critical mass to support new cultural activities that could be magnets for other businesses).

These suggestions can be adapted by a company doing its own IA, potentially tying in with market analyses.

Wilmoth, Jarboe, and Sashkin (1984–85) present another approach to scale down assessment efforts to adapt them to the narrower range of issues that typically confront industrial firms as they contemplate technological changes. They restrict attention to a very limited number of impact categories. The strategy is to define the change under study, and then list, first, all the positive consequences of the change and, second, all the negative consequences. These are then split into direct and indirect effects and then into obvious versus hidden effects. To complete the procedure, after detailing the categories, work back up the tree seeking new insights and higher-order impacts (reflecting the interaction of direct impacts).

15.2.4   Problem Formulation and Bounding

As noted in the 10 steps (Exhibit 15.1), problem definition is the first step in IA. It is one of the few steps that is never omitted, even if the IA is done cursorily or implicitly. Khakee (1985) noted that the results of a futures study depend heavily on the way the problem is formulated. Armstrong and Harman (1977) suggested spending approximately 20 percent of an IA study in setting boundaries.

Even then, bounds can be further adjusted if information emerges that suggests revision. Certainly the manager should have a reasonable sense of the study focus at the outset, but bounding should be an ongoing activity that is refined as one understands more about the topic. Major considerations in bounding a study are listed in Exhibit 15.4.

■ **Exhibit 15.4 Factors to Consider in Bounding an Assessment (Based on Berg, 1975).**

1. Form of the study (Are there better alternative studies?)
2. Time horizon(s)
3. Spatial extent—internal to a company or other organization, local (what jurisdictions?), regional, national, or international
4. Institutions to be considered (possibly, those affecting policy, those likely to use the study, and/or those impacted by the development in question)
5. Technology and range of applications
6. Impact sectors for in-depth treatment
7. Policy options for in-depth consideration
8. Intended study users   ■

Sometimes, the "take-home lesson" from an IA largely resides in the conceptualization and bounding of the problem. For example, an assessment of the impacts of office automation on employment decided that the essence of the impact was a balance between enhanced productivity, which acts to suppress employment, and increased demand (e.g., for computer-aided analyses) per unit of output, which acts to boost employment (Porter, 1987a). In another example, a TA of life-extending technologies hit its "eureka," after about a year of groping, in distinguishing two forms of life extension: (1) increasing the maximum lifespan, attained by only the tail of a normal population distribution, or (2) squaring off the distribution so that a much larger percentage of the population attains longer life, without an increase in maximum lifespan.

Are there special methods for these early stages of an assessment? Porter and colleagues (1980) suggest the possibility of conducting a microassessment—a "quick and dirty," but relatively comprehensive, run through all steps of the assessment (see Exhibit 15.2). The output would clarify the possible courses of action for the full assessment based on a rough analysis of the technological development(s), impacts, and policy options.

Roper (1988) points out how difficult the initial stages of an assessment are—it is the point at which we know the least about the topic, yet the consequences of project decisions (bounding) are most critical. Initial learning activities (sometimes characterized as "wallowing") must move toward closure, yet be sufficiently broad as not to miss critical elements. Roper recommends use of the Nominal Group Process (see Chapter 11) because it produces a rich list of factors considered vital to the issue by stakeholders and a preliminary agenda of those that are most important in a relatively short time.

15.3   IMPACT IDENTIFICATION

Prior to impact identification, the IA process needs to establish *baseline conditions*—that is, the current situation and the likely future situation in the absence of the development in question. In broad scope assessments, the baseline would be established through technological and societal context description. In more narrowly drawn assessments, the focus is more localized and, often, is shorter term, so baseline conditions may be relatively obvious.

Impact identification looks for likely changes from baseline conditions due to the development being considered. There are two main approaches to identifying such impacts:

1. *Scanning techniques* identify potential impacts by investigating the full range of candidate impacts in a single, direct step.
2. *Tracing techniques* construct structural relationships between development actions and impacts, and among impacts, creating a causal trail in which impacts become causes of higher-order effects.

Either or both approaches can be useful depending on the characteristics of the IA task.

15.3.1   Scanning Techniques

Scanning methods search the impact field to minimize the probability that significant impacts will be overlooked. The simplest approach is to use a *checklist*—an a priori listing of "all" candidate impacts. This list may be relatively brief, at a high level of abstraction, or it may be highly detailed and relatively concrete.

A different approach is to list all the *parties affected* by the development. This list is scanned, and ways in which the development could affect each party are identified. In many cases, the assessors do not have a ready-made checklist. In such cases, brainstorming (Chapter 7) offers an unstructured way to identify impacts. Edelson and Olsen (1983) identified impacts by *interviewing* over 50 parties at interest to the subject of their assessment. The interviews elicited 20 potential impact areas for further study.

In another example, Lough and White (1988) set out to identify impacts from decommissioning nuclear power plants. They began with a literature search and refined this by means of a Delphi approach. Seventeen participants represented different parties at interest (consulting firms, electric utilities, public utility commissions, and the federal government). The process yielded 19 potential impacts of varying significance. The assessors felt that use of all these impacts to rank alternative decommissioning strategies would confuse the issue. Therefore, they reduced the set to four critical impacts: (1) cost, (2) occupational radiation exposure, (3) institutional impacts (personnel requirement, regulatory and liability obligations), and (4) public attitude impacts.

The *matrix* is a logical extension of the checklist. A classic example comes from Leopold and colleagues (1971) who listed 100 possible actions on the environment and 88 potential impacts. Forming a matrix with these data resulted in 8,800 impact cells! For almost all purposes this is too detailed to be useful. As illustrated previously by Lough and White (1988), the aim is to identify significant impacts—not to create information overload. Toward this end, the crossing of a stakeholder list with a list of technology/development activities can usefully capture the distribution of impacts. Wolf (1983), in searching out sociopolitical risks from energy generation, lists stage of the fuel cycle (technology activities) against alternative energy sources to identify impacts. Crossing impacts with impacts helps identify second-order effects as well (see "Cross-Impact Analysis" in Chapter 12).

15.3.2 Tracing Techniques

Tracing methods emphasize structural relationships. These may be expressed as a formal model or as a chain of causes and effects. Of special interest are tree techniques. *Relevance trees*, treated qualitatively, graphically depict the linkages between various members of sets of elements, moving from level to level via some relationship. Such a tree begins with the technology/development activity to be assessed. The next level consists of the direct impacts. A third level traces the second-order impacts caused by the direct impacts.

To complete such a tree requires that a series of "what if" questions must be answered for each node at each level. The assessor must not become seduced by the treeing game and end up, as happened in one TA, with a 30-page impact tree of no practical value. To prevent this, the assessor must balance salience—restricting analysis to the most important branches at each level—and completeness—avoiding premature closure that misses important impacts.

A tree can be developed semiquantitatively by assigning subjective probabilities of occurrence to each impact. The timing and/or severity of each impact can also be estimated, depending on study needs. These efforts can help determine which impacts are most significant to narrow the list for detailed study (Figure 8.2 shows such a matrix).

A breakthrough insight can sometimes result from thinking of impacts through a structured approach and a holistic perspective. Harman (1983) reports on an assessment of solar energy development impacts. The analysts attacked the problem with three approaches. A large energy model helped assess market penetration possibilities under various price and subsidy assumptions. Second, a broad IA addressed implications for various stakeholder groups. Third, the solar option was considered in the context of related, broad social issues. This third level analysis effected a shift in study focus from examining effects of a specific technological intervention to a more fundamental question of the major priorities in U.S. society. For instance, consideration of the collective impacts of U.S. energy policy (rather than the incremental differences among policy options) pointed toward unacceptable future conditions. This creative interplay between impact identification, alternative perspectives, and study bounding reflects the best in IA.

15.3.3   Narrowing the Impact Set and Estimation of Effects

Many early IAs were severely criticized for cataloging too many impacts with too little significance. Restricting the impact set for in-depth analysis expedites impact analysis. As per the example from Lough and White (1988), restricting the impact set also aids impact evaluation. In their assessment of nuclear decommissioning options, inclusion of a larger set of impacts would have clouded the comparison of those options.

Mechanisms to narrow an impact set rely on expert judgment (see Chapter 11). The range of possibilities extends from casual judgment through formal compilations of stakeholder opinions (such as surveys, the Delphi process, and meetings).

Estimation of effects also can help prioritize impact areas. Dimensions to consider include: extent of impact, probability of occurrence, dynamics (timing), and importance (magnitude, salience) of the impact (Mihai, 1984–85). Techniques include ranking impacts as to importance, simple weighting mechanisms, and threshold schemes (for example, including only those impacts that could conceivably reverse a policy choice or only those attaining a limiting case benefit/cost level). The AHP technique described in Chapter 18, and supported in the TOOLKIT, provides a powerful way to focus on the most important impacts.

15.4   CONCLUSION

Sometimes asking the right questions is just as important as coming up with the right answers. IA, and impact identification in particular, have this character. The technology manager who neglects to consider the effects of technological change is gambling recklessly. The manager who systematically asks about the effects, including possible unintended and delayed effects, is well on his or her way to dealing with them effectively.

This chapter provides a framework to pose those questions and to answer them in terms of possible impacts. Chapters 16 and 17 point out ways to analyze those impacts, moving the answers toward specifying the probability and magnitude of the impacts.

EXERCISES

15.1   Focus on a given technological development. Select a topic of current assessment interest or choose one of the following:

- Construction of a 200-mile-per-hour restricted rail corridor from Washington, D.C. to Boston
- Elimination of local automobile dealerships by direct factory ordering, through a video/computer network, and custom production
- Direct international data base access for enterprises and individuals in the Soviet Union
- Spread of nuclear weapons possession to 50 or more nations

Develop a checklist or matrix for impact identification tailored to your topic. Illustrate its application.

15.2 Focus on the same technological development as in Evercise 15.1. Develop a suitable treeing structure for impact identification tailored to your topic. Illustrate its application.

15.3 For this same development, propose an approach and operational criteria to narrow the impacts identified to the most critical subset. Apply this approach to the illustrative impact set generated in Exercises 15.1 and/or 15.2.

15.4 Identify the principal impacts of the voice-activated word processor ("Talkwriter"). Use both scanning and tracing techniques. Explain the techniques used and why the results are different. Determine which impacts should be analyzed in depth and why.

15.5 Assess the environmental impacts of a combination 430-MW solar/fossil fuel power plant (see Exhibit 15.5). This hybrid technology uses a fluid-cooled solar receiver atop a tower. Surrounding the tower is a circular field of heliostats (mirrors) that direct the sun's energy onto the receiver surface. This energy is conducted by molten sodium fluid to be used in steam production to drive turbines that generate electricity. Coal is burned to heat the working fluid at night or when demand outweighs solar contributions. The plant is designed for a 30-year life. It takes up 12.3 km^2 (mainly taken up by a 2-km radius field of 60,000 mirrors). The central tower is 3,300 meters high. Water consumption is 1.2×10^{14} m^3/year. Coal consumption at full load is 200,000 kg per hour. Coal storage is 250×10^6 kg. Assume the plant is to be built in the southern California desert, 100 miles from Los Angeles. Identify environmental impacts likely from this plant. Then, narrow this set to identify those few impacts that are most significant.

■ **Exhibit 15.5 Impact Identification and Screening for a New Power Plant (Based on Hunsaker et al., 1983).**

Initial screening for the solar/fossil fuel power plant identified potential impacts on

- air quality
- water use and quality
- geology and soils
- land use
- hydrology
- occupational health & safety

- meteorology
- vegetation & wildlife
- solid waste disposal

- socioeconomics
- aesthetics

This set was further narrowed to the following determined to be of major significance:

Land Use: Large area taken; probably requiring exclusion or access limitation to limit dust and aerosol deposition on the mirrors (this could serve a secondary use as a wildlife reserve); probably not a major concern in desert locations where considerable open land exists (could compete with other uses of open space, such as off-road recreational vehicles); zoning changes might be required. Other land pressures result from urban development to accommodate workers during the five-year construction phase.

Air Quality: Major concern is whether air pollution from the plant operations would compromise the solar components. A numerical model translates atmospheric emissions into ambient concentrations. A special model of aerosols on the surface of heliostats was developed to relate emissions of aerosols from a variety of sources (such as coal handling and combustion, vehicular travel) to accumulations on the mirrors. Dust storms and crop dusting in the vicinity increase particulates. Potential attenuation of insulation by pollutants suspended in the air was evaluated with a different model. Results suggest that air pollution could reduce annual solar energy collection by 40 percent.

Hydrology: Plant construction would alter surface water runoff patterns, especially if the site were paved. Mitigation measures include not paving the site and planting suitable vegetation. This is considered a significant issue due to the sensitivity of deserts.

Geology and Soils: Grading and leveling of the site would remove vegetation and destroy the "desert crust" of packed pebbles cemented with salts, gypsum, limes, and silicates, which retards erosion and surface water runoff. Suggest minimizing the amount of grading and service roads.

Aesthetics: Significant visual impacts on nearby residents and travelers from solar reflection.

General: Key impacts are air quality and land use. Additional factors to watch include availability of quality water, presence of rare or endangered species in the locale, and socioeconomic impacts during the construction phase. ■

____16
ANALYSIS OF
THE IMPACTS
OF TECHNOLOGIES

OVERVIEW

This chapter discusses the assessment of the various categories of impacts (except economic which are treated in Chapter 17), and it introduces the reader to generic approaches to impact assessment.

16.1 INTRODUCTION

This treatment is by no means exhaustive. Volumes have been written on assessing various types of impacts. This chapter presents a generic treatment of the analysis of the major types of impacts resulting from technological change. The next chapter will deal with benefit/cost analysis—an important consideration in determining whether to proceed with a technology related development—and risk assessment, which has become critical in areas that affect human health and safety (such as nuclear energy and toxic wastes). The emphasis here is on technology-driven impacts considered from the perspective of the technology manager. It is not intended as a treatment of the environmental impact analysis, which is mandated by law for many developments in the United States and in other countries. The impacts of/impacts on distinction introduced in the last chapter will be used throughout this chapter as well.

The following topical impact areas are considered:

- Technological (scientific)
- Economic (see Chapters 14 and 17)

- Institutional (organizational)
- Social (behavioral)
- Cultural (values)
- Political (legal)
- International
- Environmental
- Health related

The following sections note the types of impacts in each category and introduce methods to deal with these.

16.2 IMPACTS ON TECHNOLOGY

It is fitting to begin with technology, since that is this book's major focus. However, since we are analyzing impacts *of* technology, our treatment will consider impacts *of* technology *on* technology. Change in one technology begets changes in other technologies. The three generic types of such impacts are vertical, horizontal, and integrative.

16.2.1 Vertical Impacts

Vertical changes relate to the natural development and succession processes within a given family of technologies. A fine example of this is the progress from vacuum tube to the transistor to the integrated circuit, with subsequent larger-scale integration. The stages of technology growth (Exhibit 4.3) suggest a path of technological progress through the various stages of the innovation process. For instance, observing high temperature superconductivity (HTSC) in a laboratory environment, at the very least, suggests the possibility of building HTSC devices of significant commercial application, such as wire and thin film (see Exhibit 16.1 for additional technological/business considerations relating to HTSC).

Technological change propagates spatially and temporally. An innovation will produce effects felt in widening circles. The likelihood of adoption of an innovation in a given time period declines with distance from the source, in the absence of other forces (one model suggests exponential decay— see Sirinaovakul et al., 1988). Adoption depends on whether active efforts are made to propagate the innovation (by government agencies or producers, for example) or to impede it (by restricting access to proprietary knowledge, for instance). It also depends on the industrial infrastructure of the industry (manufacturing or service), firm size, capital requirements and availability, and how the innovation provides competitive advantage. International diffusion appears to be increasingly rapid, sometimes faster than diffusion within the country where the invention occurred (witness Japanese firms bringing a technology, such as the home VCR, to market first, even though it was invented in the United States).

16.2.2 Horizontal Impacts

Horizontal impacts result when advances in one technology affect other technologies. The nature of these interactions can vary greatly—enabling or creating a demand for some new capabilities or inhibiting or precluding others. Advances in automotive technology led to the widespread use of automobiles, thereby creating a demand for technologies to abate the pollution caused by automobiles. Likewise, breakthroughs in the internal combustion engine in the first part of the twentieth century inhibited the development of electric and steam-powered cars. Mapping the set of technologies that relate to a target technology (see, for example, Figure 8.1) is a good first step. The analyst then must probe the relationships through in-depth understanding of development processes and interactions among these technologies.

■ **Exhibit 16.1   Potential Consequences of High Temperature Superconductivity (HTSC) for Canadian Industry (Based on Bailetti, Callahan, and Elliott, 1988).**

Interview data from 11 Canadian corporations with a business interest in superconductivity fit with Tushman and Anderson's (1986) conceptualization of technological discontinuities. These can reflect in *product* discontinuities (new products with sharply improved characteristics) and/or in *process* discontinuities (radical improvements in how unchanged products are produced). Such discontinuities may either enhance or destroy current industrial competencies.

HTSC appears likely to lead to product discontinuities. HTSC will offer opportunities (two examples are power transmission and high-power motors) by lifting technical constraints (using the same two examples, resistance and heat in power transmission; heat, size, and weight in high-power motors), thereby offering advantages (energy savings and simpler and cheaper designs, respectively) that point toward market opportunities (energy plants and industrial opportunities, respectively).

The resulting technological discontinuities will be potentially competency destroying for Canadian producers of wire and transmission cable, electric motors and generators, and microcircuit components. HTSC advances will alter product performance in these areas radically, while Canadian producers will be hard-pressed to handle the new materials, change product design, and invest in new and unfamiliar manufacturing processes. They suffer comparative disadvantages vis-à-vis U.S. and Japanese producers regarding R&D access, qualified personnel, and government incentives for early commercialization.

On the other hand, HTSC should prove competency enhancing for Canadian producers of ceramics and electronic equipment. They can build on current knowledge bases and proven adaptive abilities.   ■

■ Exhibit 16.2 Patent Analysis

Patent records provide information about many aspects of technological change. Battelle (of Richland, Washington) and Computer Horizons, Inc. (of Cherry Hill, New Jersey) have led the way in creative analysis of computerized patent records. Various measures are possible:

- *Activity:* For a selected technological area (among the hundreds specified by the U.S. Patent Office) general measures (such as total patenting rate or the number of firms engaging in patenting) can be tracked to assess how rapidly technology is changing in the area. Or, specific to a firm's interests, the patent activity of key competitors, suppliers, or customers can be tracked. Are other firms getting ahead? Is a key supplier decreasing activity in an area important to our firm? Is a customer beginning to patent in an area in which it purchases products from our firm?

- *Immediacy:* If most patents cited in recent patent applications are themselves recent, rapid technological advancement should be anticipated.

- *Dominance:* Who is citing whose patents? If a certain firm's patents are heavily cited, that firm would appear to be the forerunner in the area. In considering a possible acquisition, this indicator can be used to verify that the target firm's R&D is influential, or it can be used to map how the target firm's efforts complement those of the acquiring firm.

- *Technological Linkage:* What other technological areas are heavily cited by patents in the target area? Conversely, what other areas cite patents in the target area? For example, fuel injection technology, at some point, begins to cite computer technologies. At one time, fuel injection patents were referenced by aircraft developments, not automotive. Co-citation analysis tracks the extent to which patents are jointly cited by other patent applications. This can pinpoint the integration of two technologies to serve new applications.

Patent analysis requires access to a computerized comprehensive data base and is inherently limited by that data base. For instance, a U.S. patent base will not fully reflect worldwide developments. Industrial practices also vary greatly—in some technological areas patenting may be pursued aggressively, while in others it may be neglected in favor of other strategies (e.g., keeping information confidential by trade secret). ■

Technological opportunities and bottlenecks illustrate areas of high potential for horizontal impacts. A technological opportunity is presented when economic demand is present or supporting technologies are available for a particular development. The development by Ampex of the studio video cassette recorder presented an opportunity for its use as a medium of domestic entertainment, an opportunity that other firms were quick to seize. A technological bottleneck occurs when a single missing innovation prevents the advance of technology on a broad front. Battery technology is the current bottleneck of the electric car.

Spin-off technologies are one important horizontal technological impact. Luchsinger and Van Blois (1988) presented spin-offs from military-developed technology:

- *Energy:* Commercial nuclear power deriving from submarine reactor developments
- *Nutrition:* Radar magnetrons recast as microwave ovens
- *Environment:* Military satellite sensors spawning civilian applications and transforming weather forecasting
- *Sports:* Graphite composites for aerospace working their way into fishing rods and tennis rackets

Patent analysis provides one tool to identify emerging horizontal and vertical relations among technologies (see Exhibit 16.2).

16.2.3 Integrative Impacts

Integrative impacts refer to those that blend technological change with associated contextual changes. The technology delivery system (TDS—see Chapter 2) can help identify the interplay among institutions involved with given technologies. This may be particularly useful in identifying who will have to do what to translate gains in one technology to changes in another. Institutional interfaces affect the diffusion of knowledge and the resulting impacts.

It may be desirable to estimate impacts on business *productivity*. To understand the competitive implications of changing technology, a company should consider the following factors based on Sumanth (1988a):

- *Customer-related factors:* Product quality; product reliability; price competitiveness; product loyalty
- *Market-related factors:* Market share; company image
- *Process-related factors:* Product life cycles; production efficiency
- *Employee-related factors:* Employee dislocation; training requirements; job satisfaction
- *Vendor-related factors:* Vendor capabilities; vendor loyalty
- *Owner-related factors:* Profits; long-term company viability

Exhibit 16.1 illustrates how HTSC technological change might translate into strategic business implications. The technological roadmap notion introduced in Chapter 8 can also be applied to explore the integrative implications of a technological change for a company.

16.3 INSTITUTIONAL/ORGANIZATIONAL IMPACTS

For the purposes of IA, institutions and organizations are groups of individuals whose collective activities are involved in the implementation of a technology. Institutions may be formal (such as government agencies or corporations) or informal (such as groups of co-workers, or community groups opposing a development). Institutions may pursue goals explicitly or implicitly.

Institutional *structures* reflect patterns of relationships. *Processes* are the behaviors that take place within a structural context. Together, structures and processes combine to determine the operation of a *technology delivery system*. Depicting a TDS for a focal technology identifies the salient institutions for further analysis. For example, Figure 2.2 sketches a TDS for microcomputers in developing countries.

Technological change can impact the institutions that comprise the technological enterprise—*internal organizational change*—and/or other institutions that interact with the enterprise organization—*external institutional impacts*.

16.3.1 Internal Organizational Changes

Internal organizational change associated with technological change is a major concern in technology assessment. Change in the technology of a product can alter demands on the organization. For instance, as AT&T shifts from production of copper wire to optical fibers, it necessarily requires different suppliers and different worker skills. Those differences are likely to exert pressure for managerial change and reorganization. Change in the technology of a process (such as manufacturing method, service delivery mechanism) is even more likely to produce pressure for organizational change to secure maximum benefit from the new technology.

Organizational issues play a critical role in the success or failure of a technological change. Impact analysis can help anticipate needed changes in the organization. For instance, one author participated in a project to implant a flow process production "cell" in the midst of a batch production bearings factory. Those planning the project did not anticipate what turned out to be a critical change—instituting around-the-clock operations for the cell. Some 40 months into this 18-month project, the technology was working fine, but production goals were not achieved. The main reasons were that training the operators proved much more involved than was anticipated; three-fourths of the initial group of operators refused to operate on the new work schedules and went back to traditional batch operations in the plant; supervisors did not grasp the potential of the new production process; and workers were not motivated to push for higher production and quality goals. When these organizational issues were later resolved, production skyrocketed and quality surpassed the project goals.

16.3.2 External Institutional Changes

External institutions influence the technological enterprise (for instance, dissolution of the Congressional Joint Committee on Atomic Energy made nuclear power more vulnerable to its opponents) and are influenced by the technological enterprise (for example, accidents at Three-Mile Island and Chernobyl pressured state public service commissions to be extremely cautious).

The success of a technological change involves both impacts of and impacts on the institutions involved in the TDS. A useful exercise is to diagnose existing or potential mismatches between the requirements of the technology and the organizational capabilities. For example, Fried, Armijo, and Trejo (1984) show how these factors should determine what approaches Mexican communities use to exploit available natural resources. They estimate three factors to detect imbalances:

1. Technological complexity (What is required to accomplish the tasks?)
2. Management capabilities of the local institutions (Can the institutions manage those task requirements?)
3. Political capabilities (Do the institutions engaged have the power necessary to meet the task requirements?)

They conclude in the case of these rural Mexican communities that managerial and political factors are critical to selecting successful technoeconomic activities.

16.4 SOCIAL IMPACTS

This section separates general social impact approaches from those tied closely to the direct economic effects of technological developments.

16.4.1 Social Impact Assessment

Social impacts are those that alter the day-to-day quality of life (Burdge, 1987). Major new technologies and substantial construction projects cause social impacts. This section emphasizes those effects bearing on social groups, particularly in the lifespace of the family and local community where these effects are typically most pronounced (Little and Krannich, 1988).

Social impact assessment (SIA) is quite highly developed; however, it is largely an offshoot of environmental impact assessment. The following questions aid in assessing the major social impacts of technology-based, industrial developments:

1. What social impacts are significant?
2. What are the character of and relationships among these impacts?
3. How does one analyze these impacts?

The first two questions were discussed in the previous chapter; they can be approached by mapping a TDS. The third question can be answered by using appropriate data and methods. Analysis of secondary data, surveys, and observation methods may prove useful.

Secondary data analyses make use of data already collected for other purposes. Depending on availability, these data provide answers for many baseline questions.

Likely sources include:

- Census data (population, housing, transportation, economic characteristics)
- Local and regional sources (departments of economic development, employment, public health, environmental quality, historical and archeological commissions)
- Planning commissions
- Regional water resource councils
- School districts
- Agricultural extension centers
- Public and university libraries
- Local newspaper files

Survey approaches include expert opinion surveys and resident surveys. Surveys are often used when perceptual and attitudinal information on possible impacts and patterns of service use are needed. Phone, in-person, and mail surveys may be appropriate, depending on the information needs. Sampling is usually a critical issue (refer to Chapter 11).

Observation methods include participant observation. This is effective in studying the ongoing nature of social activities. The researcher usually must live in an area for an extended period of time and care must be taken in assessing the generalizability of the data obtained (for instance, observers are likely to be contacted by those with strong positions on the issues). In some cases, unobtrusive methods may be useful. Possibilities include videotaping of activities, content analysis of daily newspapers, and other opportunistic gatherings of pertinent information. The danger in these methods lies in unsystematic data accumulation that leads to biased interpretations.

16.4.2 Socioeconomic Impact Assessment

New technology, especially in the form of developments such as new plants, affects the community. The driving force for a whole complex of impacts of great concern to the community, and hence to the developer, is new employment. Figure 16.1 provides a general socioeconomic impact model that suggests how employment impacts population, which in turn changes demand for community services and affects public sector finances.

Employment impacts begin with the jobs provided directly by the development. Estimation of additional indirect employment may involve employment multipliers (see Exhibit 16.3), export base regional models, or input-output analyses (see Chapter 14). Additional concerns involve the timing and geographic distribution (within a region) of employment and other economic impacts. Lovejoy (1983) suggests refining those models to consider the development's labor characteristics together with characteristics of the local citizens (such as unionization, occupation, and years of education).

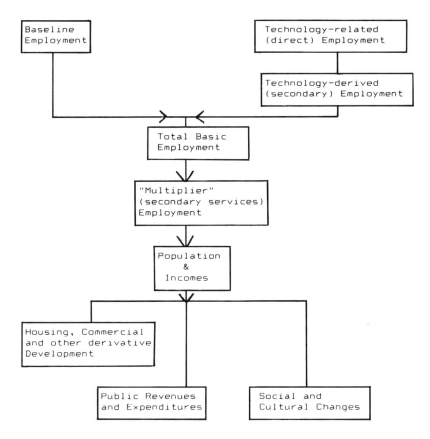

Figure 16.1 Socioeconomic impacts of development (based on Moore and Gilmore, 1983).

■ Exhibit 16.3 Multipliers

Multipliers are useful in estimating total changes in employment, population, or income. Consider a new development that pays $100 in wages at time 1. If the marginal propensity of the wage earners to consume (m) is 0.8, for example, they spend $80 at time 2, which translates into income to those who provide secondary services. This cycle continues. At equilibrium, the income multiplier is

$$1/(1 - m)$$

For $m = 0.8$, this means that the direct $100 in wages leads to a total increase in income to the region of $500. Employment multipliers of 3 to 4 are reasonable. ■

Employment projections may be of interest. Elaborating the following tautology accounts for the technological changes in the nature of work (Rumberger and Levin, 1985):

$$\text{Employment} = \text{Output} \times \frac{\text{Labor}}{\text{Output}} \times \frac{\text{Employment}}{\text{Labor}}$$

Future output reflects demand for sector goods or services; labor required per unit of output reflects future productivity, as enhanced by changes in technology; and employment per labor requirement allows for part-time or subcontracted work possibilities. The model may need to be adjusted to acknowledge changing practices as, for instance, technological changes translate into product or service enhancements instead of increased productivity (see Porter, 1987a).

Baseline *population* estimates usually will be taken from secondary sources. Project-induced population changes derive from employment projections (see Exhibit 16.3). Demographic profiles for workers and their families form the basis for computing migration. The simplest techniques multiply the average family size per worker by the number of in-migrating workers. More detailed projections take into account data (as available) on age, sex, marital status, and children of workers (Leistritz and Murdock, 1981).

Service demand projections derive from the population and population change projections. Housing, commercial services, and public services may be of concern. Leistritz and Murdock (1981) note that projection of service needs must consider: previous service levels and perceptions of their adequacy; quality and distribution; and service delivery systems.

Fiscal impact assessment projects the changes in costs and revenues of the various governmental units involved in response to a development project. Timing and amount of capital costs and operating costs of providing services are essential to policy formulation. The time stream of revenues expected needs to be estimated, taking into account: property taxes, income and sales taxes, production and severance taxes, population-related taxes, and intergovernmental transfers. Jurisdictional issues are often vital as industrial developments may locate in one jurisdiction while new workers choose to live elsewhere.

The company planning a new facility, proposing major expansion of an existing facility, or intending to close a plant needs to consider these issues. Rather than defending itself as stakeholders uncover impacts, the company, having forecast and assessed such impacts, can take a proactive stance. It can present carefully reasoned projections and propose strategies to take best advantage of those projections, thereby defusing opposition before it begins. This is not to claim that IA eliminates differences among stakeholders, but it can provide a fair base for cooperative design of mutually acceptable solutions.

16.5   CULTURAL AND BEHAVIORAL IMPACTS

This section discusses two types of individual-level effects that produce important impacts external to and/or internal to the organization.

16.5.1   Impacts of and on Values

Cultural impacts involve human values. *Values* are conceptions of desirable states of affairs that guide judgments across specific objects and situations toward more ultimate endstates (Enk and Hornick, 1983). IA deals with values and value changes as (1) criteria by which people e-"valu"-ate the desirability of technological developments; (2) changing contextual factors that can influence technological developments (impacts on); or (3) affected by a technology (impacts of).

People hold various types of values. Canan and Hennessy (1982), in assessing the desirability of various energy development options for the Hawaiian island of Molokai, sought to identify prevalent values. They used the Galileo method (Woelfel and Fink, 1980) to prepare a value map of closely held values (such as pace of life, education, Hawaiian culture) with respect to energy options (that is, electricity self-sufficiency). This method began with open-ended interviews of a small, diverse set of residents. A broad sample of residents was surveyed on the values identified as important in the first round. Principal component analysis (a form of factor analysis) was used to map the value concepts (for example, to what extent do people who hold a high regard for Hawaiian culture also subscribe to energy self-sufficiency?). In this case, values of the decision maker were also mapped to identify important differences from the residents. The main policy implications of this analysis were that electricity self-sufficiency could be favorably associated with jobs and living off the land but not with energy to be exploited by others (such as tourism, development).

Although not as deeply rooted as values, *attitudes* toward a given development can exert great force. For example one community might oppose a factory that another community welcomes. A vital early step in development planning is to find out prevailing attitudes that bear on the development. Some form of survey—ranging from circumspect, informal observation to more formal, direct inquiry—makes sense. A recommended follow-up step involves appropriate participation by affected parties in determining features of the development, insofar as is feasible. Just as with internal organizational acceptance of change, participation can significantly lower community resistance to technological development (Edelstein, 1988). Such participation processes need to be guided by knowledgeable, effective leaders.

Values change. Consider how your values differ from those of your parents—and those of your children, if you have exercised the value of having children. Values concerning sexual behavior, abortion, and the number of children per family have changed sharply over the past few decades throughout the world. These values differ within and among nations.

Changes in values affect technological development. For instance, marketing efforts accentuate sexual mores, and family size determines potential customers in various age groups. Moreover, technological choices made today impact upon future generations—people who will hold different values than we do. Long-term IA should be accompanied by efforts to gauge what values the affected society will hold at that time—a precarious assignment!

Values can change in a number of ways (Rescher, 1969), including:

- Acquisition or abandonment
- Increasing or decreasing importance/emphasis
- Increasing or decreasing standards for a value
- Widening or narrowing subscription to a value

The assessor may want to estimate types of value change taking place, their direction, their rate of change, and how widespread they are. It may be possible to identify related values likely to shift in conjunction with basic value changes. Taken together, this information allows one to attempt to forecast values.

For IA purposes, patterns of societal value changes will usually be based upon others' analyses (secondary data). These derive from repeated surveys to identify value trends (see Elgin and Mitchell, 1977) or content analyses. Another choice is for the assessor to package alternative sets of future (or present) values in the form of scenarios (see Arthur D. Little, 1975).

Certain significant technological changes also can impact values. Such changes may be broad in scope (affecting an entire society) or may be restricted to a specific group (such as the workers in an industry).

Major value changes attributable to technology are likely to be long-term and cumulative—resulting from a complex of technological and social influences. Tongue-in-cheek, Joe Coates (1971) has traced an increased acceptance of divorce to the "sixth-order" consequences of television (Exhibit 3.1). Anticipating changes (in family values, for example) as a function of technological changes is extremely difficult. Alternative value changes may be portrayed through scenarios (see Chapter 13).

John Clippinger (1983) conducted a fascinating field study to determine the psychological impacts of changing technology in the logging industry in northern New England. Open-ended interviews identified key beliefs that were then systematically examined through questionnaires addressed to traditional (horse) loggers, independently organized loggers, and highly mechanized company loggers. The following differences were attributable to the changed work technology:

- High-tech loggers think in terms of free time and vacations; traditional loggers do not.
- High-tech loggers compartmentalize their personal from their professional lives more.
- Loggers' lives differed on the job, and differences permeated family, community, and market relations.

Both management and labor would seem to have interests in assessing the likely impacts of increasing automation in given industries. Studying the patterns resulting from analogous changes in other industries, or from early changes in the focal industry, appears to be the best guide to possible psychological/value/cultural impacts on workers. Unfortunately, such studies are relatively rare.

16.5.2 Impacts on Behavior

Behavioral impacts concern changes induced in individuals due to technological changes. Two critical groups of individuals are *workers* within the organization and *users* of the products or services that the organization provides. (Impacts on workers are addressed by Clippinger, 1979.)

Analysis of user needs can help ensure a favorable impact profile and a profitable product/development. For instance, a large Air Force project seeks to produce a Designer's Associate to help those who design aircraft. Rather than develop this technology based on technical capabilities and the beliefs of the developers, the developers chose first to ask the designers what aids they presently use, for what functions, and what they would like to have in the future.

To conduct such a needs assessment, the assessor should

- Identify the intended user population
- Interview (open-ended) a few knowledgeable users to determine key issues, terminology, and how to access this user community
- Interview (focused) a representative sample of users to prioritize functional needs, willingness to pay for features, and taboos.

A good opportunity to follow up exists; these same interviewees can later be asked to review the conclusions about their needs and the organization's planned response to those needs.

16.6 POLITICAL/LEGAL IMPACTS

Political considerations certainly appear within organizations. However, this section concerns broader external political and legal impacts.

16.6.1 Political Impacts

Politics translates values (Section 16.5), through institutions (Section 16.3), into action representing the power of the body politic. Both political impacts on and impacts of technological development are based on implicit or explicit assessment of the potential impacts of the development. Those who would promote a development need to estimate the reactions of others to it.

Constructing a TDS is an excellent first step to map out the institutional interest in a new technology or project. This provides a basis for analysis of what it will take to gain acceptance of the development (impacts on):

- Who are the key stakeholders? How can they exert power over the development? What do they want from it?
- What are the key "leverage points" (critical decisions) for the development?
- How should one work with the various stakeholders? Is their active participation in planning the development feasible?

The "impacts of" refer to how political power will be altered by development of the new technology (for example, consider how the birth control pill and access to abortion have affected the progress of the women's movement).

Role-playing is an approach to consider in political analysis. Rather than having the assessment team strive to analyze the way various interests will respond at critical decision points in the development, the developmental sequence is "played" through. Such play may involve assessors "role-playing" different stakeholder interests, or it may engage representatives of those interests in the play. Play can entail constructing one or more scenarios that engage people in a sequence of development decisions or a singular crisis. Or, it may involve the use of an interactive computer simulation depicting key development features. This can be done by modifying an available simulation (see Gibbs, 1974) or by creating a new one (potentially a major task). Role-playing contrasts with other analytical methods in its relatively free-form interactions, emulation of actual conflicts, and the intent to develop insights into the actual political processes. It can foster communication among stakeholders and allow them to try out creative resolutions of complex situations.

Political factors also may be of interest at the national level. Impact assessors may wish to obtain (not construct) *political risk analyses* for countries in which a development is being planned. These draw upon trend analyses of various economic and political stability indicators and/or expert opinions about political stability (Mumpower, Livingston, and Lee, 1987). Firms, such as Frost and Sullivan (in New York) monitor and forecast political stability for a number of countries.

16.6.2 Legal Analysis

Legal analysis addresses the interaction between the law and the technological development in question. Assessors must be aware that the law affects technology (for example, national sovereignty laws restrict the use of remote sensing) and vice versa (for example, advancements in satellite technology spur development of relevant space rights sovereignty principles). The law is evolving in conjunction with changes in such technologies as photocopying, software, biomedical technologies, and genetic engineering.

Existing and potential laws, administrative regulations, and court cases must be evaluated because they could influence the course of a development by imposing conditions (requirements, fees), modifying market opportunities, or changing decision-making structures.

Legislators respond to political pressures by enacting new laws. For example, recent concerns with American technological competitiveness have led to restricting Japanese opportunities to market in the United States; permitting cooperative R&D efforts by American firms in microelectronics (such as MCC—the Microelectronics and Computer Technology Corporation—and Sematech—the Semiconductor Manufacturing Technology Institute) and other areas; and tailoring high-definition television (HDTV) standards to favor American industry.

Legal liability deserves special note. The jeopardy of a lawsuit can stifle technological implementation. As lawsuits mounted, costs of insurance rose, which destroyed the American private airplane industry. In contrast, federal limitation on liability from nuclear accident under the Price-Anderson Act facilitated initiation of the American nuclear power industry. Bases for legal liability include (Changnon et al., 1977):

1. *Trespass*: Intentional harm involving physical invasion (or intent to invade) of real property (for instance, toxic waste carried in water runoff).
2. *Negligence*: Careless harm as defendant fails to conform to standards of carefulness.
3. *Abnormally dangerous activity*: Liability without fault from uncommon, high-risk actions.
4. *Nuisance*: Substantial invasion of rights of others that outbalances social value of purpose of the conduct.

Management must confront and resolve any legal issues associated with introducing new technology.

16.7 INTERNATIONAL IMPACTS

The international dimension becomes increasingly important as the world shrinks toward a true global village. Certain impacts of technological development transcend national boundaries. As Senator Albert Gore, Jr. noted in a Congressional hearing on July 16, 1987:

> This hearing [on global climate change] reflects a growing understanding in Congress that some of the greatest threats to our environment are global in nature and that a comprehensive national and international effort must be undertaken to address these problems. Our political system is not used to confronting problems of such global scale.

This section treats two categories of international concerns: (1) international considerations in technological development and (2) impacts of development that transcend national boundaries.

International considerations in development vary across institutions. Companies engaging in the development of new technologies must determine where to conduct R&D, where to produce, and where to market. Each of these choices requires balancing of short-term profits, long-term capabilities, risks, and public relations.

National leaders have special concerns about development patterns. Developing countries must balance economic growth with environmental protection. They must determine what technologies to purchase abroad (and deal with such considerations as foreign debt) and which to develop themselves. Such development can involve a variety of arrangements with multinational companies (such as joint ventures with local companies and/or government units, requirements to use indigenous suppliers, or special export considerations). Developed nations evidence concerns about employment opportunities, long-term economic strength, and environmental impacts.

International organizations are increasing their efforts to assess technological developments. The World Bank made headlines by boosting its efforts to perform environmental impact assessment (EIA) and social impact assessment (SIA) before going ahead on major projects. The European Economic Community coordinates EIA procedures. The United Nations Environment Program promotes environmental quality; the United Nations Centre for Science and Technology for Development has instituted an advance technology alert system to foster international cooperation in technology assessment. Coordination of such efforts and cataloging of available data bases are sorely needed.

16.8 ENVIRONMENTAL IMPACTS

Environmental impact assessment (EIA) refers to the IA process, mandated by law in many jurisdictions, that typically leads to an environmental impact report and, if warranted, a more thorough environmental impact statement (EIS). Impacts on the natural environment, the focus of this section, constitute only a part of EIA, which can encompass the full spectrum of possible impacts. These issues are most important in the case of construction projects. This section is a brief overview of a complex area.

16.8.1 Environmental Impact Assessment Methods

As a generalization, environmental impacts associated with information age technologies are less severe than those associated with more intrusive, industrial era developments. To ascertain the possible importance of environmental impacts, however, consider three levels of analysis: (1) screening, (2) sketch methods, and (3) detailed methods.

General screening methods are presented in Chapter 15 (checklists, matrices, impact trees). Available maps, especially computerized, may be of great use in identifying potentially important environmental impacts. For example, overlay of a highway project on a map that maintains ecological land classification data (such as terrain, soils, habitat, vegetation, land use) can quickly identify potential problem areas (Roe, 1988).

Sketch methods are relatively general and less quantitatively precise than detailed methods. For instance, to gauge biotic disruption, aerial photography (visual or infrared) is a sketch method; a more detailed method would be species count surveys. The resource demands of detailed methods may be orders of magnitude greater, and, in a technology forecasting context, only rarely will these be appropriate. Therefore, detailed methods will not be covered here; the assessor with needs for detailed analysis must refer to volumes directed at the specific topic area of concern (for example, for details on water quality modeling for which the models continue to increase in complexity, see Stans, 1986).

Detailed environmental modeling uses various approaches. At its simplest, ecosystems may be considered as common-sense enumeration of the facets of a system (living and nonliving) subject to impact by a development. This can be supported by computerized data atlases of area resources, maintained for land use planning (see Shields, Rosenthal, and Holz, 1986; Kelly et al., 1978). Should no significant impacts be identified, this may suffice for impact analysis. To model the course and severity of project impacts, general models (such as air or water quality) may be applied to the specific situation or simpler models may be built from scratch, specific to the issues at hand. However, the scope of environmental modeling extends beyond our capability to even categorize the many different models.

16.8.2 Topical Areas

The major areas of environmental concern, the importance of which will vary from situation to situation, include ecosystems, land use, water quality and quantity, air quality, noise, and radiation. These will be very briefly introduced.

Ecosystem assessment depends on baseline data. This implies the use of some form of survey (as quantitative as possible) to project the differences attributable to the technological development in question. Ecosystem impacts of a dam project, for instance, include terrestrial and aquatic changes.

Major *land* impacts include land use, solid waste, erosion, and pollution (such as pesticides). Current land conditions are a major consideration—a given disturbance may be readily accepted in an industrialized area, but be violently rejected in a parkland. According to the U.S. EPA Region X (1973), projects that produce solid waste, either during construction or due to operations, should assess

1. Quantities and composition of solid wastes
2. Hazardous wastes produced (see Section 16.9 of this book)
3. Long-term wasteload, from the project plus other anticipated sources (such as population growth)
4. Plans for storage, collection, and disposal of the different types of wastes that will be generated
5. Consideration of alternatives (such as recycling)

Special concerns have emerged with respect to *toxic wastes*. Landfill design and operation must especially protect groundwater and surface water quality. Leachate (water percolating through to carry with it soluble and suspended substances) processes demand sophisticated analyses. Edelstein (1988) provides an informative assessment, reaching into the psychosocial impacts of hazardous waste sites in communities.

Water quality and quantity impact assessment must consider legal requirements (Federal Water Pollution Control Act, 1972 Amendments; Western U.S. Water Compacts). Quality standards vary by the body of water and its use. Quality concerns include:

- Hydrologic characteristics (flows, drainage patterns)
- Physical properties (temperature, turbulence, solubilities)
- Chemical constituents (dissolved oxygen, dissolved solids, nutrients, toxic chemicals, pH)
- Biological constituents (algae, bacteria, weeds, fish)

Many detailed models have been developed to estimate these characteristics (see Warner et al., 1974). Water quality depends on many factors. For instance, reservoir characteristics are influenced by inflow water quality, climate, reservoir size and shape, and management (influencing flow patterns and biological controls). Baseline data may be available from appropriate governmental agencies.

Air quality impacts parallel water impacts in terms of the growth of national and international attention, leading to increasingly stringent standards and real successes at quantitative modeling. The 1977 amendments to the U.S. Clean Air Act list 28 categories of pollution sources for which standards apply. The five major pollutants have been

- Hydrocarbons (emitted mainly from partial combustion of fossil fuels)
- Carbon monoxide (largely from incomplete combustion of hydrocarbons in automobiles)
- Nitrogen oxides (from high temperature combustion processes, as in automobile engines or fossil-fuel power plants)
- Sulfur oxides (from burning of fossil fuels with sulfur impurities)
- Particulates (a wide range of solid or liquid particles emitted during combustion, from grinding, or from earth moving)

These primary pollutants may combine to generate secondary pollutants (especially smog from nitrogen oxides and hydrocarbons reacting in the presence of sunlight). Certain other hazardous toxicants may be present in air, including asbestos, arsenic, lead, mercury, and radioactive wastes. Direct health effects need to be evaluated (see Section 16.9).

The 1990 Amendments to the Clean Air Act toughen provisions for hydrocarbons and nitrogen oxides and extend the antismog provisions to control ground-level ozone. On the other hand, new provisions seek to protect atmospheric level ozone that shields us from ultraviolet rays by restricting the use of ozone-depleting chemicals, especially the hydrochlorofluorocarbons. Sulfur dioxide and nitrogen oxides from smokestacks are to be halved to control acid rain. The list of toxic air pollutants to be regulated by the U.S. EPA has expanded from 7 to 189. This legislation also reflects SIA in that it provides compensation for workers displaced from their jobs by the tightened restrictions.

Numerous models are available to calculate air pollution concentrations from both point and nonpoint sources. Modeling the reactive chemicals requires combination of chemical formulas, under atmospheric conditions, with diffusion models (Cook, 1976). Agencies may provide background air data. Various mitigation measures may improve the air quality impacts of certain developments. These include altering fuels used, using technological pollution controls, and restricting hours of operation.

Noise — unwanted sound — is most associated with surface and air transportation, human activity, construction, and industry. Noise concerns center on community annoyance more than on physiological or structural damages. Noise levels must conform to standards and are successfully modeled. In the United States, federal agencies are responsible for noise standards in their domains of concern (for instance, Housing and Urban Development publishes "Noise Assessment Guidelines").

Noise measured in decibels (dBA) weights loudness according to hearing. Relative noise levels are also important. For instance, an outside noise level no more than 60 dBA for 90 percent of the time will protect 50 percent of sleepers from awakening. However, an increase of 10 dBA (a logrithmatic measure) over preexisting levels is likely to generate substantial complaints.

Noise assessment begins with data on preexisting noise levels. It should consider both construction and operation phases of a development. Noise sources should be described and noise contours estimated by time of day and duration (by using analogies or computer models, for instance). Criteria for acceptable noise levels should be established by considering existing and proposed nearby land uses, possibly involving community participants. Noise impacts can be mitigated by reducing noise at its source, locating away from sensitive populations, providing insulation, or removing affected individuals (for example, by relocating airport neighbors).

Another area requiring special attention is radiation. The U.S. Nuclear Regulatory Commission (NRC) has established special EIA guidelines. Prospective licensees must prepare an environmental report.

16.9 HEALTH IMPACTS

Health hazards include:

- Failure of large-scale technological systems
- Discrete, small-scale accidents

- Low-level, delayed-effect hazards (e.g., cancers)
- Increases in infectious or degenerative disease rates

These can affect a workplace (occupational hazards) or the community at large.

The World Health Organization has produced an extensive series of over 30 publications detailing health criteria for various environmental health factors. These include sources and exposure levels, health risk groups and prediction of how they will be affected by exposure, definitions of significance and acceptability of health impacts, and identification of mitigation measures. These sources serve the environmental health impact assessment (EHIA) process well in their specificity, detailing criteria for such diverse factors as lead, oxides of nitrogen, noise, radiofrequency and microwaves, and health risks during pregnancy. EHIA of industrial developments entails risk assessment (see Chapter 17). A quantitative risk assessment concerned with environmental health hazards may entail:

1. Quantitative plant description
2. Identification of failure cases (from historical parallels, and technical audit of plant design and operations) and estimation of primary failure frequencies (fault tree analysis)
3. Prediction of discharges, and modeling of their dispersion
4. Calculation of hazards at various distances and preparation of risk contour plots

Most EHIA methods are simple (such as checklists), although the underlying sciences (epidemiology and toxicology) of linking environmental health factors to human morbidity are complicated (Martin, 1986).

16.10 CONCLUSION

This chapter has quickly scanned an enormous range of impact categories. The manager or engineer must consider whether each category poses concerns with respect to a changing technology. If so, detailed analysis is in order. IA requires the impacts of the various categories to be integrated as well. These impacts couple strongly; techniques such as cross-impact analysis (see Chapter 12) can help analyze such interactions. Impact analysis should feed back important considerations into the technology planning process.

EXERCISES

16.1 Obtain a study of a new technology prepared by the Office of Technology Assessment (U.S. Congress, Washington, D.C. 20015). Determine what impact categories were emphasized. What methods were used to analyze the impacts in each category? Which of these impacts raise the greatest concerns for a technology manager promoting the growth of the technology under study? Specify actions that a manager might take to facilitate successful development.

16.2 Impact identification can progress quite far on brainstorming; however, impact analysis requires hard data. Bridging this data gap often proves difficult. Use Exercises 15.1, 15.2, and 15.3 to continue the development of impact analyses.

(A) Specify a restricted set of potentially key impacts (up to 6).

(B) Identify sound data sources for each.

(C) Check out those sources and obtain necessary data; explain limitations and what might be done to overcome these.

(D) Continue with analysis of those impact areas where sufficient data have been obtained (follow the guidance of the appropriate sections of this chapter and consult additional sources as appropriate).

___17

BENEFIT/COST AND RISK ANALYSIS

OVERVIEW

Technology managers must choose among alternative projects in allocating scarce resources. Benefit/cost analysis provides one vital tool to help decision makers, in the public and private sectors, make these allocations. This chapter introduces the concepts and tools to estimate the costs and benefits of new technologies. Furthermore, because projects pose different chances of success or failure, the chapter extends benefit/cost considerations to incorporate uncertainty through the use of risk analysis.

17.1 INTRODUCTION: OPPORTUNITY COSTS AND CHOICES—AGAIN

Scarce resources mean that choices must be made. Since resources are limited, it is not possible to tackle every challenge, even every desirable technological challenge. Thus managers must make decisions. To do this wisely, they need tools to discipline the allocation of resources. For example, a new product that combines hardware and software components might seem attractive, but what opportunities must the firm give up to develop it? Perhaps concentrating on marginal improvements to existing products will reduce quality problems, satisfy customers, and, within a year, provide funds to meet financial obligations to bankers and investors. On the other hand, taking some risk now to develop a new product may allow the company to leapfrog competition and gain a dominant market position in three years. What should the manager decide?

Analytical tools never will completely answer such hard questions. There always will be uncertainty. However, benefit/cost and risk analyses can bring discipline to the decision-making process, credibility to requests for financial support, and a framework within which to assess the choices that must be made.

Society also makes choices. Some opportunity costs of these choices are reflected in costs facing firms and will be considered by decision makers in the firms. Other costs and benefits are external to firms and will not be considered in business decisions unless society takes steps to have them considered. From a purely economic point of view, it is ideal for all social costs and benefits of a project to be weighed. This means that risks external to the firm also must be taken into account. Technological advances have added much to the quality of life in industrial nations. However, certain technologies (the pesticide DDT, for example) have caused severe societal impacts. As a result, some efforts to introduce new technology have been frustrated because the public and institutions were not convinced that social costs and risks had been adequately considered. The techniques of benefit–cost and risk analysis described in this chapter, used in conjunction with the impact assessment (IA) procedures in Chapters 15, 16, and 18, will enhance the manager's ability to present a fair case for an appropriate innovation and help it become a reality.

17.2 BENEFIT/COST ANALYSIS—WITHIN THE FIRM

Before undertaking any project, a firm must justify the resources that will be committed to it. Investment in the development of a new technology may return future profits. But how does the manager know if returns will be large enough? How long should he or she be willing to continue committing money before profits begin? Will bankers or financial people within the company approve?

The problem for the decision maker is that money has *time value*—a dollar today is worth more than a dollar tomorrow. For example, one could put $100 in almost any bank and watch it grow to at least $105 in a year. It follows that $105 delivered one year from today could be said to be worth as much as $100 today. Similarly, $110.25 in two years is equivalent to $100 today, as is $115.76 in three years, because compounding interest at 5 percent annually would produce these amounts. Therefore, at 5 percent per year interest, the present value of $105 in one year is $100 today; $100 is also the present value of $110.25 in two years or $115.76 in three years. In general, the *net present value* (*NPV*) of returns in some future year t is given by

$$NPV = \frac{R_t - C_t}{(1 + i)^t} \tag{17.1}$$

where i is the interest rate, R_t the revenue, and C_t the cost in year t. This simple formula is the basis of all capital budgeting decisions for which money must be spent today to secure expected returns tomorrow. By looking at the time pattern of expected costs (outlays) and returns, the most attractive projects can be determined. In doing this, the rate of interest (i) used in the calculations is known as the *discount rate*. The answer (*NPV*) sometimes is called the *discounted present value*.

For example, assume that a firm could pursue three mutually exclusive technologies A, B, and C that have the pattern of net benefits $(R_t - C_t)$ shown in Table 17.1. Suppose Project A has a rather high initial outlay, say $1 million, with relatively low annual operating costs. Further, suppose that one reason for these low operating costs is that the process produces environmentally harmful by-products that are not currently regulated. In the first year, Project A will attract $500,000 of revenue with only $100,000 of costs for material, labor, taxes, and other operating expenses. When the $1 million initial outlay is included, net benefits are −$600,000 for the first year. The next two years will produce $750,000 in revenues and $150,000 in outlays to meet product demand. In the fourth year, more restrictive environmental regulations are expected and major renovation will be needed. Outlays will exceed revenues by $200,000 that year, but the subsequent year again will bring net benefits of $300,000. The last year also will bring some revenue, but the plant's operating costs will be too high to allow it to stay in business. Since demolition will require the disposal of toxic wastes, the cost of retiring the facility will be very high. In fact, that cost will result in a net outlay of $700,000.

Project B does not have the severe environmental problems of Project A, and initial capital investment is not as large. Annual operating costs, however, are much higher. The first year results in net outlays of $400,000; net benefits in the next three years $220,000, $250,000 and $200,000, respectively. In the fifth year, major modifications will be needed making net benefits −$300,000. However, this work will extend the life of the technology for one more year and produce $300,000 of net benefits in the last year of operation.

Project C will require the highest initial outlay, but the facility is durable and has low environmental impacts. The process is costly and the expected net returns are modest but consistent throughout the project's life, since no major modifications or renovations will be needed.

How does a manager select from among these three alternatives? If the time value of money were not important, the most profitable project would be Project C: the sum of its net benefits is $340,000, compared to $270,000 for B, and $0 for A. If the popular, but dangerously simplistic, notion of payback period were applied, Project A would appear best because it returns investment in the second year. Project B requires almost three years to return investment and Project C takes

TABLE 17.1   Hypothetical Benefits for Projects A, B, and C (in thousands)

Year	Project A	Project B	Project C
1	$−600	$−400	$−700
2	600	220	190
3	600	250	220
4	−200	200	220
5	300	−300	210
6	−700	300	200

over four years. The best approach, however, is to examine the sum of the present values of each year's net benefits for each project—that is, apply Equation 17.1 to each entry in Table 17.1 and calculate the total present value for each project.

Finding present values can be tedious, but spreadsheet software will help do it quickly. (see, for example, the @NPV(x,range) function in LOTUS 1-2-3). Table 17.2 shows the results of present value calculations for the net benefits of each project in Table 17.1 for interest rates ranging from 0 percent to 50 percent per year.

Table 17.2 makes it clear that the answer to which of the three alternatives is best depends upon the discount rate. Figure 17.1 graphically displays these results and shows that from 0 percent to about 5 percent Project C is the best alternative; Projects B and A become preferable as the discount rate increases.

Choosing the discount rate for this analysis obviously is critical. Conceptually, the rate should be determined by the firm's cost of capital, which in turn depends upon its financial structure, tax laws, credit conditions, credit ratings, and other variables. As a practical matter, the discount rate will be specified by corporate finance officers if corporate funds are involved. For a small firm or individual, the rate probably will be the interest rate on the loans needed for the project. If the firm does not have to borrow the funds, the rate will be determined by the best alternative rate available on the funds (that is, the opportunity cost). Later discussions will show how this discount rate can be adjusted for various levels of risk implied by the alternative projects.

Before considering the effects of uncertainty on the example, it is worthwhile to briefly consider one other criterion for choosing projects—the *internal rate of return*. This rate often is used as a filter to eliminate less profitable projects from

TABLE 17.2 Present Value of Net Benefits for Projects A, B, and C (in thousands)

Discount Rate	Project A	Project B	Project C
0%	$ 0.0	$ 270.0	$ 340.0
1	10.3	251.6	306.4
2	19.2	234.3	274.7
3	26.9	217.9	244.9
4	33.6	202.5	216.9
5	39.3	187.9	190.5
6	44.0	174.1	165.6
7	48.0	161.0	142.2
8	51.2	148.6	120.1
9	53.8	136.8	99.2
10	55.7	125.7	79.5
..	......	......	....
20	53.6	40.5	−66.6
..	......	......	....
50	−17.0	−68.5	−228.4

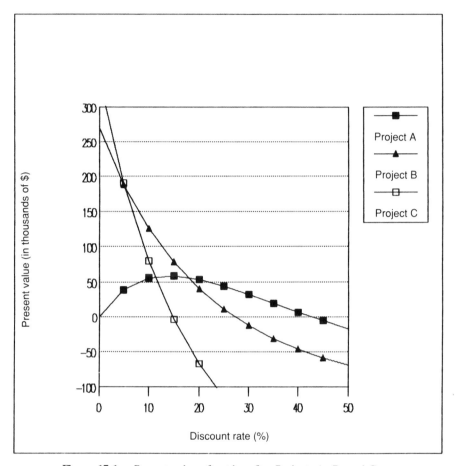

Figure 17.1 Present value of net benefits: Projects A, B, and C.

consideration. The internal rate of return is defined as the discount rate that will make the sum of the discounted present values of net benefits for the project zero. Using this definition, Figure 17.1 shows that the internal rate of return for Project A is either 0 percent or 42.7 percent. It also shows internal rates of return for Projects B and C of about 27.2 percent and 14.7 percent, respectively. Most spreadsheets or business calculators can be used to generate internal rates of return.

The problems with the internal rate of return as the criterion for choosing a project are apparent. First, for projects in which future returns may be positive or negative, the internal rate of return is not unique. Project A, for example, has an internal rate of return of either 0 percent or 42.7 percent. A zero rate of return is terrible and suggests that the project should be scrapped. On the other hand, an internal rate of return of 42 percent is clearly superior to the other two projects and most other investment opportunities. Thus this criterion gives widely differing conclusions about the desirability of Project A. Second, the result is sensitive to the rate of return selected. Note, for instance, that for Projects B and C the internal rate of return favors Project B; however, at low rates of interest (below 5 percent), Project C is more attractive.

17.3   ACCOUNTING FOR RISK—WITHIN THE FIRM

Thus far our analysis has assumed that the costs and benefits of alternative projects are known with certainty. However, managers rarely have perfect foresight about what will happen. At best, they have a somewhat subjective, probabilistic assessment of the likelihood of various outcomes. Obviously, introducing risk can change the choice among alternative projects. For instance, if the expected present values of two projects are equal, the one more certain to produce at least the expected present value is preferable. Depending upon the risk preferences of the decision maker, projects with high risk may be rejected in favor of others for which profits will almost surely be lower, but losses are unlikely.

The following distinction between risk and uncertainty is important:

Risk: An action can result in more than one outcome, depending on external conditions with known probabilities of occurrence (for instance, every weather forecaster in town agrees that there is a 50 percent chance of rain or a 50 percent chance of sun tomorrow).

Uncertainty: An action can result in more than one outcome, depending on external conditions of unknown probability (for instance, the weather forecasters disagree on the chance of rain tomorrow—one says 10 percent chance; another says 50 percent; a third says 90 percent—their average is 50 percent, but there is considerable uncertainty in this estimate).

Risk will often concern chances of injury, especially human injury. Uncertain probabilities can arise due to many causes, including: inherently random processes, measurement error, and model imperfections (limited understanding of complex processes). Exhibit 17.1 sketches one approach to deal with risk and uncertainty.

Another way to deal with risk is to use higher discount rates for riskier projects (Bierman and Hass, 1973). For example, a 20-year U.S. Treasury bond will yield about 8 percent per year in interest. The corporate bonds of a company with a high level of debt are much riskier. In fact, these securities are sometimes called "junk bonds." Issuers of junk bonds may have to pay as much as 15 percent per year or more. The additional 7 percent is a reward for the added risk of junk bonds over those of the U.S. government. A similar *risk premium* is often used to adjust project assessments. Table 17.2 shows that increasing the discount rate lowers the present value. Hence, if the discount rate of one project is increased significantly to account for its risk, then other projects may be chosen due to their more attractive risk-adjusted present values.

Although this approach to introducing risk into the analysis is simple, it has some serious drawbacks. For instance, adding a risk adjustment to the discount rate reduces present value and thus the relative attractiveness of a project. However,

■ **Exhibit 17.1 Risk as Expected Value**

Returns, or benefits, and sometimes costs as well, are not certain. One way to take this into account is to make explicit risk estimates and use these to calculate expected values.

For instance, suppose your company is considering construction of a new liquefied natural gas processing facility. Experts provide the following estimates:

- Annual chances of an accident are one in one million.
- Annual chances of sabotage are three in one million.
- The costs of either an accident or sabotage are valued at $1 billion.

These estimates translate into an expected cost of

$$\text{Expected value} = \text{Likelihood} \times \text{Magnitude}$$

$$= [(1 \times 10^{-6}) + (3 \times 10^{-6})] \times [\$1 \times 10^{9}]$$

$$= \$4,000$$

This annual cost can be combined with other costs in determining NPV, for instance. Equation 17.1 entails considerable calculation (unless one uses a computer program or spreadsheet). A shortcut form of the equation can be used when the annual costs (C) and benefits (B) are the same each year, after the initial costs (C_0) and benefits (B_0), are taken into account:

$$NPV = B_0 - C_0 + (B - C)(F)$$

where

$$F = \frac{(1 + d)^n - 1}{(d)(1 + d)^n}$$

$$d = \text{the discount rate (or interest rate)}$$

$$n = \text{the development's life in years}$$

This form assumes no salvage benefits or costs at the end of the development's life.

Assume that there are no initial benefits and that $C_0 = \$5$ million. Assume also that annual benefits are $1 million, annual costs are $400,000, the facility life is 20 years, and $d = 10$ percent.

$$F = \frac{(1 + 0.1)^{20} - 1}{(0.1)(1 + 0.1)^{20}}$$

$$= \frac{6.73 - 1}{(0.1)(6.73)}$$

$$= 8.51$$

$$NPV = -\$5,000,000 + (\$1,000,000 - \$400,000 - \$4,000)(8.51)$$

$$= -\$5,000,000 + \$5,071,960$$

$$= \$71,960$$

Thus the development's NPV is positive, but only modestly so. Does such a small return warrant such a big risk ($1 billion)? Insurance might be considered to reduce the risk at somewhat higher cost (for instance, perhaps an insurer would charge $10,000 annually instead of the expected value of $4,000), or other options of reducing risk might be explored (for instance, perhaps hiring a security force could reduce the chances of sabotage greatly).

Another inherent factor to note is uncertainty. It is likely that the risk estimate for sabotage, for instance, is far from certain. Bounds on the risk estimate (analogous to the confidence intervals discussed in Chapters 9 and 10) could be determined. NPV calculations could then be carried out under extreme scenarios with various facility life, discount rate, and risk estimates to provide management with a set of well-rounded benefit/cost/risk estimations. ■

---

the effects of this adjustment are compounded with time. For instance, the present value of $100 in one year at a 10 percent discount rate is $90.91, as compared to $87.72 for a risk-adjusted discount rate of, say, 14 percent. However, after 20 years, the present value for the higher discount rate will be less than one-half that for the lower rate. Such a compounding of risk adjustment is severe and unlikely to be appropriate for many projects. Another problem with simple discount-rate adjustment is that it implicitly assumes the independence of returns in each time period. In reality, however, early success might greatly increase the certainty of net benefits in subsequent years, while early losses could completely destroy the project's future.

One solution to these problems is to adopt a scenario strategy analogous to that discussed in Chapter 13. This approach will complicate Table 17.1, since instead of one column of net benefits for each project, there will be several. Project A might have one column that portrays terrible losses, another that depicts terrifically high benefits, and several others (including the one in Table 17.1) between these extremes. Perhaps the stream of benefits depicted in Table 17.1 is the most likely (that is, has the highest probability) and lower probabilities can be assigned to the other outcomes. Projects B and C could have similar or quite different outcome distributions. For example, sometimes there may be almost equal probabilities of great success or terrible failure and almost no chance of a middle ground. The array of possibilities depends upon the nature of the projects and the judgment of the manager who makes the decisions.

Figures 17.2 to !7.4 present the results of an analysis of alternative outcomes. The horizontal axis of each figure shows the present values of net benefits that might result; the vertical axis shows the probability of each value. Each scenario uses the same cost-of-money discount rate—that is, the return possible on a risk-free investment. For example, when comparing projects over a time horizon of six years, it might be appropriate to use the current annual interest rate paid by government securities with a six-year maturity. Or, if the firm has debts, the interest rate paid on those could be used, since avoiding these costs is an alternative to investing in a new project. The discount rate chosen is intended to introduce the appropriate figure for the time value of money.

The benefits displayed by the scenarios in Figures 17.2 to 17.4 depict both the pattern of returns and the effects of the different probabilities of occurrence. This representation should be viewed as a probability distribution. The procedures for generating such distributions can range from simply using the judgment of one

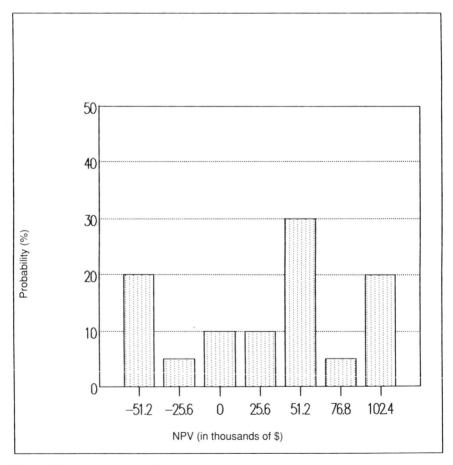

Figure 17.2 Investment analysis under uncertainty—Project A (8 percent discount rate).

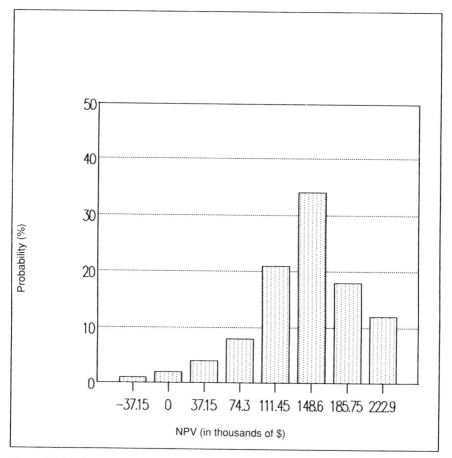

Figure 17.3 Investment analysis under uncertainty—Project B (8 percent discount rate).

analyst to applying complex computer simulations and collections of expert opinions to determine the likelihood of the various outcomes. This approach provides a much richer basis for decision making than merely using present values of the most likely outcome. It also is more realistic because no one is really certain about the future.

The choice of the best project using the scenario approach, however, is no longer so clear. Under conditions of certainty, selection of a discount rate identifies the highest present value and thus indicates a clear direction for the decision maker. With uncertainty, however, somewhat arbitrary values must be applied, and even then the results can be ambiguous. Although there may be cases in which the array of outcomes for one project clearly dominate those of others, it is more likely that the best choice will depend on the degree to which the manager is committed to risk aversion. Usually, projects with the highest potential rewards also carry high risks. Different individuals may make equally rational choices based on differing

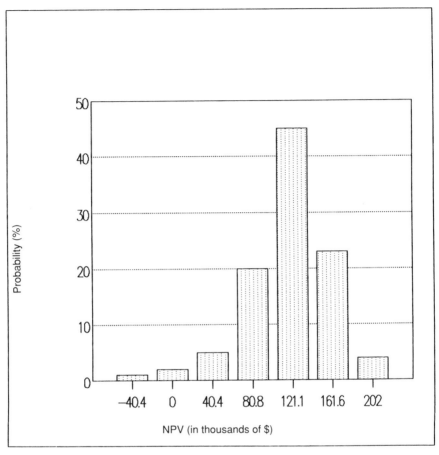

Figure 17.4   Investment analysis under uncertainty—Project C (8 percent discount rate).

preferences for risk and return. As the discussion broadens to include the social dimensions of new technology, the quest for an unambiguously good choice will become even harder. Note that even within the firm there is less than complete objectivity in decision making under uncertainty.

The firm may undertake specific risk assessments for a number of purposes (Rowe, 1989): (1) to determine the risks associated with a new development or a new market; (2) in conjunction with resource development (such as a new mine); (3) for operations analysis to ensure safety and cost-effectiveness of processes and systems; (4) as a requirement to obtain a permit (such as under the Resource Conservation and Recovery Act); and (5) to preclude or defend against legal actions (regarding product safety, for instance). Section 17.5 addresses an array of risk concerns from a societal perspective; the technology manager within a firm must be sensitive to these broader issues as well.

17.4 BENEFIT/COST ANALYSIS—SOCIETY'S STAKE AND THE MANAGER'S RESPONSE

Earlier it was noted that the decisions of managers are an important part of the appropriate allocation of society's resources. Economic systems are set up to provide signals and incentives for individuals to do things that benefit the world in which they live as well as themselves. However, no system includes all the signals and incentives required. Sometimes, this means resources are misallocated. Other times, society will impose additional restrictions on individuals through legal sanctions, regulations, taxes, or other constraints. A technology manager who wishes to commercialize a new technology must understand both its potential effects on society and the possible responses to those effects. A broad application of benefit/cost techniques along with other tools presented in this text can help the manager successfully understand this complex reality.

Since the individual or firm that develops a new technology is part of society, the benefits and costs from a decision to proceed are benefits and costs to society as well. In fact, under many circumstances, the correct private decision also is the correct social one. That is why societies decentralize such decisions rather than making them through a central authority. The private assessment of costs and benefits is not sufficient. New technologies frequently imply benefits and costs well beyond the firm. Sometimes these broader benefits are a justification for public subsidy of private developments; at other times, the broader social costs lead to regulation, control, or even to prohibition of new technology. Social benefit/cost analysis, within the broader context of societal impact assessment, provides information for such determinations.

When the word "private" is not attached, benefit/cost analysis generally means social benefit/cost analysis. The beginnings of benefit/cost analysis in the United States can be traced to water projects of the U.S. Army Corps of Engineers. According to Thompson (1980), although the River and Harbor Act of 1902 and the Flood Control Act of 1936 mandated consideration of costs and benefits, it was the Planning, Programming, Budgeting System (PPBS) work of the 1960s that brought widespread public application of the concept. Many advocate benefit/cost analysis as a politically neutral tool for analyzing expenditures or regulations; however, others view it as an illegitimate and essentially arbitrary application of quantitative analysis to questions that are largely qualitative. These disagreements cannot be resolved here. The following discussion relates social benefit/cost issues to the technology manager.

Mishan (1973) presents social benefit/cost analysis in a manner analogous to the discussion of private investment given in the preceding sections. For example, if there are three projects A, B, and C, how can decision makers decide which should be pursued? Our discussion considered revenues and outlays for a firm and showed how the decision could be based on a comparison of discounted present values. However, private decisions often produce spillover benefits and costs not considered in the firm's decision making, which must be included in social benefit/cost analysis. The difference for society is in the range of social benefits and opportunity costs that need to be considered—in essence, all impacts of the development on all parties. From a social point of view, the present value of social benefits should exceed that of social costs—that is, benefit/cost ratios should be greater than one.

The starting point for social benefit/cost analysis is *Pareto optimality*. Simply stated, this concept says that if at least someone can be made better off without making someone else worse off, then a project is unambiguously good. Even in cases where someone might be worse off, the criterion still applies if the gains of "winners" are large enough to compensate "losers" sufficiently to make them indifferent to the change. To ignore an opportunity to make Pareto improvements is to waste resources. However, two questions must be considered: Can benefits and costs be measured? If so, how?

Many simplistic approaches have been used in benefit/cost analysis. In fact, that is one reason why the concept is so controversial. Good benefit/cost analysis requires painstaking identification and measurement of increases in consumer surpluses and the rents for labor and other resource owners.

Figure 17.5 shows the familiar market demand curve of elementary economic texts. The horizontal axis represents the quantity of a product purchased (Q); the vertical axis shows the price paid (P). The demand curve shows the highest price consumers are willing to pay for various quantities of the product or, alternately, the maximum quantity they will buy at various prices. Suppose the market price of a product is P_{m1} and the quantity purchased is Q_{m1}. If the price were lower, more could be sold. If the price were higher, less than Q_{m1} would be purchased. Note, even at this price, there are consumers who would be willing to pay more. Since the price is P_{m1}, these consumers are paying less than they would be willing to for all but the very last unit. The difference between what these consumers would be willing to pay and what they have to pay is called *consumer surplus*. The value of the consumer surplus in this case is shown by the cross-hatched triangular area above the line at P_{m1}.

A new technology could increase the consumer surplus and thus produce a social benefit. For example, a new production process might lower the costs of the product and reduce the price to P_{m2}. This would raise the quantity purchased to Q_{m2}, since more consumers could afford and/or would be willing to pay for the product. The gain in the consumer surplus from introduction of the new technology is represented by the double-cross-hatched rectangle and triangle in Figure 17.5. These represent, respectively, the contribution of the lower price for those who were already buying and the consumers' surplus for new purchasers. These are gains from the new technology that are independent of the increased profits of the firms who adopt the technology in their production processes.

The benefits to labor from the new innovation also result from what they are willing to do and what the market dictates. Figure 17.6 shows a supply curve for labor. The higher the wage, the more labor will be offered to firms in this market. W_{m1} and L_{m1} represent the market wage and the amount of worker time that is used at that wage. At wages below W_{m1}, there are still workers who would be willing

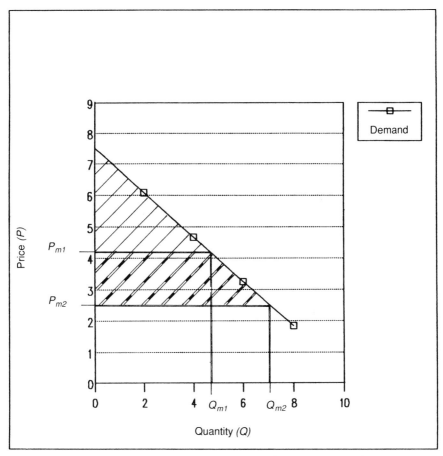

Figure 17.5 Market demand curve showing consumer surplus.

to work. Thus at W_{m1}, workers are receiving more than the minimum required. The difference between what they are paid and what they would demand for their labor is called *rent* or *producers' surplus*. Just as consumers need not always pay what they are willing to pay, producers, like labor, can get more than the minimum they would accept. The amount of the producers' surplus is represented by the cross-hatched area in Figure 17.6.

If technology increases the demand for factors like labor, then the wages in the market may move up to W_{m2}. This will increase the producers' surplus that workers receive by the amount of the double-cross-hatched area shown in Figure 17.6. A similar sort of analysis could be done for other resource owners, such as those who control land or natural resources. Sometimes suppliers of the goods and services that go into the product affected by the new technology also are examined for changes in surplus. If you add up the changes in producers' surplus for all the

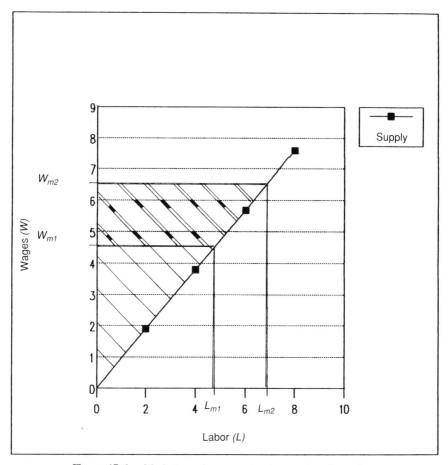

Figure 17.6   Market supply curve showing producer's surplus.

individuals and businesses involved in producing Q_{m1} and Q_{m2}, the sum gives an indication of social benefits from the new technology on the production side.

This discussion has proceeded as though the effects of new technology always are social benefits and that the surpluses produced are easily identified and quantified. Obviously, technological changes have social costs, and the costs and benefits often are hard to measure. For example, some workers may not be needed in the new production process and may find themselves unemployed. How does one calculate that social cost? If the workers can find work in another industry, the loss to them will be determined by the difference in wages. If they are unemployed, presumably their increased leisure time will have some value. Answering these questions will involve judgment about hypothetical opportunities and workers' values. Such speculations permit the analyst, intentionally or unintentionally, to introduce his or her personal values, and these will affect the outcome. Exhibit 17.2 presents important measurement considerations.

■ **Exhibit 17.2   Measurement of Social Benefits and Costs (Based on Sassone and Schaffer, 1978).**

Distribution: Data should be compiled to allow separate tabulation of the benefits and the costs for different stakeholders (city versus state or taxpayers versus indigent).

Nonmonetized Effects: Data on noneconomic effects should be tabulated in suitable quantitative units if feasible (for instance, "acres" of land to be developed or "number" of housing units displaced); if this is not feasible, qualitative measures should be used.

Benefits Gained versus Costs Saved: In many cases, categorization of effects is quite arbitrary. If an impact can reasonably be considered in either category, this should be noted so that subsequent analyses can show things both ways.

These measures facilitate multiple analyses. For instance, regional benefit/cost comparisons can indicate if certain jurisdictions are receiving proportionately more of the benefits and paying less of the costs. This information can help decision makers derive compensation schemes. Nonmonetary effects can be cumulated to augment dollar comparisons among alternative projects. If a certain project's NPV is most attractive, then careful review of the nonmonetized impacts is in order. Categorization of effects as benefits gained or costs saved particularly affects the benefit/cost ratio (increasing the numerator is quite different from decreasing the denominator). Computation of multiple economic criteria, along with use of multiple discount rates, offers a better basis for decision making.   ■

Another issue that has not been resolved is the time value of money. The previous section showed that future costs and benefits had to be discounted to their present values for effective decision making by firms. An analogous, but more complicated, argument applies to society. For an individual, the opportunity costs of forgoing returns from investing capital make it clear that future returns differ from present ones. This argument does not apply in such a straightforward way for society as a whole.

Some argue that the social discount rate should be lower than the private rate because many returns on public projects are not evaluated in monetary terms. Further, the risk of failure on any one project is low compared to the size of government resources (Porter et al., 1980). If a lower discount rate is appropriate for public projects, shouldn't that same rate be used for the public evaluation of private projects? The answer is probably no, but there is a great deal of controversy about the proper social discount rate.

The distribution of costs and benefits is another problem inherent in benefit/cost analysis. For example, the analysis does not imply that winners *must* compensate losers, only that they *could* if the benefit/cost ratio is one or greater. Obviously, if there is no mechanism for compensation, then an appropriate benefit/cost ratio may not be sufficient justification for a project that makes poor people even poorer. Further, for a new technology, the costs and benefits of which are extended into the future, it must be determined whether the well-being of future citizens should be discounted relative to the well-being of current ones. Since elections today cannot be won by tomorrow's voters, the answer in a democracy may seem to be yes. However, constitutional democracies do not always operate solely on the principal of majority rule, and future citizens have rights that should be taken into account.

Difficulties compound when technological decisions involve what economists call *externalities* (external effects) and *public goods or bads* (one person's consumption does not significantly diminish another's, and it is not feasible to exclude certain people from the good or isolate them from the bad). Technologies that damage the ozone layer of the atmosphere, for instance, harm everyone, even those who had no role in the adoption of the technology. Nor can even the most avid pacifists be excluded from the benefits of a strong military that keeps foreign enemies from taking their property or endangering their families. Techniques by which to estimate values of externalities and public goods are presented in Exhibit 17.3.

Sustainable development has emerged as a popular theme for the 1990s. The Brundtland Commission recommended that private and public development decisions take into account the full range of impacts over the life cycle of the development (*Environment*, 1989). For instance, an automaker contemplating whether to substitute plastic for metal body shells, should consider the following:

- Environmental costs associated with obtaining the materials (such as petroleum production and refining to make plastics versus strip mining for iron)
- Production impacts (such as relative job intensities, waste production, and toxic exposures to workers)
- Life cycles (such as expected life times, recycling mechanisms, and costs and hazards of final waste disposal)

Sustainable development is to be served by associating costs with all such impacts at the time of the design decision. Given that many of these costs are external to the firm, this implies considerable government intervention to get the firm to internalize them. It also implies expansion of the data requirements and scope of benefit/cost/risk calculations. The technology manager of the 1990s will need to adopt a broad, long-range perspective to function in an increasingly environmentally sensitive context.

17.5 ACCOUNTING FOR RISK IN SOCIAL DECISIONS—RISK ASSESSMENT

The discussion of social benefit/cost must consider the implications of societal risk. It has already been shown (Section 17.3) that managers must make judgments about risk and uncertainty before deciding about the development of a new technology.

■ **Exhibit 17.3   Techniques to Estimate Externalities and Public Goods (Based on Porter et al., 1980).**

Assume that a freeway extension is being constructed through a well-to-do urban area. The freeway provides a public good, use of the road, that has value in saving time, reducing commuter stress, and possibly decreasing air pollution. It provides a public bad in splintering the existing community. Externalities include the decrease in traffic on other roads as commuters use the new freeway and increased ambient noise to residents living near the proposed freeway.

Here are some techniques to estimate the value of these impacts:

1. Survey willingness to pay among the affected individuals. This usually requires disguising the purpose so people will not exaggerate their own interests (in this case, one might poll the neighborhood to find out how much they would pay, in increased property taxes, to have noise barriers built).

2. Draw analogy to private goods (perhaps, consideration of various levels of tolls for commuters would be feasible to assess value or even to convert this public good into a private one).

3. Use gaming (see Chapter 12) in which stakeholders resolve various scenarios. Again, there is danger that stakeholders will overstate their true valuations (perhaps the parties concerned with the freeway could play through alternative scenarios and compensation schemes).

4. Use public referenda (allow voters to express their preferences among a set of alternative development and pricing schemes).

Willingness to pay is difficult to assess reliably. Furthermore, it is hard to separate this from the distributional question of ability to pay. It seems even less likely that we can properly evaluate and discount all future social costs and benefits, including external ones. For this freeway example how can the interests of future commuters (in the year 2025) be identified and represented?   ■

Society also must consider the impacts of new technology. To the degree that risks are knowable, society must weigh those risks to determine if and how to control, prohibit, or promote new technologies. Business prospects for both new and old technologies often are determined by these social decisions. Moreover, the moral and legal implications of a mistake in technology can mean public disgrace and/or financial ruin for even a large company. The environmental disaster at Bhopal, the terrible results of the Dalcon Shield birth control device, and the tragedies resulting from the production and use of asbestos are just a few examples of what can happen. It is clear that knowledge of risk assessment methodology and its limitations is crucial for technology managers.

Our sense of what risk involves is colored by the media's "hazard of the week" coverage—literally, national news devotes widespread attention to 40 or 50 hazards each year (Kates et al., 1977). These hazards may be natural or technological. Examples include: nuclear power plant accidents, dam collapse, toxic waste exposure, speeding bullet trains, hang gliding, gene splicing, new drugs, medfly control programs, surgery, and the spread of AIDS.

Usually, the risk involved concerns human health. Often, that is simplified to the risk of some number of people dying, occasionally considering other aspects (such as injuries). Sometimes human mortality is linked to economics through such measures as "cost per life expectancy gain," which vary enormously for public safety programs. For instance, to gain 20 years of added life expectancy costs $50 through expanded immunization in Indonesia, $13,000 through cervical cancer screening in the United States, $440,000 through kidney dialysis, or $20,000,000 via mine safety programs (Cohen, 1980).

There are other risks as well. Assessment of environmental risks is vital to dealing with the greenhouse effect, for instance. Political risk assessment is systematically carried out on behalf of business clients who need to decide whether to invest in various countries. At the limit, risk assessment becomes synonymous with impact assessment in that any impact is likely to entail risk and uncertainty.

Risk assessment generally involves three steps: risk identification, risk estimation, and risk evaluation (Shrader-Frechette, 1985). Consider a health risk issue. Scientific domains, especially toxicology and epidemiology, identify the risk, relying heavily on biostatistical techniques. Further work in these domains may involve establishing dose-response relationships and estimating the dose received by particular populations. Risk evaluation then stretches beyond the scientific domains to determine what level of risk is acceptable. Methods such as benefit/cost/risk analysis, revealed or expressed preferences, or comparison with natural standards may be used (Shrader-Frechette, 1985).

Methods used in risk assessment range from detailed technical analyses to sweeping subjective judgments. Probabilistic risk assessment within a power plant, for instance, requires engineers to define all elements of the system subject to failure, calculate fault trees that track possibilities of failure, and model and test system performance. At the other extreme, governments and businesses try to anticipate public reaction to possible events. Reaction depends on an amplification sequence wherein information about an event is interpreted by various parties, impressions of the potential event generate further discussion, and resultant public perceptions exert real effects in the form of lost sales, increased regulation, litigation, and investor flight (Kasperson et al., 1988).

This chapter does not present risk assessment methods in depth; however, refer to Covello, Menkes, and Mumpower, (1986) and Glickman and Gough (1990) for a discussion of these methods. The following subsections outline what is involved in the key area of health risk assessment and explore three underlying issues critical to risk assessment: risk perception, implementation of decisions involving risk, and the interplay between monetary and basic values in risk evaluation.

17.5.1 A Four-Part Framework for Health Risk Assessment

Risk assessment evaluates all types of losses to people and the world they live in. However, human health risks take priority. Paustenbach and Keenan (1988) described four parts to a health risk assessment:

1. Hazard identification
2. Dose-response assessment
3. Exposure assessment
4. Risk characterization

Hazard identification seeks to determine whether a product, by-product, or process is carcinogenic (causes cancer), a developmental toxicant (causes birth defects), a reproductive toxicant (reduces the possibility of pregnancy), or whether it produces other adverse health effects. Identification often is accomplished through experimentation with laboratory animals. While this is almost certainly preferable to experimenting with people, there are reasons for caution. For example, the fact that chemicals carcinogenic to humans also cause cancer in laboratory rats does not imply that the converse is always true. An experiment may show that a substance causes malignant tumors in rats, but no human may be affected in that way by that substance. In spite of this, the Delaney Amendment to the U.S. Federal Food, Drug and Cosmetic Act says that the Food and Drug Administration cannot consider health benefits of food additives if there is any evidence that they might be carcinogenic.

Several years ago a drug called thalidomide was banned in the United States but was sold in European countries. Some women in those countries bore children with terrible birth defects caused by the drug. This focused attention on developmental toxicants that can inhibit the normal development of a fetus. Unfortunately, many common substances can affect prenatal development. For example, pregnant women are told not to drink alcoholic beverages or to take even over-the-counter drugs. The unborn are just not equipped to process chemicals. Thus managers must be very careful about substances to which pregnant women may be exposed. This applies, not only to product use, but also production activities. In fact, some jobs may not be safe for women in their childbearing years because of dangers to which they can be exposed before they realize they are pregnant.

Managers also must be conscious of effects on reproductive capability. Both men and women have been made sterile by occupational exposure. Consumer products also have produced such effects. However, a technology manager must not only be concerned about effects on humans. If a product or process affects the ability of endangered species to reproduce, then the government may take action. For example, a major complaint about the insecticide DDT was its effects on the eggs of birds.

There are, of course, many other kinds of adverse health effects. Each year, for example, new toys are marketed that can injure or even kill the children who play with them. Side effects of drugs and other chemicals alone or in combination can make people sick. The health effects of long-term exposure to video display terminals have been questioned, and various production procedures can lead to injury and/or physical or mental disabilities. Although these impacts may take years to develop, it is important for the technology manager to recognize that such problems are his or her responsibility.

Dose-response assessment examines the amount of the substance necessary to produce adverse effects. This often is expressed in terms of the risk-specific dose or RSD (the dose necessary to produce risk of one in a million, for instance). For chemical substances, the assessment is performed in laboratories using animals. Generally the doses administered are much higher than humans are likely to encounter; this way the margin of error favors protection of health. Investigators would like to define a safety *threshold* below which substances are not harmful. Then regulators can provide complete protection by banning exposures above the threshold. Unfortunately, there is not always a threshold below which there is no risk; therefore, the question of the appropriate relationship between exposure and risk arises. A conservative approach is to use a linear relationship. This implies that risk increases with exposure at the same rate at low concentrations or high ones. Some criticize this approach and suggest the weight of evidence be applied rather than statistical tests with limited data. For example, the U.S. Environmental Protection Agency (EPA) found that occupational exposure to levels of dioxin well above the RSD associated with a one-in-a-million risk actually caused no increased risk of cancer (Paustenbach and Keenan, 1988). Such contradictory evidence might be used to raise the RSD. On the other hand, Finkel (1989) notes that being conservative might be appropriate because the predicted risk levels based on statistical results might actually underestimate the probability of adverse effects.

Exposure assessment is another consideration in risk assessment. Who is likely to come in contact with a product or by-product? To answer such a question, the manager must know where the substance goes, what happens to it over time, and the characteristics of the people, plants, and animals that will be affected. Some chemicals degrade and can be assimilated by the environment. Others build up cumulative impacts through the food chain. There was a time when business could send waste products to a landfill and forget them. That time is past. Now disposal of even seemingly harmless substances must be done according to strict regulations. Moreover, the waste producer remains ultimately responsible, even if someone is hired to dispose of it in accordance with regulations. It is clear that technology managers must carefully assess the risks of exposure to products and by-products of their technologies.

The manager also must realize that regulators have tended to be conservative in estimating exposure risks. Exposure assessments, for instance, often assume that the maximally exposed individual (MEI) is located at the worst possible location. If there also is an assumption that the MEI has high susceptibility, then the implications for risk appear much higher than for the average person in the exposed area. Making several worst-case assumptions can compound the estimated risk, and a manager might feel that the assessment is unfairly weighted against technology.

Risk characterization is the most important and difficult part of the risk assessment. It uses information from the dose-response and exposure assessments to describe what can happen. In the United States, the Centers for Disease Control (CDC) plays an important part in developing information for risk assessments, although its protocol specifically forbids the CDC from quantifying risks (U.S. EPA, 1985, p. 36). The EPA and most firms would prefer some quantification so that judgments can be made about the management of risk. However, it is hard to make quantitative judgments about health without seeming insensitive, since society's values imply that human beings are individuals not numbers.

17.5.2 Perceived Risk

Another problem with quantifying the risks of technological change is that people respond to risks in very complex ways. Certain individuals seek risk recreation, such as hang gliding, skydiving, or caving (Machlis and Rosa, 1990); yet those same individuals may be unwilling to accept the much lower risks of commercial air travel. Or those who do not wear seatbelts may be totally opposed to a hazardous waste disposal facility in their neighborhood, even though the projected risk to them from the facility is much lower than not wearing a seatbelt.

In a classic paper, Chauncey Starr (1969) compared various risks and reached the following conclusions:

1. The "public is willing to accept voluntary risks roughly 1,000 times greater than 'involuntary' risks."
2. The risk of death from disease is a "psychological yardstick" for the acceptability of risk.
3. The acceptability of a risk is roughly proportional to the cube of the real or imagined benefits of the activity.

More recently, Starr and Whipple (1984) compared financial risks of the type described in an earlier section with the health and safety risks that surround technology. They noted that information on risks is often based on "fuzzy estimates." At best, the analysts can provide upper and lower bounds for risks and these may diverge widely when uncertainties are high. Even so, these estimates still involve many judgmental decisions and, hence, opportunities for bias. Exhibit 17.4 identifies a number of perceptual factors that affect risk judgments.

Recognition of the issues in risk perception is vital, but it does not give the technology manager an overall assessment of how people will react. That assessment is vital to determine an organization's decision to proceed with a development and, if proceeding, how to accommodate those concerns. Two approaches can provide hard evidence on perceived risk:

1. *Revealed Preferences*: Assessors seek analogous cases and statistical data on human responses (such as nearby property values when a similar plant was introduced into a similar community, or the accepted level of acciden-

---

■ **Exhibit 17.4   Selected Perceptual Factors Affecting Risk Evaluation**

Probability Squeeze: People tend to overestimate risk from low-probability events (for instance, death from nuclear power plant accident) and underestimate risk from high-probability events (for instance, chance of getting heart disease from smoking).

Sense of Control: Personal willingness to tolerate risk skyrockets when exposure is voluntary or controllable (as per Starr's observations).

Dread: A number of factors cluster together as the opposite of Sense of Control, including: catastrophic, uncontrollable, not equitable, high risk to future generations.

The Unknown: This composite factor reflects unobservable effects, unfamiliar risks, and delayed effects. When a technology combines Dread with The Unknown characteristics, perceived risk is greatest (Slovic, Fischoff, and Lichtenstein, 1986).

Omission over Commission: Government agencies, for example, lean against innovation as the public encourages them not to take chances with an individual's health (for instance, the U.S. FDA leans toward avoiding introduction of a harmful new drug at the expense of missing opportunities for that drug to reduce health hazards). The legal system pushes companies in this same overly conservative direction.

Economics Be Damned: Safety and health measures taken by government often bear little relationship to benefit/cost tradeoffs (recall the disparities among safety and health programs in terms of dollars per 20-year life expectancy gains). Americans do not like to confront lives for dollars choices.   ■

---

tal deaths associated with a similar technology). Human behavior reflects acceptable balances between the advantages and disadvantages of the many technologies already in use.

2. *Expressed Preferences*: Assessors survey the affected people directly to determine their perceptions. Public hearings, voting, and other means may be used to elicit public attitudes.

The first approach draws on actual behavior but requires extrapolation to the current situation. The value assumptions that underlie revealed preferences have also been questioned (Shrader-Frechette, 1985). The second approach relates directly to the current situation but relies on expressed opinions, requiring extrapolation to real behavior. Attitudinal data gathering may require the techniques introduced in Exhibit 17.3.

Other approaches may also help evaluate perceived risks (Buss, Craik, and Dake, 1986)—implied preferences (such as regulations and legal rulings applied to similar technologies), risk comparisons (such as relative risks of nuclear versus coal-fired power plants), and expert judgments. Expert judgments, however, often deviate from broadly perceived risks. Slovic, Fischoff, and Lichtenstein (1981) had various informed publics rate the riskiness of a list of activities and technologies. League of Women Voters, college students, and active club members tended to agree highly among themselves, but poorly with experts (who tend to agree well with technical estimates of annual fatalities). The glaring discrepancy in the list was for nuclear power—rated #1, #1, and #8 by the three informed lay groups, but only #20 by the experts.

17.5.3 Risky Decisions

Accurate estimates of risk are difficult, and risks can be misestimated for new technologies. A helpful general distinction has emerged between two types of hazards (Hohenemser et al., 1986):

Energy Hazards: Release energy, typically with short-duration consequences (often less than a minute) on those immediately exposed with minimal risks other than human mortality;

Materials Hazards: Release materials, usually over times of more than a week, exerting delayed consequences that act at the molecular level, affecting future generations and potentially impacting human and nonhuman systems significantly.

Hohenemser and colleagues (1986) synthesize the psychometric studies of perceived risk to arrive at a way to characterize the nature of the risk involved with various technologies (see Table 17.3). They translate this characterization into a "triage" decision strategy: give extraordinary attention to multiple extreme hazards; provide distinctive effort for each entity representing a single extreme hazard; and take ordered, routine response for the remainder. The technology manager needs to maintain such a perspective because heated concerns can arise over what may be rather mundane risks. Table 17.3 directs attention to extreme societal hazards. The energy versus materials hazard distinction can help managers make judgments about steps to take to counter the routine hazards.

TABLE 17.3 Classes of Risk

Class	Instances
1. Multiple extreme hazards	Recombinant DNA; Dam failure
2. Hazards extreme in one factor	
a. Intentional biocides	Bacterial resistance to antibiotics; water chlorination
b. Persistent teratogens	Uranium mining
c. Rare catastrophes	Commercial airplane crashes
d. Common killers	Automobile accidents
e. Diffuse global threats	Atmospheric ozone depletion
3. Hazards (extreme in no factor)	Skateboards, aspirin

Source: Hohenemser et al., 1986.

Researchers try to carefully control for all apparent variables but can never completely succeed. For example, there have been recent studies that related cancer risks to diesel engines. Schwing, Evans, and Shreck (1983) criticized the data and model uncertainties of these studies and then noted that because of their greater energy efficiency diesel-powered vehicles could be 25 percent heavier and thus significantly safer for occupants. As this example shows, there are a great many dimensions to the issue of risk, and the analyst must view the issue broadly.

Conservative risk decisions may bear an additional cost—loss of better risk information to be gained from trying a new technology. Sequential decisions particularly should take this into account (consider the example in Exhibit 17.5).

■ Exhibit 17.5 Uncertainty and Sequential Risky Decisions (Based on Zeckhauser and Viscusi, 1990).

Two alternative medical treatments are available for two patients, to be treated one at a time. Treatment A is known (no uncertainty) to cure 50 percent (considerable risk) of patients. Treatment B is still experimental (considerable uncertainty) but is judged to have equal chance of being a success (curing 90 percent of patients—implying low risk) or a failure (curing no one). This uncertainty poses a dilemma. Give a value (magnitude) of 0 to failure and 1 to success. For the first patient, Treatment B is not an attractive option:

$$\text{expected value (EV)} = \text{likelihood} \times \text{magnitude}$$

$$= (0.5)(0) + (0.5)(0.9)(1)$$

$$= 0.45$$

This is less than the expected value from Treatment A:

$$(0.5)(0) + (0.5)(1) = 0.5$$

For the set of patients (two in this simple example), however, Treatment B is best. Give Treatment B to the first patient. Half the time the treatment fails. If so, give Treatment A to the other patient for an expected total payoff of

$$[0 + 0.5(1)] = 0.5$$

The other half of the time, Treatment B is a sound treatment. This will result in a cure 90 percent of the time, to be followed by giving Treatment B to the other patient, for an expected payoff of

$$[1 + 0.9(1)] = 1.9$$

In 10 percent of the trials where Treatment B is sound, it will not result in a cure. Observing this, the second patient would be given Treatment A, for an expected payoff of

$$[0 + 0.5(1)] = 0.5$$

Overall this yields:

$$EV = (0.5)(0.5) + 0.5[(0.9)(1.9) + (0.1)(0.5)]$$

$$= 1.13$$

The average EV for each patient is $1.13/2 = 0.565$ —which is considerably better than Treatment A's 0.5. However, this splits out as 0.45 for the first patient and 0.68 for the second. In this hypothetical illustration, one would like to be the second patient treated!

This example illustrates the interplay of risk and uncertainty in decision making. Moreover, it suggests the social value of experimenting with new technologies to reduce the uncertainty as to their results but the difficulties in carrying out such experimentation. ∎

Starr and Whipple (1984) also note that beyond the "fuzziness" of study results lies the political dimension of decisions about technology. The issue may not be the benefits, risks, and costs, but rather who reaps the benefits and who bears the risks and costs. Furthermore, there is the problem of the "myth of abundance" described by Freeman and Portnoy (1989):

> Even when it knows better, the public likes to be told that its government is working to eliminate all environmentally transmitted risks. Sensing this, politicians shy away from analytical approaches based on the premise that resources are finite and priorities have to be set. (p. 3)

The perception of risk by impacted communities typically shows less variation than the scientific and technical estimates of risk. The problem of relating the best estimates of risk to the perception of risks by the impacted public has not been adequately solved. In operational decisions by the lay public, it is the perception of risk that drives decisions. The acceptability of perceived risk is the most important go–no go criterion. One of the most critical issues in risk assessment today is relating estimated risk to perception of risk and acceptability of risk. Here the issue is one of combining apples, oranges, and watermelons into a new fruit. Such a fruit has yet to emerge.

In their public communications, businesses also try to minimize the risks and emphasize the benefits of products to counter the "no risk is acceptable" attitude they perceive in society. Discussions of risks related to technology frequently become polarized, with each side relying on experts to support their view. The public has responded by taking an increasingly skeptical view of experts. Shrader-Frechette (1988) argues that decision makers often dismiss the layperson's view as being based on ignorance or irrationality. However, the public may perceive that prospective benefits are low and that both benefits and risks will be distributed unfairly. Therefore, it will not support implementing a particular technology, even when the expert view is that the benefits are high enough to justify risks. Shrader-Frechette (1988) goes on to argue that decisions about technology involve both producer risk and consumer risk and that the proper approach is to minimize the latter. This leads to a conservative posture about new technology because it suggests that it is better to err on the side of protecting the public than on providing increases in welfare.

17.5.4 Assessing Risks: Monetary and Basic Values

Another problem in dealing with the risks of technology is that balancing monetary variables against basic values often is considered unacceptable. A whole literature has emerged on the value of life (see Exhibit 17.6). Given that the major focus of risk lies on human mortality, analysts seek a dollar value to complete the benefit/cost calculations. Keeney (1986) notes that various revealed preference approaches have been tried, yet none approach a consensus agreement on how to proceed.

It is also important to examine to what extent societal risk preferences hold across cultures (see Douglas and Wildavsky, 1982). For instance, our society would find it totally unacceptable for a mother to talk about how much money she would require to expose her baby to danger. However, it is clear that we *implicitly* make such decisions all the time. Consider the interaction between societal norms and risk in the following example from Zeckhauser and Viscusi (1990):

■ Exhibit 17.6   The Value of Life

What is a human life worth? Methods to make such judgments and the resulting estimates vary widely. Kahn (1986) gathered a number of estimates based on wage/risk tradeoffs by workers in the United States and the United Kingdom.

The key to the estimations can be illustrated by this example. If a worker wants an additional $800 per year to compensate for an increased chance of dying that year of 1 in 10,000, then the value of life is calculated as $8,000,000. Overall, one person in 10,000 will die and each of the 10,000 will be paid $800 to bear this risk, for a total payment of $8,000,000.

Kahn balanced eight labor market studies, considering their leanings toward overestimating or underestimating the value of life, to arrive at a best estimate of about $8,000,000 in 1984 dollars for an American life. Three questionnaire surveys also support an estimate of about this magnitude—a value far greater than that typically used in policy analyses. Kahn concludes, therefore, that policy analysts underestimate the value of life and that too many risks are being taken.■

You are baby-sitting. The baby is sleeping, and you need to drive off on a 10-minute errand. Do you leave the baby alone in the house while you run the errand? Doubtful — that would flout a clear norm in American society. Instead, you take the baby, thereby exposing the baby to greater risks of a car crash than of any home risks. (p. 560)

Antilock braking systems on cars provide another example. Although these systems are a great safety feature, not everyone will buy them as an extra-cost option on the family car. Nevertheless, it seems wrong to *explicitly* trade off risks to human life against economic benefits. This becomes particularly apparent when a value is placed on human life to permit a quantitative approach to decision making analogous to that a firm uses for investment decisions.

Freeman and Portnoy (1989) conclude that the analysis of ventures with environmental risks should explicitly consider all risk dimensions (such as their voluntary or involuntary nature) and the distribution of the costs and benefits among elements of society. Mishan (1973) suggested that risk could be introduced into social investments analysis by using subjective probabilities and presenting arrays of outcomes (for instance, the most likely together with lower and upper bounds). This is a step in the right direction. The analysis of technology from a social point of view requires a broader perspective. Mason (1988) maintained that proper economic analysis should consider all implications of the changes that may flow from a new technology. Benefit/cost analysis needs to be integrated with impact assessment as described earlier.

17.6 CONCLUSION

The choice of projects under conditions of uncertainty is not simple; it involves judgment and values, particularly in regard to risk avoidance. Proper consideration of whether a technology should be developed must include an impact assessment that evaluates potential risks to the environment and to the public. The manager who makes decisions about technology needs to take everything into account. Therefore, the proper procedure may be to first perform the internal investment analysis. If this is positive, then there should be a social benefit/cost analysis that includes risk assessment. Even when these are completed, an impact assessment will be required to develop information about social, political, economic, and environmental effects. Anticipating all the impacts of technology and society's reactions to them may slow the decision-making process, but in the long run it is the best way to assess the most likely outcome and properly allocate the firm's and society's resources.

EXERCISES

17.1 Suppose you have a terrific idea for a new software package to be developed by the firm for which you are working. List all the various sources of opportunity costs associated with the project to develop this new software. How would you decide conceptually if the project was worthwhile?

17.2 What is the present value of $100 per year for 10 years if the relevant rate of interest (that is, the discount rate) is 10 percent?

17.3 Two new technologies are available for development in your firm. The pattern of net cash flow over the next three years is known with certainty to be as follows:

	Year 0	Year 1	Year 2	Year 3
Technology A	$-18,000	$ 9,000	$ 9,000	$ 9,000
Technology B	$-29,000	$12,000	$12,000	$20,000

Assuming there are no other factors to consider, which project should be undertaken? Is the size of discount rate important to your answer? Why did you use the value you used?

17.4 Suppose a risk element is introduced into Exercise 17.3. The best judgments of the worst case, most likely case, and best case scenarios for each technology are as follows:

	Probability	Year 0	Year 1	Year 2	Year 3
Technology A	0.3	$-20,000	$-3,000	$ 5,000	$ 7,000
	0.5	$-18,000	$ 9,000	$ 9,000	$ 9,000
	0.2	$-14,000	$ 9,500	$10,000	$11,000
Technology B	0.25	$-33,000	$ 10,000	$ 9,000	$ 8,000
	0.6	$-29,000	$ 12,000	$12,000	$20,000
	0.15	$-25,000	$ 14,000	$22,000	$32,000

Does the uncertainty change the selection process to pick one technology for development? If the discount rate is 10 percent, which one would you choose and why? Is there a unique answer for all decision makers?

17.5 The total market demand per year for a product Q is given by

$$Q = 10,000,000 - 8,000P$$

where P is the price per unit. If the price is $500, the firms in this industry will sell a total of 6 million units. How much consumer surplus would be gained if the industry adopted a new technology that lowered costs and allowed the market price to fall to $40 per unit?

17.6 The technological change in the product in Exercise 17.5 will have an effect on labor markets. Worker compensation in the industry that makes the product is $10 per hour, and the labor required per unit before the change in technology was 20 hours. Because of the change in technology, the labor required drops to 10 hours per unit. If each worker works 2,080 hours per year, how many workers will lose their jobs in this industry? Would employment increase in any other industries? How would you determine if there was a net gain or loss of producers' surplus for all workers as a group?

17.7 A supply curve of labor for an industry is given by

$$L = -10,000 + 20,000W$$

where L is number of workers and W is the hourly wage rate. Because of a technological change, the number of workers hired in this market rises from 70,000 to 90,000. What is the change in the workers' surplus (producers' surplus)?

17.8 The development of the automobile was an extremely important technological change. Give three examples of external costs of automobile production and use. Were there any external benefits over previous technology for transportation?

17.9 Suppose you work for a pharmaceutical company that believes it may have an effective drug to counteract the effects of being exposed to the AIDS virus. Unfortunately, the drug may also cause birth defects, and it will be very expensive to produce. List the steps you would take in deciding whether to develop, produce, and market the drug.

17.10 Pick a hazardous substance with which you are familiar (such as lead, asbestos, PCBs, herbicides, etc.), and show how the steps of risk assessment can be applied.

17.11 List the involuntary risks that you encounter on a day-to-day basis. Can you identify factors that determine your personal level of willingness to accept them?

17.12 You have been assigned the task of locating a hazardous waste incinerator. What steps would you take to make this project as advantageous as possible for your company?

17.13 Chapter 13 described the use of scenarios to handle multidimensional issues. Show how scenarios might be applied to introduce the results of societal risk assessment into the internal decision to develop a specific technology within a private firm.

____18

EVALUATION OF TECHNOLOGIES AND THEIR IMPACTS

OVERVIEW

This chapter deals with the process of valuing the impacts of technologies and assigning a relative value to technological alternatives. It discusses the criteria for evaluation (including those relating to the selection of alternatives) and the measures used to operationalize the criteria. Multiple objective approaches to selecting nondominated alternatives and clarifying the value preferences of decision makers are introduced. The potential roles of participation and mediation in impact evaluation are discussed as well.

18.1 INTRODUCTION

As developed so far in this book, the impacts of a single technological development, or the comparative analysis of a number of alternative technologies, must be evaluated. This evaluation supports making "go–no go" decisions, adoption decisions, or taking other actions to deal with the technologies.

This chapter confronts the issue of evaluation prospectively. Information that relates to the future is evaluated in the present; hence it is uncertain. In that respect prospective evaluation differs from evaluation of past or ongoing projects and programs that have generated data about their performance.

Evaluation is the process of assigning value; it requires criteria and measures for them. *Criteria* reflect the values held by the "e-valu-ators," or the parties whose judgment they are trying to address. *Measures* reflect the degree to which criteria are met.

There are at least four basic questions to be asked in any evaluation:

1. What is to be evaluated?
2. Who is to be involved in the evaluation and what roles do they play?
3. What criteria are to be used in the evaluation and how are they weighted?
4. How are the criteria to be measured?

The answer to the first of these questions is the technologies whose impacts are being evaluated. (A different answer would focus evaluation on the forecast or assessment study per se. The interrogation questions of Exhibit 4.2 provide a good framework to evaluate a study.) The answer to the second question depends on the organizational/institutional environment within which the evaluation is being conducted. Sometimes, there is a single decision maker; more often, decision-making authority is diffused among multiple parties; quite frequently, stakeholders will hold distinctly different values.

It is important to clarify the values of the parties at interest. Kenneth Hammond and his colleagues point out the advantages in separating differences due to "cognitive conflict" (misunderstanding of a situation) from those due to "motivational conflict" (incompatible criteria). The technique of *policy capture* (Hammond and Adelman, 1976) provides one way to identify and characterize the values and criteria of various stakeholders (see Exhibit 18.1). Clarification of values, however achieved, may help establish evaluation criteria. It also may help each side understand the other, possibly leading toward win/win results in which stakeholders work out mutually acceptable tradeoffs.

■ Exhibit 18.1 Policy Capture

The county landfill is almost full. Siting a new landfill, incineration, and a recycling proposal each generate heated debate. Suppose the two dominant considerations are costs (C) and environmental protection (E). A number of specific scenarios are devised that implicitly cover the full range of possible levels for C and for E. Stakeholders are asked to participate in a policy capture exercise in which they give preference scores for each scenario (on a 1 to 100 scale). This process yields the following data on each stakeholder's preference regarding each scenario:

	Levels		Stakeholder A	Stakeholder B
Scenario	C	E	Preference	Preference
1	30	51	99	10
2	94	72	5	40
3	78	87	40	90
4	60	75	35	60
5	12	23	70	20

(This could be extended over additional scenarios, stakeholders, or considerations.)

A multiple regression program is then used to calculate the weightings that the stakeholders have implicitly given C and E (by statistically associating the C and E values with the preference values over the set of scenarios). (Chapter 9 discusses how to perform regression calculations.) This information can capture the extent to which each party values C and E, for instance:

$$\text{Stakeholder A's preference function} = 77.5 - 1.5C + 0.9E$$

$$\text{Stakeholder B's preference function} = -23.7 - 0.2C + 1.3E$$

In other words, Stakeholder A prefers low-cost alternatives (the negative coefficient, -1.5, indicates that A downgrades alternatives with high cost). Stakeholder A secondarily factors in high environmental protection; Stakeholder B emphasizes environmental protection and only slightly considers low cost.

Two-dimensional plotting of the scenarios against the C and E axes can further clarify choices. For instance, some of the options may dominate others—Scenario 3 offers better E at lower C than Scenario 2. This could simplify choices by showing that the only reason for favoring a dominated choice would be personal interests, especially "NIMBY"—not in my backyard!

There are pitfalls in applying policy capture, including failure to include all pertinent factors, sensitivity to presentation, time demanded of participants, representative sampling concerns, and nonlinearities (Mitchell et al., 1975; Crews and Johnson, 1975). ∎

18.2 CRITERIA

Values underlie the criteria for technology evaluation. The policy capture example (Exhibit 18.1) trades off cost with environmental protection. More generally, Gastil (1977) suggests four essential values that may come to bear on a situation (see Figure 18.1):

1. *Utility*—The greatest net social good
2. *Equity*—The evenness with which those social goods are distributed. For example, comparison of cost/benefit ratios may show Technology A superior to Technology B; however, distributional justice demands to know who gets the benefits and who pays the costs. Technology B may be preferred if it distributes fewer goods more equitably.
3. *Transcendence*—Nonmaterial (spiritual) values that people hold dear. For instance, the ancient Greeks devoted much of their available resources to learning, architecture, and the arts; medieval Europe devoted resources to monasteries; and modern America devoted resources to space exploration. These higher (transcendent) human attainments come at the expense of utility—the man on the street might be materially better off if the United States had never

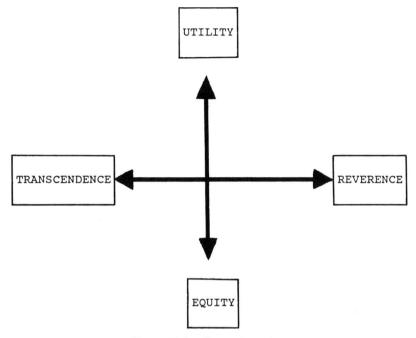

Figure 18.1 Competing values.

invested in the space program. However, that "one great leap for mankind" in going to the moon has enormous transcendent value.

4. *Reverence* — Another nonmaterial value maintains the sacredness of certain considerations. Reverence may oppose eliminating an endangered species, thereby denying the material utility in exploiting a natural forest. Similarly, reverence may lower the utility of a straight highway in favor of respecting an Indian burial ground.

The approaches to follow do not assume that different stakeholders' values or criteria can always be traded off commensurably. All considerations should not be forced to use a single measure, say dollars. More basically, Kenneth Arrow (1963) has demonstrated the logical impossibility of a shared (social) utility function (expressing the weighting of the criteria for deciding) that incorporates the very different preferences of all individuals concerned. Thus in most cases it will be necessary to deal with both multiple deciders and multiple sets of decision criteria.

Selection of criteria critically influences any evaluation, since any answer depends upon the question. A few evaluations hinge upon a single criterion. Choice of technology for a low-cost component of a complex military system might be based unidimensionally on relative reliability, for instance. More frequently, there are multiple criteria. These may relate to a common objective. A company designing a new computer faces tough choices as it strives to keep costs down and performance up. Many technology evaluations may entail multiple criteria that reach beyond economic utilitarian formulas to address equity and nonmaterial criteria. In such cases one should aim for a multidimensional portrayal of how well the various criteria are met; formal evaluation methods may be supplemented or supplanted by participative processes that play a major role in decision making.

In technical cases in which nonmaterial values are not involved, *requirements analysis* may serve as the vehicle for criteria selection. Requirements serve as criteria, although not necessarily as the complete set of criteria. For instance, an Air Force project seeks to design a "pilot's associate" to help fighter pilots fly better. A logical starting point is to ask pilots what they do now, what activities are most troublesome, and what they might like from such an associate. Such inputs must be integrated with an initial estimate of what could be developed with the resources and time available. A good follow-up would be to mock up one or more prototypes to get pilots' reactions to something more concrete. Finally, the requirements analysis should provide explicit performance criteria for the technology. Additional criteria (such as low cost) will round out the evaluation profile for technology choices to be made.

Impact assessment often generates a large set of potential criteria for evaluation. Impact evaluation may need to focus on a *reduced set*, those weighted most heavily, as key decision criteria. Multiple evaluation criteria, often differing for different stakeholders, are likely to remain, even after attempts to reduce the set. The analytic hierarchy process (see Section 18.5) may prove helpful in clarifying the relative importance of different criteria. Multiple objective methods (see Section 18.6) provide another tool to deal with complex sets of criteria.

18.3 ALTERNATIVES

There are some instances when the alternatives to be evaluated are not clearly specified. Conflicting interests may disagree over the alternatives, as well as the criteria for evaluation. Evaluations involving multiple parties are inherently delicate. Determining the set of alternatives to be considered is often a political decision. The following ground rules can help avoid too large a set of alternatives that unduly increase the evaluation workload and could confuse the issue:

- Exclude clearly inferior alternatives
- Eliminate alternatives that are technically or economically infeasible
- Establish certain a priori minimal standards for alternatives to be considered
- Seek to configure alternatives to be comparable in scale
- Try to have roughly comparable levels of information available on all alternatives.

Peter Nijkamp (1986) distinguishes several relevant characteristics of alternatives that affect the course of evaluation:

- Discrete versus continuous alternatives (determining what price to set on a new technology would be an example of continuous alternatives)
- Concurrent versus sequential alternatives (deferred choices pose different alternative considerations—time value of payoffs, risks involved, etc.)
- Mutually exclusive versus nonexclusive alternatives (nonexclusive choices complicate evaluation)
- Static versus dynamic alternatives (evaluations under dynamic conditions where alternatives evolve over time raise considerations about process and criteria)

Lawrence Susskind (1983) points out further complexities: How far into the process do you go before additional options are added? How diverse a set of options do you consider in one evaluation process? How do you package hybrid options?

18.4 MEASURES

Impact evaluation requires explicit attention to the types of measures to be used and how these are to be manipulated.

18.4.1 Types and Levels of Measures

Types of measures vary on a few critical dimensions. Units with which variables are measured also often differ. In the simplest case, all the criteria are measured in the same units (say, dollars). However, different units are required for different variables (such as dollars; quads of energy).

There are four levels of measurement. *Nominal measures* occur when a name or label is attached (for example, the numbers football players wear). *Ordinal measures* relate to a ranking without clearly defined intervals between the entities being ranked (for example, chocolate ice cream tastes better than coffee ice cream to a certain individual). In *interval measures*, the intervals between the entities being ranked are clearly defined. *Ratio measures* are interval measures with a defined zero point.

Interval and ratio measures are the most attractive. However, they are not always available, especially for subjective measurements. Ordinal measures are often the best available. These can be handled in several ways.

Two important approaches to the creation of ordinal scales are rating and ranking. *Ratings* compare against some standard(s), whereas *rankings* give relative indications among a set of alternatives. McConnell and Khalil (1988) presented a sample technology rating scale:

5—Excellent technology for this attribute

3—Average

1—Poor

0—Technology does not possess this attribute

Mason (1986) used another scale in matching community strengths with corporate needs:

+5—Abundance of the feature

0—Moderate availability

−5—Poor availability

(Specification of intermediate value meanings may enhance inter-rater comparability.)

Scales range in precision from binomial (0 or 1; yes or no) to as fine a gradation as desired (for example, from 1 to 1,000). Scales with an odd number of values (5-point scales) allow raters to opt for a neutral or middle position. An even number of values forces raters to express a leaning.

An interesting rating scale results from creating an interval scaling that may be based on subjective, component judgments. The futures forgone (FF) index (Freeman, Frey, and Quint, 1982) was used to compare alternatives for 106 discrete land regimes in the United States (potential natural vegetation communities—PNCs), for each of 10 activity categories (e.g., wood harvest, tree life forms). For each PNC, the FF index was calculated as:

$$FF = \frac{\text{Base year total} - \text{Projected year total}}{\text{Projected year total}} \tag{18.1}$$

Projected year totals could derive either from quantitative trend projections or from subjective expert estimates.

Ranking also can be done with various scales. Very simply, rankers may be asked to judge one alternative higher or lower than another (with or without an option to say "the same"). Refinements can take many forms. Sharif and Sundararajan (1984) use the more precise analytic hierarchy process scaling to compare technological alternatives. This model is discussed in Section 18.5 and supported in the TOOLKIT.

Sometimes it is important to measure stochastic (probabilistic) information separately. The Futures Group (1975) devised an impact likelihood versus desirability matrix (see Figure 18.2). This two-dimensional array of information allows participants in the evaluation to sort out reasons for their relative enthusiasm for an alternative. For instance, number 13's lack of enthusiasm for the alternative mapped is based on preference, while number 11's misgivings relate to likelihood.

It is often helpful to normalize measures across criteria. Yet care must be taken to note that weightings, if any, need to be applied and that treating ordinal measures as if they could be added and multiplied meaningfully is subject to the caveat that the resulting numbers are meant to show no more than a relative valuation. If all measures are dimensionless and on similar scales, nothing needs to be done.

	Very desirable	Desirable	Neutral	Undesirable	Very undesirable	No opinion
Almost certain				3		
Very likely		7 8	16	9	13	
As probable as not			12	10	4	
Very unlikely			1 2 12			
Almost impossible						6
No opinion						

Participant codes

1. Private industry
2. Private industry
3. Government — administrative
4. Environment and research
5. Public utility
6. Government — administrative
7. Government — legislative
8. Environment and research
9. Public utility
10. Government — legislative
11. Public utility
12. Private industry
13. Environment and research
14. Private industry
15. Private industry
16. Government — administrative

Figure 18.2 Impact likelihood versus desirability matrix. (*Source:* The Futures Group, 1975, p. 369.)

However, if measures differ, they should be made comparable. Standardization may prove suitable:

$$\text{Standard score} = \frac{\text{Raw score} - \text{Sample mean}}{\text{Sample standard deviation}} \qquad (18.2)$$

Another useful technique is to sum the scores, say, for all the alternatives on a given criterion and then divide each score by this sum. This results in decimal values that sum to 1. Converting these to percentages by multiplying by 100 may yield an informative measure of relative performance.

Nominal measures can sometimes be useful, but they should not be subjected to statistical manipulations, such as computing correlations. In some cases nominal variables may be converted to sets of binomial variables for which certain statistics are justifiable. In the case of ordinal measures, numerical manipulations may yield relative and indicative results without giving absolute information. Appropriate statistical manipulations include computation of rank order correlations and nonparametric inference tests.

18.4.2 Measurement Inputs and Combinations

Measures must be associated with each criterion to determine the extent to which the criterion is met. Combining these measures requires knowing their levels and types as discussed in the previous section.

When it is necessary to rank alternatives using human judgment, *pairwise comparison* is a useful approach to obtain valid judgments. A given alternative is compared with one other (using whatever ranking scale), then with a second alternative, and so forth. This simplifies the judgments required but is very demanding that is, $(n - 1)!$ judgments for a set of n alternatives, for each criterion considered. A matrix of pairwise comparisons can be constructed and consolidated to an ordering of the factors (Sharif and Sundararajan, 1984).

Interpretive structural modeling (ISM) is an approach to simplify the generation of pairwise comparisons and convey the results graphically. ISM computer programs can facilitate judgments by assuming transitive relationships (if you prefer A to B and you prefer B to C, the program will assume you prefer A to C, and save you making that judgment). Relationships can also be portrayed using directed graphs produced by the program (see Watson, 1978).

Nijkamp (1986) points out that evaluation is not primarily aimed at identifying the optimal solution. Rather, the purpose is to rationalize the decision process by presenting and making explicit information on criteria, alternatives, interest conflicts, and so forth. For instance, rank ordering the alternatives separately for each of the criteria will help people perceive the tradeoffs involved. Dominance among alternatives across criteria can be noted.

Nevertheless, it is often of interest to consolidate measures across criteria to compare alternatives. Suppose a design team has narrowed a choice to three alternative technologies (A, B, and C). Suppose that the selection criteria boil down to three (D - dollars; F - performance on the target function; and R - reliability). Imagine that Table 18.1 reflects the design team's consensus as to: (1) the relative *weights* that should be assigned to each criterion and (2) the measure of how well each technology fulfills each criterion.

The calculations with Table 18.1 give a linear additive weighting model. Virtually identical calculations masquerade under labels such as weighted scoring model, decision matrix, relevance trees, and attribute trees. Simpler ways to combine criteria and alternatives are possible (such as equally weighting all criteria or binary scoring in which an alternative does or does not meet minimal requirements for

TABLE 18.1   Weighted Decision Matrix

Weight[a]	Criteria	Alternatives		
		A	B	C
0.3	D (Dollars)	0.2	0.5	0.3
0.6	F (Functionality)	0.6	0.2	0.2
0.1	R (Reliability)	0.1	0.4	0.5

Linear additive calculations yield total scores for each alternative of:

A $= 0.3(0.2) + 0.6(0.6) = 0.1(0.1) = 0.43$

B $= 0.3(0.5) + 0.6(0.2) + 0.1(0.4) = 0.31$

C $= 0.3(0.3) + 0.6(0.2) + 0.1(0.5) = \underline{0.26}$

$$1.00$$

[a] Relative weights assigned to the criteria should sum to 1.0. In this example, the performance of all the alternatives on each criterion sum to 1.0; this is one way—not the only way—to ensure that criteria are not subtly weighted by differing scoring patterns on each.

each criterion)—but these seem to use the available information less fully at no great computational savings.

The linear additive weighting model facilitates *sensitivity analysis* (see Appendix C). For instance, a stakeholder could check the data in Table 18.1 to see that changing the weights for reliability to 0.3 and functionality to 0.4 results in Alternative B being favored.

18.5 THE ANALYTIC HIERARCHY PROCESS (AHP)

AHP was created by Thomas Saaty to structure complex judgments (Saaty, 1980; Saaty and Kearns, 1985). It does this through four basic stages:

1. Systematizing the judgments into a hierarchy or tree
2. Performing elemental, pairwise comparisons
3. Synthesizing those pairwise judgments to arrive at overall judgments
4. Checking that the judgments combined are reasonably consistent with each other

The AHP process is hierarchical. As an illustration, assume that your objective, the highest level of the decision tree, is to get a good job. Suppose you break this objective down into three criteria, constituting a second level of the tree. Assume these criteria are salary, location, and opportunity to advance. The importance of these three criteria relative to each other can be determined using the AHP procedure. Furthermore, suppose that five alternative jobs, each in a different city, are being considered. These can be compared for each of the three criteria using the AHP procedure at this level (yielding local priorities for the set of elements on the second level immediately above). AHP's hierarchy is structured from the top down, much like relevance trees. It also relates closely to the mission statements and support graphs introduced in Chapter 3.

Once the decision hierarchy is specified, you can turn to the judgments to be made. People can judge between two items more easily than they can make composite judgments of multiple items all at once. Therefore, AHP uses pairwise comparison among each relevant pair of items as the basic judgments. Other techniques, such as interpretive structural modeling (ISM), also use pairwise comparisons. However, in contrast to ISM's dichotomous judgment, AHP employs a nine-point scale (Table 18.2). Saaty (1980) documents the superiority of this scale over alternatives.

Consider a sample judgment as to the relative importance of two of the good job criteria of the example just introduced. How much more important is salary than location? Referring to Table 18.2, suppose you feel that salary is more important, meriting a ranking of 4. AHP will fill in the complementary judgment of location compared to salary with the reciprocal value—1/4. Concerning the three criteria of salary, location, and opportunity, you will be required to make two more judgments

TABLE 18.2   Relative Importance Scale

Importance[a]	Definition[b]	Explanation
1	Equal importance	Alternatives contribute identically to the objective
3	Weak dominance	Experience and judgment slightly favor one alternative over the other
5	Strong dominance	Experience and judgment strongly favor one alternative over the other
7	Demonstrated dominance	One alternative's dominance over the other is demonstrated in practice
9	Absolute dominance	Evidence favoring one alternative over the other is affirmed to the highest possible order

Source: Based on Saaty, 1980.

[a] The intermediate values of 2, 4, 6, 8 reflect compromise between two judgments.

[b] If element A is judged relatively dominant (more important, preferred, more likely, etc.) to element B (rated a 7), the judgment of B relative to A will take the reciprocal value (in this case, 1/7)

(salary versus opportunity and location versus opportunity). AHP will fill in the complements. The result will be a 3-by-3 matrix (with 1's on the diagonal—that is, salary is equally important to salary). The same items appear as the rows and as the columns of the matrix—salary, location, and opportunity.

The next AHP stage is to synthesize the judgments within a given matrix (for local priorities) and then across matrices (global priorities). The idea is quite simple—to collapse the set of separate judgments into a properly weighted overall judgment. Calculation involves matrix mathematics but need not be a direct concern.[1] The TOOLKIT, or more elaborate programs such as EXPERT CHOICE,[2] provide these weighted priorities for each matrix.

[1] Precise solution requires calculation of eigenvectors for each matrix, which are then normalized to unity. However, this can be approximated. Conceptually, an easy approximation is to normalize the elements in each column (add them up and divide by their sum) and then average each row of these normalized elements to estimate the priority vector. The TOOLKIT uses another approximation. It calculates the geometric mean of the elements in each row, normalizes these, and then averages each row to estimate the priority vector. If all judgments are consistent, these estimates will be precise (Saaty, 1982); if not, the judgments are subjective and mathematical precision will not make these exact anyway.

[2] This attractive AHP program is available from Expert Choice, Inc., 4922 Ellsworth Ave., Pittsburgh, PA 15213. They can also supply paperback copies of Saaty's books at a reasonable price.

The fourth AHP stage is to check the consistency of the judgments in each matrix. Collections of pairwise judgments are apt to show inconsistencies. These may reflect crude scaling (such as A seems a little better than B; B seems a little better than C; C seems a lot better than A—the imprecision of the "little better" designation leading to considerable uncertainty). Or raters may just be flagrantly inconsistent (for instance, preferring A to B, B to C, and C to A). AHP provides a helpful indicator to signal the degree of inconsistency in a matrix of judgments. This requires extension of the synthesis calculations.

Calculation of the degree of inconsistency again requires matrix manipulations.[3] These yield the Consistency Ratio:

$$\text{Consistency Ratio} = \frac{\text{Consistency Index}}{\text{Random consistency number}}$$

(See Footnote 3 for calculation of the Consistency Index.) The random consistency number indicates an expected value if judgments were taken at random over the scale from 1/9 to 9. The random consistency number varies as a function of the size of the element set:

Matrix size	2	3	4	5	6	7	8	9	10
Random consistency #	0	0.58	0.90	1.12	1.24	1.32	1.41	1.45	1.49

Saaty suggests that the Consistency Ratio should be 10 percent or less; sometimes up to 20 percent may be tolerated.

Exhibit 18.2 discusses practical steps in conducting an AHP analysis. These integrate the four stages into a typical sequence of activities. Exhibit 18.3 illustrates the calculations involved using a basic example implemented in the TOOLKIT.

Saaty and Kearns (1985) document a range of AHP applications that illustrate the following:

- Inclusion of interdependencies among criteria and how these alter priorities in comparison to assuming independence of the criteria
- Formulation and comparison of alternative scenarios, using an example of seven scenarios for higher education in the United States through 2000, analyzed over four primary factors (economic, technological, etc.), six actors (faculty, government, etc.), and various actor objectives (four faculty objectives, six governmental objectives, etc.)

[3]Formally, solves the following matrix equation:

$$A'w' = \lambda_{\max} w'$$

where $A' = (a_{ij})$. $\lambda_{\max}$ is the largest eigenvalue of the judgment matrix A.

To approximate this calculation by hand, sum the first column of the judgment matrix and multiply this by the first component of the normalized priority vector; sum the second column and multiply by the second component; and so on. Then add the results to get $\lambda_{\max}$.

The consistency index is calculated as

$$(\lambda_{\max} - n)/(n - 1)$$

where n is the number of elements being compared.

■ **Exhibit 18.2   Steps in the Analytical Hierarchy Process (Based on Saaty and Kearns, 1985).**

1. Define the problem and what you want to know. Uncover assumptions and preconceptions reflected in the problem definition; revise the problem definition if these are not viable. Identify affected parties; check how they define the problem. Consider ways for them to participate in the AHP.

2. Structure the hierarchy from the top—that is, from the overall objective to the intermediate level(s) factors or criteria to the lowest level (usually the alternatives under consideration). Check that levels are internally consistent and complete and that relationships between levels are clear.

3. Construct one pairwise comparison matrix covering the set of elements in the lowest level for each element in the level immediately above. In complete simple hierarchies, every element in the lower level affects every element in the higher level. In other hierarchies, lower-level elements affect only certain upper-level elements, requiring construction of unique matrices.

4. Make the judgments to fill in the matrices—$n(n - 1)/2$ judgments per each $n \times n$ matrix. The analyst (or the group participating) judges whether element A dominates element B—if so, inserting the suitable whole number (see Table 18.2) in the cell at row A, column B—or, if B dominates A, inserting the whole number in row B, column A. The reciprocal is automatically inserted in the counterpart cell.

5. Calculate the Consistency Ratio for each matrix. If unsatisfactory, redo the judgments. Repeat steps 3 through 5 for all levels of the hierarchy.

6. Analyze the matrices (preferably using a computer program such as the TOOLKIT or EXPERT CHOICE) to establish local and global priorities. Check the hierarchy's consistency by multiplying each Consistency Index (see Footnote 3) by the priority of the corresponding criterion and adding them up; then compute a consistency ratio. If this is too high, redo the judgments (for instance, rephrase questions or recategorize elements). Saaty recommends that each set include no more than seven elements; larger sets can be broken down into multiple groups, repeating one element in each to use as an anchor. ■

■ **Exhibit 18.3   AHP Calculations**

Consider the following hierarchy:

Objective:	Good Job
Criteria:	Salary
	Location
	Opportunity to Advance
Alternatives:	Atlanta
	Boston
	Denver
	New York
	San Francisco

Start at the lowest level. Three matrices are required. The first will compare the five job sites on salary; the second will compare the five sites as to location; and the third will compare on advancement opportunity. Suppose the salary matrix were as follows (the TOOLKIT provides example matrices such as this):

		Salary Preferences Judgment (Row over Column Item)					Normalized Priority
No.	Name	1	2	3	4	5	
1	Atlanta	1.0	0.5	3.0	5.0	0.2	0.165
2	Boston	2.0	1.0	3.0	3.0	0.333	0.241
3	Denver	0.333	0.333	1.0	3.0	0.333	0.098
4	New York	0.2	0.2	0.333	1.0	0.2	0.046
5	San Francisco	5.0	3.0	3.0	5.0	1.0	0.449

Note that the judgments are reciprocal. For instance, San Francisco is preferred to Boston (3.0); Boston judged against San Francisco shows 0.333 (1/3).

To synthesize these judgments, the TOOLKIT uses a geometric mean approximation. For example, first multiply each value in the first row:

$$1 \times 0.5 \times 3 \times 5 \times 0.2 = 1.5$$

Next take the fifth root of 1.5, which is 1.084. Follow a similar procedure for each row. Add up the resulting values $(1.084 + 1.585 + 0.644 + 0.305 + 2.954 = 6.572)$. Divide each row's geometric mean by the sum (for example, for Row 1, $1.084/6.572 = 0.165$) to obtain the components of the normalized priority vector (listed in the last column of the table).

Consistency can be estimated as follows (the TOOLKIT uses a more exacting matrix calculation). First, sum the entries in the first column (8.533) and multiply this by the first component of the normalized priority vector (0.165). Repeat this process for the second column (5.033×0.241) and the other three columns

$(10.333 \times 0.098, 19 \times 0.046,$ and $2.066 \times 0.449)$. Add these five values to estimate $\lambda_{max} = 5.420$. Then compute the Consistency Index (CI):

$$\text{Consistency Index} = \frac{\lambda_{max} - n}{n - 1} = \frac{5.42 - 5}{5 - 1} = 0.105$$

Finally, estimate the Consistency Ratio (CR):

$$\text{Consistency Ratio} = \frac{\text{Consistency Index}}{\text{Random consistency number for } n = 5} = \frac{0.105}{1.12} = 0.094$$

This is smaller than the 0.10 guideline, so the AHP analysis can proceed in comfort.

Next complete judgments and calculations for analogous matrices comparing the five alternatives with respect to location and then with respect to opportunity.

Attention would then move up a level in the hierarchy. Another matrix would be completed to determine the relative importance of salary, location, and opportunity.

As a simple illustration, imagine that the five alternatives were judged as follows on the location criterion (Atlanta = 0.3; Boston = 0.1; Denver = 0.3; New York = 0.1; and San Francisco = 0.2) and were judged exactly equal on the opportunity criterion. Imagine further that the weights determined for salary, location, and opportunity were 0.5, 0.2, and 0.3, respectively. Global priorities would then be calculated by multiplying the local priority by the criteria weights, as follows:

	Salary		Location		Opportunity		
Atlanta	0.165 (0.5)	+	0.3 (0.2)	+	0.2 (0.3)	=	0.2025
Boston	0.241 (0.5)	+	0.1 (0.2)	+	0.2 (0.3)	=	0.2005
Denver	0.098 (0.5)	+	0.3 (0.2)	+	0.2 (0.3)	=	0.1675
New York	0.046 (0.5)	+	0.1 (0.2)	+	0.2 (0.3)	=	0.103
San Fran	0.449 (0.5)	+	0.2 (0.2)	+	0.2 (0.3)	=	0.3245

These results give an overall priority to the choices. The localized priorities are also important. They allow the decision maker or the group to see what components are most heavily weighted and even to track back to specific pairwise comparisons. This facilitates sensitivity analysis in which a particular judgment could be changed or a criteria weight could be modified to see how much difference it makes. ∎

- Forward planning, backward planning, and integrated forward/backward planning
- Considering risk by posing judgments for different time periods to yield criteria that vary over time
- Treating cost-benefit analysis by computing separate cost and benefit matrices to compare three ways to cross a river (existing ferry, tunnel, bridge); the benefit matrices incorporate three criteria (economic, social, and environmental benefits), each, in turn, subdivided (social benefits include safety, communications, community pride, for instance)

AHP provides a robust tool to gather and analyze human preferences. It can help the technology manager to prioritize R&D projects; to work through differences in perspective among design, manufacturing, and marketing within a development project; or to evaluate the effects of one or more new technologies.

18.6 MULTIPLE OBJECTIVE METHODS[4]

Multiple decision makers, applying multiple criteria, often must select from among multiple alternatives. Decision tools have been developed that help evaluate alternatives and assist in the selection of a preferred outcome in such situations. These tools have helped with problems in technology selection, health services delivery, production scheduling, inventory control, and civilian and military procurement.

Perhaps the primary distinguishing feature of contemporary work on decision aiding is the recognition that the solutions to complex problems must explicitly embrace a range of competing concerns. Such concerns give rise to multiple, conflicting, and noncommensurate criteria against which alternatives must be evaluated before eventual selection. For example, consider the evaluation of different designs for an automobile (White et al., 1986; White and White, 1988). This problem is faced by consumers before making a purchase, by manufacturers before deciding which models to produce, and by government regulatory agencies before issuing standards.

All the different automobile designs under consideration comprise the set of alternatives to be evaluated. Each design is characterized with respect to a range of different attributes, such as size, weight, structural configuration, fuel economy, styling, performance, and various safety features. These attributes determine the value or score of the car with respect to any set of criteria.

Among the many criteria for a good design, we will focus here on just two— safety and cost. These criteria are conflicting, because improvements in safety generally lead to designs that are more expensive to build and operate. They also are noncommensurate, because there exists no universally acceptable transformation between safety (as measured in terms of prevented injuries and fatalities) and cost (as measured in terms of dollars).

The recognition that decisions must balance competing goods has given rise to a body of theory and practice called *multiple objective decision analysis* (Chankong and Haimes, 1983; French et al., 1983; Goicoechea, Hansen, and Dudestein, 1982; Hansen, 1983; Hwang and Masud, 1979; Keeney and Raiffa, 1976). Two related problems are addressed. These are problems arising in *multiple objective optimization theory* (MOOT) and problems arising in *multiattribute utility theory* (MAUT). MOOT seeks to identify nondominated solutions to the problem of determining preferred alternatives. MAUT concerns the formal representation of the preference structure of the decision maker.

---

[4]This section was prepared by Preston White to whom the authors wish to express their sincere appreciation.

As an example, consider evaluation of the six alternative automobile designs listed in Table 18.3 on the basis of safety and cost alone. The safety value of each design is measured on a scale of 1 to 10, with 1 representing the safest. The annualized cost of each design, including average annual operating and maintenance costs and amortization of purchase price, is measured in dollars. Our objective is to determine the best automobile—that is, the car with the best combination of safety and cost.

The safety values and costs of the six designs are cross-plotted in the solution space in Figure 18.3. Designs closest to the origin have lesser costs and greater safety. In the solution space note that the Battlebus is preferred to the Fashionable in terms of greater safety, but that the Fashionable is preferred to the Battlebus in terms of lower cost. Thus it is impossible to choose between Battlebus and Fashionable at this stage in the analysis, without stating a preference for the tradeoff between safety and cost.

In contrast, the Commuter is preferred to the Fashionable in terms of both the objectives. If only one alternative is to be chosen, then it clearly is not Fashionable. Commuter is said to dominate Fashionable, and, for this reason, Fashionable can be eliminated from further consideration. Similarly, Battlebus dominates Aggressor, and Durable dominates Exciter. The design alternatives represented by Battlebus, Commuter, and Durable are said to be nondominated, in the sense that no other alternative is superior (or at least as good) with respect to both of the objectives.

The set of all feasible nondominated alternatives can be thought of as discrete points along a curve in the solution space called the Pareto optimal curve (or transformation curve or efficient frontier). Further ordering of the alternatives on the Pareto optimal curve cannot be achieved without the introduction of value judgments concerning the relative preference between safety and economic objectives. The Pareto optimal curve for the example problem is shown in Figure 18.4.

To proceed further, MAUT must be applied. If we could develop exact preference information using MAUT, then a family of isopreference curves could be superimposed over the Pareto optimal curve, as illustrated in Figure 18.4. Isopreference curves have the property that any two points in the solution space that lie on the same curve are equally valued. The most preferred alternative is that which has the greatest value or utility. This alternative is located in the solution space

TABLE 18.3 Data for the Automobile Example Problem

Automobile Design	Safety Value (Dimensionless)	Annualized Cost
Aggressor	4	$5,000
Battlebus	3	4,700
Commuter	5	3,500
Durable	8	2,400
Fashionable	7	4,200
Exciter	9	3,100

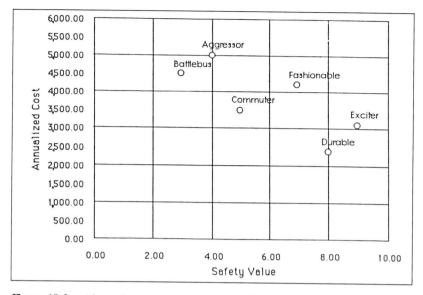

Figure 18.3 Alternatives in the solution space for the automobile design problem.

at the point of tangency of the Pareto optimal curve and the highest isopreference curve (the point representing Commuter in Figure 18.4).

Whether or not we can actually compute explicit isopreference curves, the concept illustrates the importance of determining who will make the decision for a specific problem. Different decision makers may well have the same preference

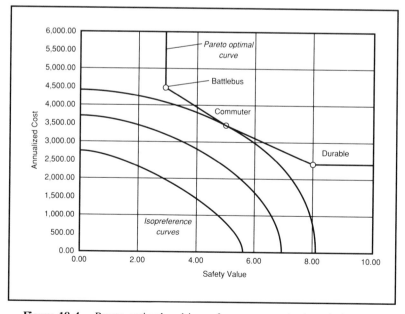

Figure 18.4 Pareto optimal and isopreference curves in the solution space.

orders with respect to each individual objective, but they are quite likely to have different isopreference curves. Different consumers, different manufacturers, and the different regulatory agencies all are likely to prefer safer and less-expensive automobiles. Nevertheless, they may disagree as to which nondominated design is the best, because they disagree regarding the appropriate tradeoffs between safety and cost.

Table 18.4 provides a taxonomy of the large variety of techniques that have been developed for multiple objective decision aiding (Deason, 1983). As an example, consider the application of a generating technique—the constraint method—to a formal multiple-objective problem.

TABLE 18.4   A Taxonomy of Multiple Objective Decision-aiding Techniques

A. Nondominated solution generating techniques
  1. Constraint method
  2. Weighting method
  3. Multiple objective dynamic programming
  4. Multiple objective simplex method
  5. Noninferior set estimation method
B. Techniques involving a priori complete elicitation of preferences
  1. Optimal weights
  2. Utility theory
  3. Policy capture
  4. Techcom method
C. Techniques involving a priori partial elicitation of preferences
  1. Lexicographic approach
  2. Goal programming
  3. ELECTRE method
  4. Compromise programming
  5. Surrogate worth tradeoff method
  6. Iterative Lagrange multiplier method
D. Techniques involving the progressive elicitation of preferences
  1. Step method
  2. Semops method
  3. Trade method
  4. Pairwise comparisons
  5. Tradeoff cutting plane method
E. Visual attribute level displays
  1. Objective achievement matrix displays
  2. Graphical displays
  3. Mapping

Based on Deason, 1983.

A two-objective decision problem can be expressed formally very simply as

$$\min_{a \in A}[f_1(a), f_2(a)]$$

where a is an individual alternative from the set of all alternatives (A), and where $f_1(a)$ and $f_2(a)$ are the objective functions for the first and second objectives, respectively. In our automobile example, the set of alternatives A is defined by the automobiles listed in the first column in Table 18.3. The first objective function (safety) is defined by the combination of the first and second columns in the table, and the second objective function (cost) is defined by the combination of the first and third columns.

As the name implies, generating techniques generate the set of nondominated solutions and assess the tradeoffs between objectives at various levels of objective accomplishment. No attempt is made to incorporate preferences. This is essential if no single decision maker can be identified for the problem. Generating methods contribute to the analysis of decision problems by reducing the set of all alternatives, feasible and infeasible, to the Pareto optimal solutions illustrated in the previous example.

In the constraint method, one objective is optimized while the remaining objectives are constrained to some specified value. This generates one point on the nondominated frontier. The constraint values are then changed, and the optimization is repeated. This generates a second point on the nondominated frontier. The entire process is repeated until the entire nondominated solution set is generated, one point at a time. An illustration follows.

Using the constraint method, solve the following problem repeatedly, with different values of $K(i)$ at each iteration i:

$$\min_{a \in A} f_1(a)$$

subject to

$$f_2(a) \leq K(i)$$

A systematic procedure for implementing the constraint method is illustrated in Table 18.5. At the first iteration, a very large value is chosen for $K(1)$. This relaxes the constraint on $f_2(a)$.

In this example, the solution to this scalar optimization problem is the safest alternative irrespective of cost ($K(1) = $ infinity):

$$a_{\text{opt}}(1) = \text{Battlebus}$$

At the next iteration,

$$K(2) = f_2(a_{\text{opt}}(1)) - 1 = f_2(\text{Battlebus}) - 1 = \$4,699$$

TABLE 18.5 Solution of the Example Two-Objective Problem Using the Constraint Method

Iteration i	$K(i)$	$a_{opt}(i)$	$f_1(a_{opt}(i))$	$f_2(a_{opt}(i))$
1	Infinity	Battlebus	3	$4,700
2	$4,699	Commuter	5	$3,500
3	$3,499	Durable	8	$2,400
4	$2,399	None feasible		

This makes Battlebus, as well as all of the more expensive alternatives (in this case, Aggressor) infeasible. This procedure is repeated until at the last iteration there are no feasible solutions.

To represent the preference structures of individual decision makers (the goal of MAUT), policy capture may also apply (see Exhibit 18.1). Clarifying the preferences of stakeholders and decision makers can lead to increased mutual understanding and can foster win-win alternatives. MOOT, MAUT, policy capture, and POSTURE (see Chapter 11) all try to clarify preferences; as such, they can provide a helpful basis for the participative approaches discussed in the following section.

18.7 PARTICIPATION AND MEDIATION

Evaluation of the impacts of technological change has a vital dimension beyond the technical analyses presented in the chapter to this point—the *process* of evaluation. In particular, this involves two questions: Who makes the evaluation? When are various parties involved in the decision? Participatory evaluative processes have much to offer, both on internal adoption of new technology (for instance, operators help formulate the changes) and on external responses to new developments (for instance, neighbors of a new plant provide meaningful input during the planning process).

18.7.1 Participation

In impact evaluation, participation without professionalism is empty while professionalism without participation is blind (Rossini and Porter, 1982). Participation is the best guarantor of acceptance of an evaluation. Without acceptance, it is virtually impossible to implement any decision effectively. Hence, participation in technology evaluation makes good sense within an organization (such as concerning the choice of a component within a new technology product) and externally (such as stakeholder involvement in assessing a controversial new facility proposed for a community).

Participation can begin at various stages of the assessment process—from involvement at the beginning of an assessment to later involvement restricted to the evaluation per se. Early involvement can increase commitment and build trust, but it consumes more resources. Later involvement uses resources more efficiently, but it can raise issues that suggest redoing earlier assessment steps at even greater cost and delay.

Participation can take many forms. This section addresses both public participation in impact assessment and internal participation in technology evaluation within an organization. Susskind (1983) lists two keys to participation: (1) defining those interests with a legitimate stake and (2) injecting additional participants into an ongoing assessment process (such as determining what conditions to set on late-joiners and whether earlier agreements are to be reopened).

Novel representation arrangements include advisory committees (Arnstein, 1975). Planning cells engage small groups, chosen to reflect the perspectives of major interests, to work intensively for a short period of time to express value preferences (Peters, 1986).

According to Redelfs and Stanke (1988), participation can fulfill various functions:

- Informing the participants (regarding the issue and/or the decision-making process)
- Informing decision makers
- Collaborative decision making (cooperatively or through adversarial processes, such as legal actions to block a development).

Bregha (1989) suggests that the benefits of competition in the assessment process require that the participants have sufficient resources and opportunity to challenge establishment information and conclusions.

18.7.2 Mediation

Substantial participation often generates conflict. Mediation, involving third parties with more or less authority, is a good method of working out conflicts to generate acceptable development plans (Susskind, 1983). Mediation is most common in large projects involving substantial disruption to a community or a region.

Susskind, McMahon, and Rolley (1987) present sequences of steps to follow during prenegotiation, negotiation, and postnegotiation phases. Special concerns include:

- Taking time to ensure that all parties understand the issues and the alternatives fully (possibly also providing training in negotiation)
- Actively directing the energies of all parties toward a consensus arrangement
- Keeping constituents abreast of negotiations as they progress
- Preempting escalation of disputes due to selective perceptions
- Developing incentives for good faith bargaining (including bounding of the concerns of "opponents," and getting formal authorities to accede to agreements to be reached)
- Devising mechanisms to bind all parties to their agreements (legal contracts)

Unfortunately, environmental mediation efforts rarely lead to successful agreements. Buckle and Thomas-Buckle (1986) studied 81 attempted mediations of environmental conflicts associated with technological development. From the perspective of the mediators, in 73 cases mediation was rejected before a second meeting, and in only 3 cases was a stable agreement implemented. On the positive side, in 40 cases, participants credited the mediation effort with helping them to improve their relationships with other parties at interest and/or understanding of the matter at dispute.

18.8 CONCLUSION

Evaluation brings together estimates about technological change with preferences. Estimates of what is likely to happen draw upon the impact assessment techniques described in Chapters 14 through 17. Determination of preferences about what should happen requires use of one or more of the approaches described in this chapter.

Evaluations of technology range from the totally informal—tacit (internal) models with no explicit measurements—to formalized computer models. This chapter has emphasized formal models, not to make choices per se, but to help rationalize the decision process by making criteria, alternatives, and estimates explicit. Such models may, at times, help integrate divergent values by clarifying positions and suggesting common grounds. Other times, they may help disaggregate values by separating criteria and helping parties recognize multiple objectives.

EXERCISES

18.1 Identify one technology/development/policy issue under considerable debate (for instance, the U.S. Strategic Defense Initiative; a policy to reduce chloroflourohydrocarbon emissions; or federal subsidies to support U.S. integrated circuit chip advanced technologies). Characterize the nature of the debate along the dimensions of Figure 18.1.

18.2 Use brainstorming or another creativity technique (such as NGP—see Chapters 7 and 11) to generate a set of alternative policies to resolve the issue. Apply suggestions from Section 18.3 to reduce the set of alternatives.

18.3 Use Nijkamp's set of characteristics (Section 18.3) to categorize these alternatives. Generate an issue that differs from this one on at least two of these characteristics.

18.4 Identify suitable measures to quantify differences among the set of alternatives (Exercise 18.2). Try to construct a futures foregone index (Section 18.4.1) for these measures.

18.5 For this same issue (Exercise 18.2), structure a policy capture process to find out how two or more interest groups weight two or more dimensions to the issue. Follow the guidance of Exhibit 18.1. Make up at least 10, preferably 20, instances that reflect combinations of high and low values along the key dimensions. (For instance, concerning SDI, these instances might involve various compromise positions on the extent of R&D and funding levels.) Use a statistical package to perform a linear regression on the responses. Plot results to depict where the instances fall and to show how the interest groups differ. (If you do not have access to representatives of different interests, assign colleagues to role play.)

18.6 Design a participation/mediation strategy to resolve differences among these interests of Exercise 18.5.

18.7 Consider Exhibit 18.4. Take the position of the Air Force program manager responsible for these two component activities to help develop pilot/aircrew automation. Present and interpret the data given to justify your budget for next year. Do this embedded in a funding scenario in which, to some extent, you compete with others for available resources. May the most persuasive win!

■ **Exhibit 18.4   The AIMTECH Study (Based on Reitman et al., 1985).**

The AIMTECH study forecasted advances in artificial intelligence technologies. The study methodology is intriguing because it combined *needs analysis* for three target areas with identification of contributing technological *milestones* and an *evaluation* scheme to help set priorities. For instance, requirements in one of the target areas, pilot/aircrew automation, included achievement of 11 milestones. Illustrated below for two technologies are the types of estimates provided for each:

Factor	Highly Parallel Programming	Large Vocabulary, Continous Speech Recognition in a Limited Domain
Probability of success	0.5–0.9	0.8
Years required	10-15	4–10
Person-years of effort required (over current funded baseline)	64	40–70
Cost ($ millions)	9.6	6–10.5

The estimates are displayed in various arrays. One display charts the minimum and maximum time requirements for each of the milestones as parallel time lines (see Figure 18.5). This gives a quick visual sense of the likely roadblocks to achieving a given target that depends on several of them.

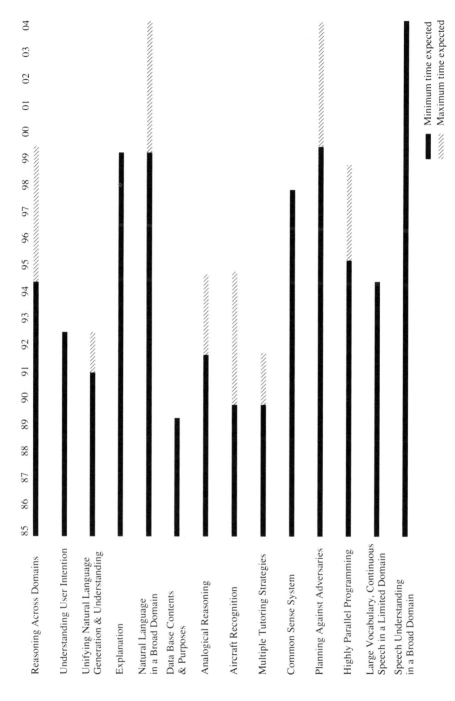

Figure 18.5 Time lines for technology milestones—AIMTECH.

Another compelling chart lists each of the milestones required to achieve a given target, which illustrates the magnitude of resources required to meet the target. These data provide the prospective technology manager with a beginning basis for developing a return on investment analysis.   ■

18.8 Try AHP (see Section 18.5). Access the TOOLKIT example problem. Adjust the matrix so as to show that Atlanta is the best city for newly graduating computer engineers. How few changes can you make in the given matrix to accomplish this, with an acceptable consistency index?

18.9 Apply AHP to a two-level decision. Establish at least three decision criteria as the top level; apply these to at least three alternatives as the bottom level. Follow the steps of Exhibit 18.2. Use the TOOLKIT to perform the computations.

18.10 In the automobile problem, the safety value and cost listed for the Aggressor and Fashionable designs are for the standard models, equipped with seatbelts. Each of these models also can be equipped with optional airbags. For the Aggressor, the airbag option improves the safety value to 2, at an additional cost of $200 a year. For the Fashionable, the airbag option improves the safety value to 5, at an additional cost of $100 a year. Use the constraint method to determine the new nondominated set of alternatives.

____19
MANAGING THE PRESENT FROM THE FUTURE

OVERVIEW

This chapter addresses two "big picture" questions: How successful are forecasts of technological change and of its impacts? How can a future perspective be applied in the present?

19.1 ALTERNATIVE TEMPORAL PERSPECTIVES

To manage technology, the manager or engineer must make decisions. If those decisions are to be made intelligently, he or she must interpret situations in light of information deemed salient. The manager's temporal perspective—past, present, or future—distinctly colors that information selection.

One traditional temporal perspective interprets the present in light of *past* experience. The title of a classic nineteenth century novel by Edward Bellamy, *Looking Backward*, tells it all. This perspective tends to lock the decision maker into the framework of past organizational goals, strategies, and experiences by drawing on evaluation of prior experiences. Its view of the future accentuates continuity, an extrapolation of that same past.

Another perspective emphasizes the *present*. This "now" orientation stresses immediate solutions to pressing problems. Picture the harried production manager totally consumed in keeping the line running to meet today's production requirements. American management has come under fire for focusing on short-term payoffs within annual budgeting cycles. Reward based on current profit and loss discourages a long-term perspective.

Another variant of the present perspective places faith in market mechanisms. The future-oriented Carter administration chose to offer major federal subsidies to instigate a synthetic oil industry now to help take care of future energy shortages. In contrast, the present-oriented Reagan administration believed that the country would fare better by avoiding long-term energy planning. Rather, when energy becomes sufficiently scarce, the price will increase, and the market will efficiently respond. Accordingly, the Reagan administration eliminated the synthetic fuels subsidy program.

Clearly, this book advocates a *future* perspective. The forecasting and assessment methods it presents offer ways to analyze the future. However, these methods utilize past, present, or future information in different ways. Trend extrapolation uses past data to anticipate future developments. If taken too literally, trends can tie the forecaster too tightly to the past, so that milestones or barriers that could be anticipated are missed.

An extreme example shows that some professionals refuse to deal with the future at all. An assessment of the prospects for widespread use of controlled environment (greenhouse) agriculture done in the 1970s involved some fine agricultural scientists. They provided excellent data showing strong trends of advances in plant genetics over time. For the purposes of the study, they were asked to project future plant yields. They refused. Their explanation was that scientists do not speculate on the future. They were locked into a past/present perspective, unwilling to even extrapolate.

Monitoring might be considered a present-oriented technique. However, it too gains a future orientation when the forecaster filters and structures the information gathered to indicate likely future developments. The best-selling book, *Megatrends* (Naisbitt, 1982), and its successor, *Megatrends 2000* (Naisbitt and Aburdene, 1989), illustrate the use of monitoring the present to predict the future.

Other techniques, such as expert opinion methods and scenarios, directly engage the future. For instance, asking an expert to predict the year that technology X will become available commercially presumes that the expert has a tacit (internal) model of the future that he or she can tap. Likewise, scenarios can be used to predict future possibilities without relying on past or present information.

After this future information is generated, it must be related back to the present. What are the actions that can be taken now (or planned for now) to address the perceived future? In the words of this chapter's title (borrowed from Smits, Rossini, and Davis, 1987), we must "manage the present from the future."

19.2 LOOKING BACK: HISTORICAL LESSONS ON TECHNOLOGICAL CHANGE

Technology forecasting purports to predict future changes in technology. As discussed in Chapter 2, theory about sociotechnical change is relatively weak. Consequently, it is advantageous to look back, empirically, on past changes in technology to identify what causes them. This section compares different types of historical evidence.

Probably the most powerful method to determine cause and effect is the scientific experiment. In it, the experimenter designs a comparison, prototypically between a group of subjects exposed to some treatment and a control group that does not receive the treatment. Ideally, the treatment is precisely meted out by the experimenter and the subjects are randomly assigned to either the treatment or control group. Under such conditions, superior performance in the treatment group can often be unambiguously attributed to the treatment. This model works well to determine the potency of new drugs, for instance.

In most situations, technological change takes place under far-less-controlled conditions. Imagine assigning one computer company randomly to develop a particular technology and another to ignore that technology. Instead, alternative study designs attempt to determine what makes some technologies change more rapidly than others, or be diffused more rapidly, or to fail in the market. The most common design is the *case study,* in which a particular development or implementation is tracked in depth (see Yin, 1989). For instance, Kidder (1981) relates how an engineering team designed and developed a minicomputer at Data General in a lively and compelling way. However, it is unknown to what extent success was due to external forces (demand peaked at just the right time, in a booming economy), internal good luck (the personal chemistry of the people on the design team clicked), or a complex of factors. Thus it is hard to generalize lessons learned to other design teams or to the task of designing new computers in other situations.

Sometimes, the historian of technological change happens upon a natural experiment—that is, a chance to compare one or more instances "with" versus "without" some technology. Suppose that you came upon two architecture and engineering (A&E) firms that had adopted two different computer-aided design (CAD) packages. A wonderful opportunity exists to compare which package works better! Yes, but with caution. Perhaps one firm chose the more powerful CAD system because its engineers were more sophisticated, or perhaps they had greater financial resources, or maybe they worked on more complex building designs—any of these features would confound interpretation of which CAD tool is better. Cook and Campbell (1979) lay out many "quasi-experimental designs," pointing out their pitfalls and their potential informativeness.

Some years ago, the National Science Foundation (NSF) funded a number of retrospective technology assessments—essentially technology case histories. They reasoned that looking back to understand the effects of past technologies would help us look forward to predict the effects of future technologies. The studies proved interesting and informative (Tarr, 1977). Hindsight showed that the first trans-Atlantic submarine telegraph cable impacted trade and international relations by speeding up communications from weeks to minutes. However, these effects could not be separated from others taking place simultaneously (Coates and Finn, 1979).

In essence, looking back is the best teacher available concerning what drives successful technological change and what the effects of changed technology are.

However, it is often a befuddling teacher. Sociotechnical change rarely provides clean baseline comparisons. Usually the technological variable intrudes into complex natural and social processes, themselves complexly interdetermined. Technological change becomes a partial cause and a partial effect of socioeconomic changes. The complexity is so great that no individual or institution can fully know or control the technological change process. Looking back on technological changes imparts humility to our efforts to look forward.

19.3 LOOKING BACK: PREVIOUS FORECASTS AND ASSESSMENTS

It is somewhat easier to learn from past forecasts and assessments. Reviewing old forecasts shows that many are too ambiguous to tell if they proved right or wrong! George Wise (1976) gathered a large number of forecasts that he could judge right or wrong and sorted these into 18 technological areas (such as computers, factory automation, new materials). He aggregated predictions for each area, reporting average percent right by area. These ranged from 18 percent right for 22 housing technology forecasts to 78 percent right for 18 new materials forecasts. The median by area was 45 percent right; the interquartile range was 38 to 51 percent. (Recall that this means that one-quarter of the areas fared worse than 38 percent right; one-quarter did better than 51 percent.) This rough estimate helps put forecasting expectations in perspective—far short of perfection (certainty) but much better than chance (blind ignorance). Or, as baseball players, forecasters would make pretty good batters, hitting around .450.

Chapter 4 pointed out that it is actually unfair to gauge a forecast just by whether it came true or not. Forecasts are conditional, and a forecast that induced an organization to avoid a predicted negative outcome by altering key factors must be judged a rousing success. Yet, there is much to be learned in evaluating technology forecasts and assessments.

Michael Scriven (1967) draws a useful distinction between two types of evaluation—formative and summative. *Formative evaluation* considers a study while it is underway. Formative procedures are often quite informal, for instance, providing early drafts of a forecast to experts and decision makers. In this way, the experts can tell the decision makers if the forecast misses a critical element, misunderstands a driving force, or is on target, while there is time to make corrections. The decision makers can tell the forecasters if the study misses a critical concern, is hard for them to understand, or is just right. Their participation in focusing the study can also enhance their acceptance of it later.

Such a formative evaluation procedure was followed effectively by Don Kash and colleagues (1973) in assessing the impacts of offshore oil and gas development. They provided early drafts, warts and all, of their impact assessment to environmentalists and to oil industry representatives. The environmentalists hated it, pointing out multiple flaws. The industry representatives hated it too, also pointing out multiple flaws. But after a couple of iterations in which the assessors took into account many suggestions from both perspectives, the final assessment was considered highly credible by all sides. It provided an effective basis for discussion for all parties, and its recommendations translated into federal legislation.

Summative evaluation reviews a study after it is done. For a forecast or assessment, this means judging a study in terms of its validity and utility (see Chapter 13); the interrogation model provides a good set of questions to ask of a study (Exhibit 4.2). It is helpful if a study specifies its objectives in advance; then one can see how well these have been achieved. For instance, a well-known system dynamics study, *The Limits to Growth* (Meadows et al., 1972), was roundly criticized in terms of the validity of this attempt to model world dynamics. Yet, the stated objective of the authors was to get world leaders to think about potential worldwide environmental disasters. The study succeeded spectacularly at drawing attention to these issues, whether or not the model itself was right.

Ideally, those who commission forecasts and assessments would perform some of them experimentally. As sketched in the previous section, learning is enhanced by performing experiments that make explicit comparisons under controlled conditions. Porter and Rossini (1977) proposed a variety of designs in which multiple forecasts or assessments would be performed in tandem. In this way, managers could learn much about how to forecast, or assess, technological change more effectively. For instance, a manager might commission a high-study-resource forecast of, say, nanotechnology (molecular scale assembly of micromachines) and a concurrent low-resource forecast of the same topic by a different organization. (It would be even nicer to compare many such parallel forecasts.) The manager could compare the high-resource and low-budget forecasts, thereby checking the main study and also getting a measure of the relative payoffs from the different scales of study effort. This knowledge could then guide future forecasting efforts.

Several authors have reviewed broad sets of forecasts or assessments (see Chapter 4 as well). William Ascher (1978) drew insightful conclusions about various forms of forecasting by reflecting on what worked well and what did not work in many actual studies. Schnaars (1989) reviewed a number of tech forecasts, largely from the 1960s or later. Exhibit 19.1 highlights some of his conclusions. Schnaars emphasizes the tendency of technology forecasters toward overoptimism. Arthur C. Clarke (1984) points out a number of instances to the contrary—instances in which forecasters lacked the vision to foresee key advances.

Hal Linstone pushes analysts toward explicit elaboration of *multiple perspectives*—technical (such as systems analysis), organizational (or societal), and personal (or individual). Each emphasizes different goals, modes of inquiry, ethical bases, planning horizons, and other characteristics (Linstone, 1989). It is critically important for technology forecasting and assessment to blend these perspectives to avoid ill-founded optimism, pessimism, or just plain myopia. Consider this illustration based on Linstone (1989):

> A technology manager faces a decision about whether to enter a new business area. She has a detailed cost/benefit analysis from her technical staff, indicating that this area is ripe for development. However, she does not make the decision based just on this technical assessment. She talks to various department heads in her company to determine the extent of support or opposition to such an expansion. Then over the

■ **Exhibit 19.1   Megamistakes (Based on Schnaars, 1989).**

There are five prominent contributors to erroneous forecasts:

1. *Fascination with the exotic:* Tech forecasters exhibit a bias toward the optimistic and a disregard for the realities of the marketplace.

2. *Enmeshed in the Zeitgeist:* Attention to technologies exhibits a mob flavor— everyone sees the same technologies as hot (devaluing expert consensus), and everyone emphasizes the same pressing societal needs.

3. *Price-performance failures:* Many technologies deliver lesser benefits at greater costs than anticipated.

4. *Shifting social trends:* Changing demographic trends and social values are not well considered; these change user desires and market opportunities.

5. *Ultimate uses unforeseen:* Rarely do forecasters anticipate applications fully. ■

weekend, she arranges to play golf with an old friend whose company is involved in the target area, but with different interests. She draws on his intuition, experience, and advice. With no particular weighting function, she then integrates these different, possibly conflicting, perspectives—technical, organizational, and personal—to arrive at the decision.

Drawing on these evaluations and our own observations, Exhibit 19.2 provides 11 recommendations. In an earlier version, Porter and Rossini (1987), included "Pray" as an explicit recommendation; this is now left implicit for the devout forecaster.

■ **Exhibit 19.2   Eleven Commandments for Technology Forecasting**

1. *Get the right technology*: Understand the technical domain sufficiently to address the essential functions at the right level of aggregation.

2. *Pick the right parameters*: Technological parameters must pertain to the decision to be made; data must be attainable.

3. *Get the context right*: Identify the institutions involved, socioeconomic influences, and critical decision points—now and in the future.

4. *Beware of core assumption drag:* Technical myopia or ideological fixation can miss qualitative changes from past patterns.

5. *Beware of the Zeitgeist*: Challenge the conventional wisdom; try out alternative perspectives; do not allow the forecast to just mirror the prevailing mood of the moment.

6. *Keep the time horizon short*: This suggests that frequent, lower-cost forecasts be conducted rather than more substantial, but less frequent, studies.

7. *Do it simply*: Invest study resources in reducing the most critical uncertainties; rarely will these respond to elaborate modeling.

8. *Use multiple approaches*: Seek convergence from diverse approaches with complementary strengths.

9. *Perform sensitivity analyses*: Deliberately estimate how the forecast would change if assumed initial conditions, influential variable levels, functional relationships, or milestones change.

10. *Provide uncertainty estimates*: Give ranges of parameter projections over time; if possible, give specific confidence estimates.

11. *Take the middle path*: Balance between far-out forecasts (that depend upon many developments with low probabilities all occurring) and too conservative assessments (often offered by committee). ∎

19.4 LOOKING BACK: FROM THE FUTURE

Once all is said and done, the key to dealing with sociotechnical change, whether as a manager, an engineer/analyst, or an impacted party is understanding that most uncertain of entities—the future. How is one to grasp the future?

The first choice might be to use a theory of sociotechnical change to predict the future. Occasionally some theory covering a limited domain (such as economic diffusion rates) will be used to project a component of the future. However, the theoretical option is essentially unavailable because adequate theory does not exist. Thus ad hoc technique must substitute for theory.

The techniques used to predict the future largely rely on data from the present and past, plus some attempts to estimate the future via direct human judgment. The technology manager must integrate the resulting information to arrive at the best estimate of the future.

Techniques for analyzing the future are not definitive. Nor can small-scale sociotechnical experimentation or other empirical studies extrapolate with assurance to large-scale systems. These activities do, however, narrow the range of possibilities for the future (a key premise of forecasting—see Chapter 3).

Imagine the decision maker as a juggler. He or she must balance: (1) the set of future alternatives of various likelihoods (provided by the forecasters); (2) the present, with its resources and constraints; and (3) organizational and personal goals (ideally with consideration of how present goals may change in the future). The interplay among these three factors determines how the future is managed.

Contrast two types of information about the future. The first extends current trends and developments—an *extrapolative* future. The second describes a desired future, in line with the goals—a *normative* future. These two viewpoints are polar types that may blend into intermediate cases.

In the extrapolative future case, the manager accesses studies of the future in the form of forecasts or impact assessments in areas over which his or her organization exerts little or no direct control. The future, as depicted in the analyses, exerts pressure on the organization's goals. Organizational goals may need to be adjusted in response to such a forecast. For example, suppose a forecast asserts a high probability of free trade (no significant taxes or barriers) between Canada and the United States in the immediate future. Your firm produces aluminum, but Canadian firms can produce aluminum more cheaply (they have cheaper energy and raw materials, lower transportation costs). Suppose your goal had been to dominate the U.S. aluminum market. You need to adjust that goal. Options might include expanding business in other markets (in the integrated European community after 1992, for instance), merging with the Canadian competition, or getting out of the raw aluminum production business in favor of machining aluminum components.

In the normative future case, the organization designs its future to achieve certain objectives. Consider the free trade with Canada illustration again, but change the premise—your firm's goal is to dominate the U.S. aluminum market, now and in the future. Toward this goal, your firm seeks ways to influence future trading conditions. Perhaps, you lobby influential congressional representatives to exclude aluminum from the free trade agreement. If you succeed, you can maintain your present goal of dominating the U.S. aluminum market. Normative forecasting identifies a desired future and points out how to attain that future.

Our actions and aspirations will impact the future. Policies and actions will invariably come to fruition in an environment different from the one in which they were conceived. Thus the perspective from which to make present decisions should be the future instead of the present. Visions of the future enter the present to serve as a principal source of information for engineering action and managerial decision making.

19.5 VISIONS

To bring the future home to the present requires images of that future. Scenarios (Chapter 13) provide such images. However, the future images need to go beyond the rather cold sense of a set of discrete, alternative scenarios. The future is very much an open system, in terms of fuzzy, penetrable boundaries in both sociotechnical space and time. Future images must accommodate uncertainty, must be adaptable, yet also must provide a "star of Bethlehem" to guide present actions. These images include both goals and contexts. *Visions* best capture the sense of these future images.

The futurist must first create one or more visions of the future; then the manager can apply such a vision to guide present decisions and actions. Vision creation variously combines observation, analysis, and intuition.

Both formal and informal *observations* of sociotechnical development patterns provide raw material to fuel a vision. These observations need to be interpreted and structured according to some framework. A conceptual framework consists of the pertinent beliefs, assumptions, and goals. These may holistically guide the overall vision of an organization. Consider, for instance, how the evaporation of the Warsaw Pact/Soviet military threat alters the conceptual framework of the U.S. Department of Defense. Long-held beliefs are undermined; new assumptions of future threats must emerge; and defense goals must adapt accordingly. In other words, a new vision is needed.

Analysis contributes to creating vision, too. You must forecast social and technological changes; the tools presented in this book apply directly. Likewise, you must assess the potential impacts of various changes; again, the tools presented herein pertain. Forecasts and assessments should address holistic, general contextual features; in addition, they must focus sharply on specific issues at hand. For instance, were IBM to consider commitment to develop a particular optical storage technology, most forecasting/assessment information should bear directly on this technical issue. However, this should be embedded in a serious consideration of changing corporate interests and the evolving international economy. As the technology manager strives to integrate the results of these analyses, a sense of action options within plausible contexts emerges.

Intuition is the third ingredient in producing a future vision. Intuition thrives on rich input; it is not well done in a vacuum. It builds from the observations and analyses just noted. It also draws on more speculative futurist writings such as science fiction. Intuition yields patterns, sometimes rather ill-shaped, certainly less specified than set scenarios. Some people seem naturally intuitive, able to generate compelling visions with relatively sparse input. Others can enhance their synthetic capabilities by applying creativity tools (see Chapter 7).

Interaction between forecasters and those who would use their forecasts is vital to generate credible visions. Analysts need to understand the worldviews, assumptions, and goals of the users to incorporate these into their observations, analyses, and intuitive syntheses. Conversely, users (managers and policymakers) need to understand what the forecasters are trying to do and how they go about doing it. The users must be convinced of the efficacy of the forecasting and assessment methods and their application to the issues at hand. They must believe in the capability and professionalism of the forecasters. Interaction can go beyond mutual understanding. For instance, forecasters may help users clarify their assumptions and focus their goals. On the other hand, users may help forecasters define the problem, bound the study, identify experts, and interpret analyses.

Visions can only be fulfilled if they are brought to bear on present decisions and actions. Making use of visions requires deliberate attention to that task. Spelling out the path from the present to a particular vision of the future helps. This can highlight incongruities between present policies and the future vision. The pathway should come replete with red flags pointing out gaps, mismatches, and contradictions that separate the future from the present. Analysts may offer suggestions on how to overcome such barriers, or they may defer to the decision makers to devise such solutions. Obviously, unless solutions are created and implemented, the future vision will not be attained.

To attain a vision, the manager must array its demands against present resources. These include money, materials, technical capabilities, and personnel qualifications. These resources enable present action. The magic in translating future visions into present action derives from the perspective. The decision maker sees the set of possible present actions with the eyes of someone in the future—that particular future captured in the vision. An analogy comes to mind in the area of sports psychology. If you imagine the play of key points in your upcoming tennis match to be successful (overhead smashes to the backhand, service aces, etc.), then in the actual play, you may achieve those visions. Put more philosophically, because the present is seen from the future, the courses of action for achieving goals will carry the present into the future. The present is only the arena to be worked—the source of resources to implement the future vision.

One component of the present should be visions of the future that can motivate and guide our actions. The outcome of forecasts and assessments should be tangible beacons for the present. Seeking to grasp the future moves us, as individuals and as institutions, into a proactive posture toward dealing with the world. We are not passive victims of sociotechnical change processes beyond our grasp. Instead, we become actors in an ongoing drama whose script is being written in considerable measure by the players themselves. This book was written to empower this perspective on change.

EXERCISES

19.1 Jot down a few phrases to describe the content of each course you are presently taking (if that is not applicable, do this for the last semester in which you took classes). Classify this material by its orientation: past, present, or future. Comment.

19.2 Browse through a local Sunday newspaper to identify one key legislative action receiving attention at each of three levels: national, state (province), or local. Classify the driving force behind each as past-, present-, or future-oriented. Comment on why this is.

19.3 Find two old technology forecasts that have had sufficient time to prove right or wrong (you might browse through back issues of a journal such as *The Futurist* at a library). Did they prove optimistic, pessimistic, on target, or just plain wrong? Consider Schnaars' and Clarke's points of view, and the eleven commandments to explore what went wrong (or what went right).

19.4 Review an organizational decision about the effectiveness of a past technology-intensive investment. What sort of study design would the given case provide—case study, quasi-experiment, true experiment, or other? What factors confound interpretation of why the investment was or was not successful?

19.5 Suppose you have just been assigned to perform a forecast of a certain technology (of your choosing).

    (A) Lay out your strategy for doing this forecast on one page.

    (B) Review that strategy and identify points at which formative evaluation could be helpful. At each point, sketch how you might obtain that evaluation in a sentence.

    (C) Describe a sensible approach to summative evaluation for your technology forecast.

(D) Sketch two visions that might emerge from that forecast to guide present decisions.

19.6 Think back over future-oriented books you have read (for instance, science fiction; *Engines of Creation* by Eric Drexler; *Player Piano* by Kurt Vonnegut). Focus on one book and identify the vision it provides and how that could translate into present actions.

THE TECHNOLOGY FORECASTING TOOLKIT (VERSION 1.00)

OVERVIEW

Many of the methods presented in this book are quite tedious to perform without the use of computer support. This appendix presents a set of computer-based tools, the Technology Forecasting TOOLKIT, that supports these methods. The Technology Forecasting TOOLKIT was designed explicitly with the student of technology forecasting in mind. The toolkit contains enough functionality to enable the student to learn and appreciate a particular method, without needing to learn many different software packages or without becoming intimidated or bewildered by the advanced features contained in many commercial systems. Although it was designed for students, the toolkit holds many features and functions of interest to both journeyman and expert technology managers.

A.1 INTRODUCTION

The text has presented numerous methods and techniques to aid the technology manager. Although conceptually straightforward, many of these methods are difficult to apply without the use of computer support. For example, even with the use of a hand-held calculator, applying the curve fitting techniques presented in Chapters 9 and 10 requires an enormous amount of both time and patience to perform the many computations required for even modest amounts of data.

Throughout the chapters, references have been made to numerous commercial computer-based tools, each of which focus on one or two of the specialized methods presented. These tools are more than acceptable when used by experts in a particular method but often overwhelm the beginner with more features and functions than are needed. Thus our goal was to develop a set of computer-based tools that would provide a novice technology manager with a platform to learn and appreciate a particular method, without beccomming bewildered by a plethora of advanced options. In other words, *the beginner should focus on the method rather than the software*.

This appendix presents the Technology Forecasting TOOLKIT, a set of computer-based tools designed to support many of the methods presented in the text while achieving this goal. The TOOLKIT presents methods and techniques in a consistent and easy-to-use manner, featuring pull-down menus, color graphics, and extensive error checking to aid the technology forecasting and management student. Although designed for the student, more experienced technology managers may find many of the provided features and functions useful in their activities.

The Technology Forecasting TOOLKIT is provided on one $5\frac{1}{4}$ inch floppy diskette that contains the TOOLKIT program as well as the necessary help and support files. Version 1.00 of the TOOLKIT supports:

Analytic hierarchy process
Creativity stimulation
Cross–impact analysis
KSIM—Kane Simulation
Project scheduling
Trend extrapolation

A.2 HARDWARE SYSTEM REQUIRED TO USE THE TECHNOLOGY FORECASTING TOOLKIT

The TOOLKIT was designed to run on a computer system meeting the following minimum requirements:

1. IBM PC/XT/AT or true compatible computer or IBM PS/2 or true compatible computer
2. Monochrome, CGA, EGA, or VGA video adapter
3. 512 Kilobytes of RAM
4. MS-DOS or PC-DOS, versions 2.1 or above

Recommended, but not required, are:

1. Hard disk drive with at least 400 Kilobytes of free disk space
2. CGA, EGA, or VGA video adapter to view graphs and plots or EGA or VGA video adapter to view graphs and plots with color

A.3   INSTALLING AND RUNNING THE TECHNOLOGY FORECASTING TOOLKIT

The TOOLKIT may be run from either the provided diskette or from your system's hard disk drive. In either case, it is recommended that you make a backup copy of the provided diskette and file the original in a safe place for protection. See your DOS manual for more information regarding making copies of diskettes.

A.3.1   Installing the Technology Forecasting TOOLKIT

If you do not wish to run the TOOLKIT from your system's hard disk drive, you may skip this section and proceed to Section A.3.2, "Running the Technology Forecasting TOOLKIT."

   If you wish to run the TOOLKIT from your system's hard disk drive (recommended for purposes of increased running speed), you will need to create a directory for it and copy the contents of the provided diskette into the new directory. This can be accomplished by following the flowchart given in Figure A.1. Proceed to the next section for details on running the TOOLKIT.

A.3.2   Running the Technology Forecasting TOOLKIT

The TOOLKIT may be run from either the provided diskette or from your system's hard disk drive. If you wish to run the toolkit from the hard disk drive and have not yet installed the software, see the previous section, A.3.1, "Installing the Technology Forecasting TOOLKIT", for installation instructions. To run the TOOLKIT, follow the flowchart given in Figure A.2.

A.4   A QUICK TUTORIAL

The Technology Forecasting TOOLKIT features pop-up and pull-down menus to help you select desired items and actions. To maneuver around the menus and to select items, you will be required to use the cursor keys ($\uparrow$, $\downarrow$, $\leftarrow$, $\rightarrow$) typically located at the right-hand side of the keyboard. For example, to select a tool from the menu of available tools (see Figure A.3), you would press either the $\uparrow$ key or the $\downarrow$ key until the tool you wish to use is highlighted. Once you have highlighted the desired tool, press Enter to start that tool.

   For purposes of this tutorial, assume that you have installed the TOOLKIT and it is now running on your system (see Section A.3 for information on running TOOLKIT). At the menu of available tools (see Figure A.3), use the $\uparrow$ and $\downarrow$ keys to highlight the Trend Extrapolation tool. To select the tool, press <Enter>. The Trend Extrapolation tool is now ready to use (see Figure A.4).

   Notice the top line of the screen. It contains the items File, Edit, View, and Curve Fit. This top line is called the *menu bar*. Each of these items represents either an action you can perform, or it leads to a menu of actions. To select one

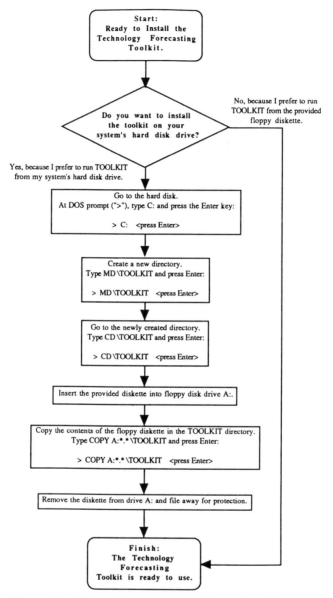

Figure A.1   Installing the Technology Forecasting TOOLKIT.

of these items from the menu bar, use the → and ← arrow keys to highlight the item and press <Enter> to select it.

On the bottom of the screen is a line containing a detailed description of the currently highlighted item. As you use the arrow keys to highlight different items, notice that the description line changes to provide you with more information on the highlighted item.

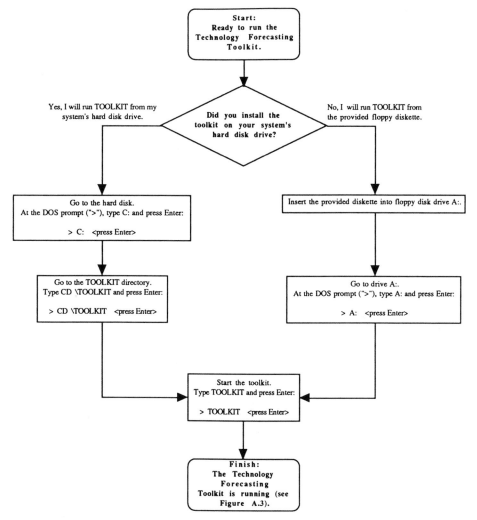

Figure A.2   Running the Technology Forecasting TOOLKIT.

Also on the bottom of the screen is the item F1: Help. If you need any help on how to use the system, need more information on a particular highlighted item, or need information on how to edit data, press the <F1> key (typically located on the left-hand side or top of the keyboard) for additional help.

As an example of using the menus, use the arrow keys to highlight the File item, and select it by pressing <Enter>. As shown in Figure A.5, a pull-down menu appears on the screen. This menu contains items that pertain to manipulating files containing trend data. These items can be selected by using the ↑ and ↓ keys and pressing <Enter>. As with the menu bar items, these items also have a more detailed description appearing on the bottom of the screen.

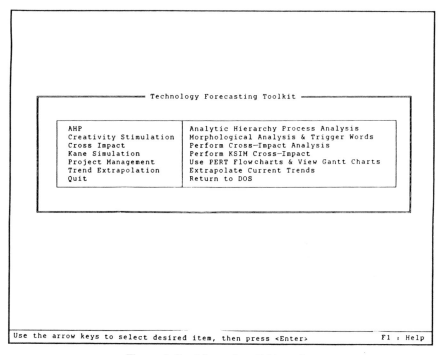

Figure A.3   Menu of available tools.

Next use the arrow keys to highlight the Open . . . menu item. Press <Enter> to bring up a list of trend files available for opening (see Figure A.6). There are two files to choose from: CATV and CATV2. Use the arrow keys to highlight the CATV2 file. Press <Enter> to open that file. A window appears, and the CATV2 trend data can be seen through the window (see Figure A.7). A window is simply an area of the screen that contains a well-defined group of information. In this case, the window contains trend data. There can be many windows on the screen at one time. For example, pressing <F1> will open another window containing help information.

By using the items located on the menu bar, you can edit the trend data, view different forms of the data, or attempt to fit various curves to the data. For example, use the arrow keys to select the View item, and press <Enter>. A menu appears containing various items to view. Use the arrow and <Enter> keys to select Plot. Assuming that your system has a graphics monitor, a data plot will appear. After viewing the plot, press any key to return to the menus.

Feel free to explore the other menus at this point. Try fitting a Gompertz curve to the data and then view the resulting plot. Try changing some of the data using Edit. View the statistics to see how well various curves fit the data.

When you are ready to quit and return to the main menu of available tools, select **File** from the menu bar. Select **Quit** to return to the main menu. If you have made changes to the data during your exploration, the system will warn you that

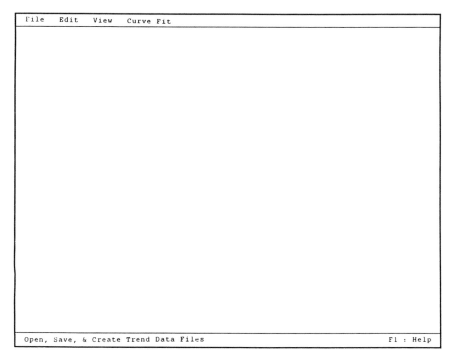

Figure A.4   Trend Extrapolation tool.

changes have been made and will ask you if you would like to save these changes. For the purposes of this tutorial, select **No, do not save the file**. Of course, when you are using TOOLKIT for your own work and assignments, you would probably choose the **Yes** option to save your work to disk.

A.5 INDIVIDUAL TOOLS AVAILABLE IN THE TECHNOLOGY FORECASTING TOOLKIT

Version 1.00 of the Technology Forecasting TOOLKIT contains six tools to support the student of technology forecasting and management: Analytic hierarchy process, Creativity stimulation, Cross-impact analysis, KSIM—Kane simulation, Project scheduling, and Trend extrapolation. Although each tool supports a different technique or method, the tools contain many similar interface features, such as pull-down and pop-up menus, common editing commands, and data file handling. This section describes these common features and explains the unique features of each tool.

1. *File handling.* Each tool contains an identical File menu on the menu bar. Using this menu, you may create or open data files, save current data, get help, print the data to a disk file, or exit the tool and return to the main menu of available tools.

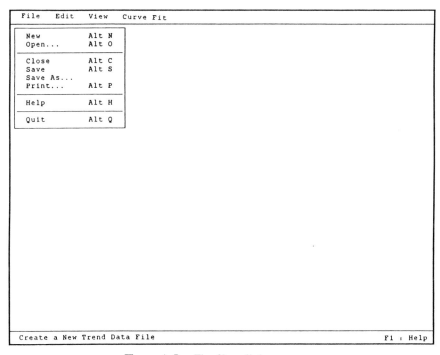

Figure A.5   The file pull-down menu.

New File—Selecting this item from the menu will create a new data file and place you in the data file editor. If there is a current data set open, it will be closed. If you have made changes to the current data set and have not yet saved the file, a warning message will appear, asking if you wish to save your changes, discard your changes, or cancel the operation.

Open File—Selecting this item from the menu will open an existing data file. If there is a current data set open, it will be closed. If you have made changes to the current data set and have not yet saved the file, a warning message will appear, asking if you wish to save your changes, discard your changes, or cancel the operation.

Close File—Selecting this item from the menu will close the current data file. If you have made changes to the current data set and have not yet saved the file, a warning message will appear, asking if you wish to save your changes, discard your changes, or cancel the operation.

Save File—Selecting this item from the menu will save the current data set. The file will be saved under the name specified at the top of the data set window. If you wish to save the file under a different name, use the **Save As . . .** item instead.

Save As . . .   File—Selecting this item from the menu will save the current data set. You will be asked to specify a new name for the file. Any eight-character

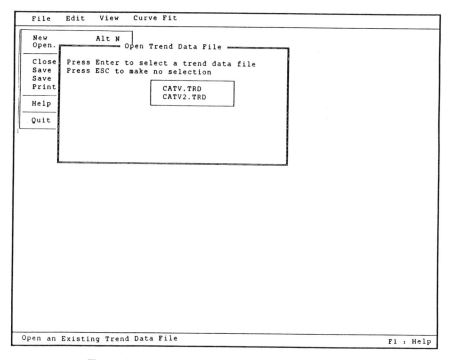

Figure A.6   Available data files for Trend Extrapolation.

DOS name is valid. If the file already exists, a warning message will appear, asking if you wish to overwrite the file or cancel the operation.

Print File—Selecting this item from the menu will print the contents of the current data file and the results of any analyses to disk. You will be asked to specify a name for the print file. Any eight-character DOS name is valid. If the file already exists, a warning message will appear, asking if you wish to overwrite the file or cancel the operation. You may then use the DOS print command to print the file to your printer.

Help—Selecting this tool from the menu will open a window containing general help for the particular tool.

Quit—Selecting this item from the menu will return you to the TOOLKIT main tool menu. If there is a current data set open, it will be closed. If you have made changes to the current data set and have not yet saved the file, a warning message will appear, asking if you wish to save your changes, discard your changes, or cancel the operations.

2. *Editing*. While editing data for individual tools, you will be entering information into fields. A field "knows" what type of information is allowed in that field and will not let you enter the wrong type of information (for example, you would not be able to enter test into a field expecting numeric information). To move from field to field, you should press the <Enter> key.

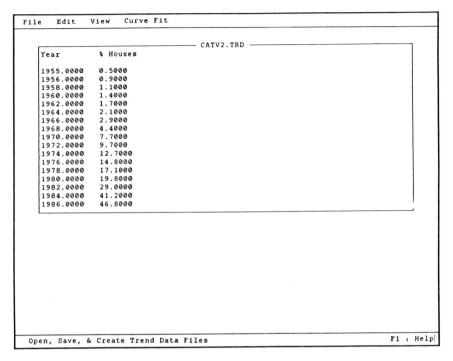

Figure A.7 Window containing trend data: CATV2.

Pressing <Enter> does not save your changes, it only moves you to the next field. To save your changes you will need to press the <CTRL> key and the <Enter> key at the same time (press and hold down <CTRL>, then press <Enter>). To abandon any changes you have made, press <Esc>. Any editing changes you had made to the data would not be reflected in the data set.

3. *Menu navigation.* All tools rely upon pull-down and pop-up menus for action selection. Use the arrow keys (↑ , ↓ , →, ←) typically located at the right-hand side of your keyboard to navigate around the menus. As you move around, the current item will be highlighted (reverse video highlighting on monochrome monitors, color changes on color monitors). To select the highlighted item from a menu, press <Enter>. Note that occasionally the arrow keys do not appear to work once TOOLKIT is first run. Check to see that the Num Lock key is not active. Typically, if this is the case, the keyboard's Num Lock light will be on and the arrow keys do not appear to have any effect on the menus. Press Num Lock once to turn off the Num Lock feature. The Num Lock light should go off, and the arrow keys should now work properly.

A.5.1 Analytic Hierarchy Process (from Chapter 18)

The Analytic Hierarchy Process tool supports the selection of a solution to a problem from a list of alternative solutions. Use this tool to state the problem, list alternative solutions, specify judgment criteria, and make pairwise comparisons between alternatives. The tool will use these judgments to compute an overall rating of alternatives and will measure your consistency in making comparison judgments. Features of note include:

1. *Computer-generated consistency measures and alternative ratings.* The computed consistency measures and ratings of alternatives are shown in the Data File window of any open data set.
2. *Editing.* Editing of problem statements, solution alternative names, judgment criteria, and judgments of priorities is done via the Edit item in the menu bar.

   Edit Names of Alternatives—Select this item to enter or modify the problem statement and the descriptions of alternative solutions to the problem. Up to seven alternatives may be entered. To delete an existing description, enter spaces into the desired field.

   Edit Priority Judgments—Select this option to enter your pairwise priority judgments between alternatives. Use the priority scale provided in the Edit window. Remember, you are entering a judgment rating of the ROW alternative over the COLUMN alternative.

A.5.2 Creativity Stimulation (from Chapter 7)

The creativity stimulation tool helps you to stimulate your creativity by allowing you to view trigger concepts and apply morphological analysis to a problem. Use this tool to view trigger concepts, define dimensions of a problem, list attributes for those dimensions, and generate alternative combinations of those attributes. Features of note include:

1. *Trigger concepts.* Select this item to view randomly selected trigger concepts. Three concepts will be presented at a time. You may repeatedly view concepts by pressing the <M> key.
2. *Editing.* Editing of dimension names and dimension attributes is done via the Edit item in the menu bar.

   Edit Names of Dimensions—Select this item to enter or modify dimension names for the given morphological analysis. Up to five dimensions may be entered. To delete an existing dimension, enter spaces into the desired field.

   Edit Dimension Attributes—Select this item to enter or modify attribute names for the dimensions of the morphological analysis. Up to 14 dimensions may be entered for each dimension. To delete an existing attribute, enter spaces into the desired field.

3. *Combining attributes*. Select this item to view randomly combined dimension attributes. One attribute from each dimension will be selected and presented. You may repeatedly view new combinations by pressing the <M> key.

A.5.3 Cross-Impact Analysis (from Chapter 12)

The cross-impact analysis tool helps perform an event-on-event cross-impact analysis on a set of events. Use this tool to define a set of events, estimate their probabilities, estimate the conditional probabilities, and execute a Monte Carlo simulation. Features of note include:

1. *Editing*. Editing of event names and their marginal probabilities, the conditional probabilities of occurrence, and the conditional probabilities of nonoccurrence are done via the Edit item in the menu bar.

   Edit Names and Marginals—Select this item to enter or modify event names and their initial marginal probabilities for the given cross-impact analysis. Up to seven events names may be entered. To delete an existing event, enter spaces into the desired field. Marginal probabilities must take a value between 0 and 1.

   Edit Occurrence Conditionals—Select this item to change the conditional probabilities of events depending upon the occurrence of other events. Conditional probabilities must take a value between 0 and 1. Remember, the number you are entering is the conditional probability of the ROW event occurring given that the COLUMN event has already occurred.

   Edit Nonoccurrence Conditionals—Select this item to change the conditional probabilities of events depending upon the nonoccurrence of other events. Conditional probabilities must take a value between 0 and 1. Remember, the number you are entering is the conditional probability of the ROW event occurring given that the COLUMN event has not already occurred.

2. *Viewing*. Viewing the conditional probabilities of occurrence and the conditional probabilities of nonoccurrence is done via the View item in the menu bar.

   View Occurrence Matrix—Select this item to view the conditional probabilities of events depending upon the occurrence of other events. Remember, the number you are viewing is the conditional probability of the ROW event occurring given that the COLUMN event has already occurred.

   View Nonoccurrence Matrix—Select this item to view the conditional probabilities of events depending upon the nonoccurrence of other events. Remember, the number you are viewing is the conditional probability of the ROW event occurring given that the COLUMN event has not already occurred.

3. *Running the simulation*. Select this item to run a Monte Carlo simulation for the current data set. You can specify the number of Monte Carlo rounds to run the simulation. During the simulation, a counter provides the number of rounds completed and the number remaining in the simulation.

A.5.4   KSIM—Kane Simulation (from Chapter 12)

The KSIM—Kane Simulation tool helps perform a trend-on-trend analysis for a set of trends. Use this too to define a set of trend variables, provide their starting values, describe trend-on-trend impacts, and define the start time, stop time, and time delta. The tool models the impacts of variables upon other variables and either plots or tabulates the results. Features of note include:

1. *Editing.* Editing of trend variable names and their initial values, the analysis time period, and the trend-on-trend impacts are done via the Edit item in the menu bar.

   Edit Names, Values, and Time Periods—Select this item to edit the names and initial values of trend variables. Also select this item to edit the start time, stop time, and time delta for the analysis. Up to seven trend variables may be entered. To delete an existing trend variable name, enter spaces into the desired field. Initial values must lie between 0 and 1.

   Edit Alpha Values—Select this option to edit the long-term trend-on-trend impacts. Impact values must be between $+3$ and -3. Remember, the number you enter is the long-term impact of the ROW variable on the COLUMN variable.

   Edit Beta Values—Select this option to edit the short-term and trend-on-trend impacts. Impact values must be between $+3$ and -3. Remember, the number you enter is the short-term impact of the ROW variable on the COLUMN variable.

2. *Viewing.* Viewing of the alpha and beta value matrices is done via the View item in the menu bar.

   View Alpha Values—Select this option to view the long-term and trend-on-trend impacts. Remember, the number you are viewing is the long-term impact of the ROW variable on the COLUMN variable.

   View Beta Values—Select this option to view the short-term trend-on-trend impacts. Remember, the number you are viewing is the short-term impact of the ROW variable on the COLUMN variable.

3. *Running the model.* Select this item to run the KSIM analysis. Analysis information can be output in either graphical or tabular form.

   Run Plot—If you are using a graphics monitor, select this item to plot the effect of trends upon other trends over the specified time period.

   Run Table—Select this item to list the effect of trends upon other trends over the specified time period.

A.5.5   Project Scheduling (from Chapter 5)

The project scheduling tool helps schedule technology management projects. Use this tool to list project tasks, specify dependencies among tasks, discover critical tasks, compute task start and end times, and view a Gantt chart and task network for the project. Features of note include:

1. *Editing*. Editing of task names and their durations, start and finish tasks, and task dependencies is done via the Edit item in the menu bar.

   Edit Tasks—Select this item to enter, edit, and/or delete project tasks and their durations. Up to 12 tasks may be entered. To delete an existing task, enter spaces into both the task number and task description fields.

   Edit Start/Finish Tasks—Select this item to enter or change the project start and finish tasks. Each project should have one start task and one finish task.

   Edit Task Dependencies—Select this item to edit dependencies between tasks. To specify a task dependency, enter a 1 into the dependency field. To specify no dependency, enter a 0 (the default). Remember, the dependency field specifies that the task number listed in the ROW is dependent upon the task number listed in the COLUMN header.

2. *Viewing*. Viewing of task dependencies, computer PERT values, Gantt charts, and the task network is done via the View item in the menu bar.

   View Dependencies—Select this item to view the task dependency matrix. A value of 1 in the dependency field specifies a dependency, while a value of 0 specifies no dependency. Remember, the dependency field specifies that the task number listed in the ROW is dependent upon the task number listed in the COLUMN header.

   View PERT Values—Select this item to view the computed PERT values and critical tasks for the project.

   View Gantt Chart—If you are using a graphics monitor, select this item to view a Gantt chart for the project.

   View Task Network—If you are using a graphics monitor, select this item to view a network representation of the task and dependency information.

A.5.6 Trend Extrapolation (from Chapters 9 and 10)

The trend extrapolation tool helps extrapolate data on trends using linear, Gompertz, Fisher-Pry, and exponential curve fitting techniques. Use this tool to enter a two-dimensional data set, fit curves to the data, plot the data, and compute statistics on the accuracy of the curves. Features of note include:

1. *Editing*. Select the Edit item from the menu bar to enter or modify trend data. Up to 17 data points may be entered. To enter a data point, enter a non-zero value into the X and Y data fields. The data limit field (used for Gompertz and Fisher-Pry fits) must be set to a value greater than the largest Y value. To delete an existing data point, enter 0 into both the X and Y fields.

2. *Viewing*. Viewing of computed curve fit equations, statistics, data plots, and user-defined extrapolated values is done via the View item in the menu bar.

   View Equations—Select this item to view the solved curve fit equations for the selected curve fit techniques (selected via the Curve Fit menu item).

   View Statistics—Select this item to view various statistics computed for the selected curve fit techniques.

View Plot—If you are using a graphics monitor, select this item to plot the data. If curve fit techniques have been selected, their respective curves will also be plotted.

View Extrapolated Data—Select this item to enter pointer for extrapolation along the selected curve fit techniques. Enter an X value for extrapolation and press <Enter> to extrapolate to points outside the current data set.

3. *Fitting curves to the data.* Various curve fit techniques can be chosen via the Curve Fit item in the menu bar.

Curve Fit – Linear—Select this item to apply a linear fit to the data set. Selecting this item again will remove the linear fit.

Curve Fit – Gompertz—Select this item to apply a Gompertz fit to the data set. Selecting this item again will remove the Gompertz fit.

Curve Fit – Fisher-Pry—Select this item to apply a Fisher-Pry fit to the data set. Selecting this item again will remove the Fisher-Pry fit.

Curve Fit – Exponential—Select this item to apply an exponential fit to the data set. Selecting this item again will remove the exponential fit.

_____Appendix B.1
F Statistic Values

v_1 = Degrees of Freedom for Numerator

v_2	α	1	2	3	4	5	6	7	8	9
1	.100	39.9	49.5	53.6	55.8	57.2	58.2	58.9	59.4	59.9
	.050	161.4	199.5	215.7	224.6	230.2	234.0	236.8	238.4	240.5
	.025	647.8	799.5	864.2	899.6	921.8	937.1	948.2	956.7	963.3
	.010	4052.2	4999.5	5403.4	5624.6	5763.6	5859.0	5928.4	5981.1	6022.5
	.001	40600.	50000.	54000.	56200.	57600.	58600.	59300.	59800.	60200.
2	.100	8.53	9.00	9.16	9.24	9.29	9.33	9.35	9.37	9.38
	.050	18.51	19.00	19.16	19.25	19.30	19.33	19.35	19.37	19.38
	.025	38.51	39.00	39.17	39.25	39.25	39.33	39.36	39.37	39.39
	.010	98.50	99.00	99.17	99.25	99.30	99.33	99.36	99.37	99.39
	.001	998.50	999.00	999.17	999.30	999.30	999.36	999.36	999.37	999.39
3	.100	5.54	5.46	5.39	5.34	5.31	5.28	5.27	5.25	5.24
	.050	10.13	9.55	9.28	9.12	9.01	8.94	8.89	8.85	8.81
	.025	17.44	16.04	15.44	15.10	14.88	14.73	14.62	14.54	14.47
	.010	34.12	30.82	29.46	28.71	28.24	27.91	27.67	27.49	27.35
	.001	167.03	148.50	141.11	137.10	134.58	132.85	131.58	130.62	129.86
4	.100	4.54	4.32	4.19	4.11	4.05	4.01	3.98	3.95	3.94
	.050	7.71	6.94	6.59	6.39	6.26	6.16	6.09	6.04	6.00
	.025	12.22	10.65	9.98	9.60	9.36	9.20	9.07	8.98	8.90
	.010	21.20	18.00	16.69	15.98	15.52	15.21	14.98	14.80	14.66
	.001	74.14	61.25	56.18	53.44	51.71	50.53	49.66	49.00	48.47
5	.100	4.06	3.78	3.62	3.52	3.45	3.40	3.37	3.34	3.32
	.050	6.61	5.79	5.41	5.19	5.05	4.95	4.88	4.82	4.77
	.025	10.01	8.43	7.76	7.39	7.15	6.98	6.85	6.76	6.68
	.010	16.26	13.27	12.06	11.39	10.97	10.67	10.46	10.29	10.16
	.001	47.18	37.12	33.20	31.09	29.75	28.83	28.16	27.65	27.24

Source: J. Banks and R. G. Heikes, Handbook of Tables and Graphs for the Industrial Engineer and Manager, ©1984 p. 610–614 Reprinted by permission of Prentice-Hall, Englewood Cliffs, New Jersey.

α = right-hand tail area. v_2 = degrees of freedom for denominator.

ν_1 = Degrees of Freedom for Numerator

ν_2	α	1	2	3	4	5	6	7	8	9
6	.100	3.78	3.46	3.29	3.18	3.11	3.05	3.01	2.98	2.96
	.050	5.99	5.14	4.76	4.53	4.39	4.28	4.21	4.15	4.10
	.025	8.81	7.26	6.60	6.23	5.99	5.82	5.70	5.60	5.52
	.010	13.75	10.92	9.78	9.15	8.75	8.47	8.26	8.10	7.98
	.001	35.51	27.00	23.70	21.92	20.80	20.03	19.46	19.03	18.69
7	.100	3.59	3.26	3.07	2.96	2.88	2.83	2.78	2.75	2.72
	.050	5.59	4.74	4.35	4.12	3.97	3.87	3.79	3.73	3.68
	.025	8.07	6.54	5.89	5.52	5.29	5.12	4.99	4.90	4.82
	.010	12.25	9.55	8.45	7.85	7.46	7.19	6.99	6.84	6.72
	.001	29.25	21.69	18.77	17.20	16.21	15.52	15.02	14.63	14.33
8	.100	3.46	3.11	2.92	2.81	2.73	2.67	2.62	2.59	2.56
	.050	5.32	4.46	4.07	3.84	3.69	3.58	3.50	3.44	3.39
	.025	7.57	6.06	5.42	5.05	4.82	4.65	4.53	4.43	4.36
	.010	11.26	8.65	7.59	7.01	6.63	6.37	6.18	6.03	5.91
	.001	25.41	18.49	15.83	14.39	13.48	12.86	12.40	12.05	11.77
9	.100	3.36	3.01	2.81	2.69	2.61	2.55	2.51	2.47	2.44
	.050	5.12	4.26	3.86	3.63	3.48	3.37	3.29	3.23	3.18
	.025	7.21	5.71	5.08	4.72	4.48	4.32	4.20	4.10	4.03
	.010	10.56	8.02	6.99	6.42	6.06	5.80	5.61	5.47	5.35
	.001	22.86	16.39	13.90	12.56	11.71	11.13	10.70	10.37	10.11
10	.100	3.29	2.92	2.73	2.61	2.52	2.46	2.41	2.38	2.35
	.050	4.96	4.10	3.71	3.48	3.33	3.22	3.14	3.07	3.02
	.025	6.94	5.46	4.83	4.47	4.24	4.07	3.95	3.85	3.78
	.010	10.04	7.56	6.55	5.99	5.64	5.39	5.20	5.06	4.94
	.001	21.04	14.91	12.55	11.28	10.48	9.93	9.52	9.20	8.96

$v_1 = $ Degrees of Freedom for Numerator

v_2	α	1	2	3	4	5	6	7	8	9
11	.100	3.23	2.86	2.66	2.54	2.45	2.39	2.34	2.30	2.27
	.050	4.84	3.98	3.59	3.36	3.20	3.09	3.01	2.95	2.90
	.025	6.72	5.26	4.63	4.28	4.04	3.88	3.76	3.66	3.59
	.010	9.65	7.21	6.22	5.67	5.32	5.07	4.89	4.74	4.63
	.001	19.69	13.81	11.56	10.35	9.58	9.05	8.66	8.35	8.12
12	.100	3.18	2.81	2.61	2.48	2.39	2.33	2.28	2.24	2.21
	.050	4.75	3.89	3.49	3.26	3.11	3.00	2.91	2.85	2.80
	.025	6.55	5.10	4.47	4.12	3.89	3.73	3.61	3.51	3.44
	.010	9.33	6.93	5.95	5.41	5.06	4.82	4.64	4.50	4.39
	.001	18.64	12.97	10.80	9.63	8.89	8.38	8.00	7.71	7.48
15	.100	3.07	2.70	2.49	2.36	2.27	2.21	2.16	2.12	2.09
	.050	4.54	3.68	3.29	3.06	2.90	2.79	2.71	2.64	2.59
	.025	6.20	4.77	4.15	3.80	3.58	3.41	3.29	3.20	3.12
	.010	8.68	6.36	5.42	4.89	4.56	4.32	4.14	4.00	3.89
	.001	16.59	11.34	9.34	8.25	7.57	7.09	6.74	6.47	6.26
18	.100	3.01	2.62	2.42	2.29	2.20	2.13	2.08	2.04	2.00
	.050	4.41	3.55	3.16	2.93	2.77	2.66	2.58	2.51	2.46
	.025	5.98	4.56	3.95	3.61	3.38	3.22	3.10	3.01	2.93
	.010	8.29	6.01	5.09	4.58	4.25	4.01	3.84	3.71	3.60
	.001	15.38	10.39	8.49	7.46	6.81	6.35	6.02	5.76	5.56
20	.100	2.97	2.59	2.38	2.25	2.16	2.09	2.04	2.00	1.96
	.050	4.35	3.49	3.10	2.87	2.71	2.60	2.51	2.45	2.39
	.025	5.87	4.46	3.86	3.51	3.29	3.13	3.01	2.91	2.84
	.010	8.10	5.85	4.94	4.43	4.10	3.87	3.70	3.56	3.46
	.001	14.82	9.95	8.10	7.10	6.46	6.02	5.69	5.44	5.24

v_1 = Degrees of Freedom for Numerator

v_2	α	1	2	3	4	5	6	7	8	9
25	.100	2.92	2.53	2.32	2.18	2.09	2.02	1.97	1.93	1.89
	.050	4.24	3.39	2.99	2.76	2.60	2.49	2.40	2.34	2.28
	.025	5.69	4.29	3.69	3.35	3.13	2.97	2.85	2.75	2.68
	.010	7.77	5.57	4.68	4.18	3.85	3.63	3.46	3.32	3.22
	.001	13.88	9.22	7.45	6.49	5.89	5.46	5.15	4.91	4.71
30	.100	2.88	2.49	2.28	2.14	2.05	1.98	1.93	1.88	1.85
	.050	4.17	3.32	2.92	2.69	2.53	2.42	2.33	2.27	2.21
	.025	5.57	4.18	3.59	3.25	3.03	2.87	2.75	2.65	2.57
	.010	7.56	5.39	4.51	4.02	3.70	3.47	3.30	3.17	3.07
	.001	13.29	8.77	7.05	6.12	5.53	5.12	4.82	4.58	4.39
40	.100	2.84	2.44	2.23	2.09	2.00	1.93	1.87	1.83	1.79
	.050	4.08	3.23	2.84	2.61	2.45	2.34	2.25	2.18	2.12
	.025	5.42	4.05	3.46	3.13	2.90	2.74	2.62	2.53	2.45
	.010	7.31	5.18	4.31	3.83	3.51	3.29	3.12	2.99	2.89
	.001	12.61	8.25	6.59	5.70	5.13	4.73	4.44	4.21	4.02
50	.100	2.81	2.41	2.20	2.06	1.97	1.90	1.84	1.80	1.76
	.050	4.03	3.18	2.79	2.56	2.40	2.29	2.20	2.13	2.07
	.025	5.34	3.97	3.39	3.05	2.83	2.67	2.55	2.46	2.38
	.010	7.17	5.06	4.20	3.72	3.41	3.19	3.02	2.89	2.78
	.001	12.22	7.96	6.34	5.46	4.90	4.51	4.22	4.00	3.82
60	.100	2.79	2.39	2.18	2.04	1.95	1.87	1.82	1.77	1.74
	.050	4.00	3.15	2.76	2.53	2.37	2.25	2.17	2.10	2.04
	.025	5.29	3.93	3.34	3.01	2.79	2.63	2.51	2.41	2.33
	.010	7.08	4.98	4.13	3.65	3.34	3.12	2.95	2.82	2.72
	.001	11.97	7.77	6.17	5.31	4.76	4.37	4.09	3.86	3.69

v_1 = Degrees of Freedom for Numerator

v_2	α	1	2	3	4	5	6	7	8	9
80	.100	2.77	2.37	2.15	2.02	1.92	1.85	1.79	1.75	1.71
	.050	3.96	3.11	2.72	2.49	2.33	2.21	2.13	2.06	2.00
	.025	5.22	3.86	3.28	2.95	2.73	2.57	2.45	2.35	2.28
	.010	6.96	4.88	4.04	3.56	3.26	3.04	2.87	2.74	2.64
	.001	11.67	7.54	5.97	5.12	4.58	4.20	3.92	3.70	3.53
90	.100	2.76	2.36	2.15	2.01	1.91	1.84	1.78	1.74	1.70
	.050	3.95	3.10	2.71	2.47	2.32	2.20	2.11	2.04	1.99
	.025	5.20	3.84	3.26	2.93	2.71	2.55	2.43	2.34	2.26
	.010	6.93	4.85	4.01	3.53	3.23	3.01	2.84	2.72	2.61
	.001	11.57	7.47	5.91	5.06	4.53	4.15	3.87	3.65	3.48
100	.100	2.76	2.36	2.14	2.00	1.91	1.83	1.78	1.73	1.69
	.050	3.94	3.09	2.70	2.46	2.31	2.19	2.10	2.03	1.97
	.025	5.18	3.83	3.25	2.92	2.70	2.54	2.42	2.32	2.24
	.010	6.90	4.82	3.98	3.51	3.21	2.99	2.82	2.69	2.59
	.001	11.50	7.41	5.86	5.02	4.48	4.11	3.83	3.61	3.44
∞	.100	2.71	2.30	2.08	1.95	1.85	1.77	1.72	1.67	1.63
	.050	3.84	3.00	2.61	2.37	2.21	2.10	2.01	1.94	1.88
	.025	5.02	3.69	3.12	2.79	2.57	2.41	2.29	2.19	2.11
	.010	6.64	4.61	3.77	3.32	3.02	2.80	2.64	2.51	2.41
	.001	10.81	6.85	5.43	4.61	4.15	3.78	3.50	3.29	3.12

v_1 = Degrees of Freedom for Numerator

v_2	α	10	15	20	25	30	50	75	100	∞
1	.100	60.2	61.2	61.7	62.1	62.3	62.7	62.9	63.0	63.3
	.050	241.9	245.9	248.0	249.3	250.1	251.8	258.8	253.0	280.7
	.025	968.6	984.9	993.1	998.1	1001.4	1008.1	1011.5	1013.2	1018.0
	.010	6055.8	6157.3	6208.7	6329.8	6260.6	6302.5	6323.6	6334.1	6366.0
	.001	60600.	61600.	62100.	62400.	62500.	63000.	63200.	63300.	63700.
2	.100	9.39	9.42	9.44	9.45	9.46	9.47	9.48	9.48	9.49
	.050	19.40	19.43	19.45	19.46	19.46	19.48	19.48	19.49	19.51
	.025	39.40	39.43	39.45	39.46	39.46	39.48	39.48	39.49	39.50
	.010	99.40	99.43	99.45	99.46	99.47	99.48	99.49	99.49	99.64
	.001	999.40	999.43	999.45	999.46	99.47	999.48	999.49	999.49	999.49
3	.100	5.23	5.20	5.18	5.17	5.17	5.15	5.15	5.14	5.13
	.050	8.79	8.70	8.66	8.63	8.62	8.58	8.56	8.55	8.53
	.025	14.42	14.25	14.17	14.12	14.08	14.01	13.97	13.96	13.90
	.010	27.23	26.87	26.69	26.58	26.50	26.35	26.28	26.24	26.14
	.001	129.25	127.37	126.42	125.84	125.45	124.66	124.27	124.07	124.00
4	.100	3.92	3.87	3.84	3.83	3.82	3.80	3.78	3.78	3.76
	.050	5.96	5.86	5.80	5.77	5.75	5.70	5.68	5.66	5.63
	.025	8.84	8.66	8.56	8.50	8.46	8.38	8.34	8.32	8.27
	.010	14.55	14.20	14.02	13.91	13.84	13.69	13.61	13.58	13.47
	.001	48.05	46.76	46.10	45.70	45.43	44.88	44.61	44.47	44.46
5	.100	3.30	3.24	3.21	3.19	3.17	3.15	3.13	3.13	3.11
	.050	4.74	4.62	4.56	4.52	4.50	4.44	4.42	4.41	4.37
	.025	6.62	6.43	6.33	6.27	6.23	6.14	6.10	6.08	6.02
	.010	10.05	9.72	9.55	9.45	9.38	9.24	9.17	9.13	9.04
	.001	26.92	25.91	25.39	25.08	24.87	24.44	24.22	24.12	23.98

v_1 = Degrees of Freedom for Numerator

v_2	α	10	15	20	25	30	50	75	100	∞
6	.100	2.94	2.87	2.84	2.81	2.80	2.77	2.75	2.75	2.72
	.050	4.06	3.94	3.87	3.83	3.81	3.75	3.73	3.71	3.67
	.025	5.46	5.27	5.17	5.11	5.07	4.98	4.94	4.92	4.85
	.010	7.87	7.56	7.40	7.30	7.23	7.09	7.02	6.99	6.88
	.001	18.41	17.56	17.12	16.85	16.67	16.31	16.12	16.03	15.92
7	.100	2.70	2.63	2.59	2.57	2.56	2.52	2.51	2.50	2.47
	.050	3.64	3.51	3.44	3.40	3.38	3.32	3.29	3.27	3.23
	.025	4.76	4.57	4.47	4.40	4.36	4.28	4.23	4.21	4.15
	.010	6.62	6.31	6.16	6.06	5.99	5.86	5.79	5.75	5.66
	.001	14.08	13.32	12.93	12.69	12.53	12.20	12.04	11.95	11.72
8	.100	2.54	2.46	2.42	2.40	2.38	2.35	2.33	2.32	2.29
	.050	3.35	3.22	3.15	3.11	3.08	3.02	2.99	2.97	2.93
	.025	4.30	4.10	4.00	3.94	3.89	3.81	3.76	3.74	3.67
	.010	5.81	5.52	5.36	5.26	5.20	5.07	5.00	4.96	4.87
	.001	11.54	10.84	10.48	10.26	10.11	9.80	9.65	9.57	9.41
9	.100	2.42	2.34	2.30	2.27	2.25	2.22	2.20	2.19	2.16
	.050	3.14	3.01	2.94	2.89	2.86	2.80	2.77	2.76	2.71
	.025	3.96	3.77	3.67	3.60	3.56	3.47	3.43	3.40	3.33
	.010	5.26	4.96	4.81	4.71	4.65	4.52	4.45	4.41	4.31
	.001	9.89	9.24	8.90	8.69	8.55	8.26	8.11	8.04	7.93
10	.100	2.32	2.24	2.20	2.17	2.16	2.12	2.10	2.09	2.06
	.050	2.98	2.85	2.77	2.73	2.70	2.64	2.60	2.59	2.54
	.025	3.72	3.52	3.42	3.35	3.31	3.22	3.18	3.15	3.08
	.010	4.85	4.56	4.41	4.31	4.25	4.12	4.05	4.01	3.91
	.001	8.75	8.13	7.80	7.60	7.47	7.19	7.05	6.98	6.92

$v_1 =$ Degrees of Freedom for Numerator

v_2	α	10	15	20	25	30	50	75	100	∞
11	.100	2.25	2.17	2.12	2.10	2.08	2.04	2.02	2.01	1.97
	.050	2.85	2.72	2.65	2.60	2.57	2.51	2.47	2.46	2.40
	.025	3.53	3.33	3.23	3.16	3.12	3.03	2.98	2.96	2.88
	.010	4.54	4.25	4.10	4.01	3.94	3.81	3.74	3.71	3.61
	.001	7.92	7.32	7.01	6.81	6.68	6.42	6.28	6.21	6.05
12	.100	2.19	2.10	2.06	2.03	2.01	1.97	1.95	1.94	1.90
	.050	2.75	2.62	2.54	2.50	2.47	2.40	2.37	2.35	2.30
	.025	3.37	3.18	3.07	3.01	2.96	2.87	2.82	2.80	2.73
	.010	4.30	4.01	3.86	3.76	3.70	3.57	3.50	3.47	3.36
	.001	7.29	6.71	6.40	6.22	6.09	5.83	5.70	5.63	5.48
15	.100	2.06	1.97	1.92	1.89	1.87	1.83	1.80	1.79	1.76
	.050	2.54	2.40	2.33	2.28	2.25	2.18	2.14	2.12	2.07
	.025	3.06	2.86	2.76	2.69	2.64	2.55	2.50	2.47	2.40
	.010	3.80	3.52	3.37	3.28	3.21	3.08	3.01	2.98	2.87
	.001	6.08	5.54	5.25	5.07	4.95	4.70	4.57	4.51	4.35
18	.100	1.98	1.89	1.84	1.80	1.78	1.74	1.71	1.70	1.66
	.050	2.41	2.27	2.19	2.14	2.11	2.04	2.00	1.98	1.92
	.025	2.87	2.67	2.56	2.49	2.44	2.35	2.30	2.27	2.19
	.010	3.51	3.23	3.08	2.98	2.92	2.78	2.71	2.68	2.57
	.001	5.39	4.87	4.59	4.42	4.30	4.06	3.93	3.87	3.70
20	.100	1.94	1.84	1.79	1.76	1.74	1.69	1.66	1.65	1.61
	.050	2.35	2.20	2.12	2.07	2.04	1.97	1.93	1.91	1.84
	.025	2.77	2.57	2.46	2.40	2.35	2.25	2.20	2.17	2.09
	.010	3.37	3.09	2.94	2.84	2.78	2.64	2.57	2.54	2.42
	.001	5.08	4.56	4.29	4.12	4.00	3.77	3.64	3.58	3.42

v_1 = Degrees of Freedom for Numerator

v_2	α	10	15	20	25	30	50	75	100	∞
25	.100	1.87	1.77	1.72	1.68	1.66	1.61	1.58	1.56	1.52
	.050	2.24	2.09	2.01	1.96	1.92	1.84	1.80	1.78	1.71
	.025	2.61	2.41	2.30	2.23	2.18	2.08	2.02	2.00	1.91
	.010	3.13	2.85	2.70	2.60	2.54	2.40	2.33	2.29	2.17
	.001	4.56	4.06	3.79	3.63	3.52	3.28	3.15	3.09	2.92
30	.100	1.82	1.72	1.67	1.63	1.61	1.55	1.52	1.51	1.46
	.050	2.16	2.01	1.93	1.88	1.84	1.76	1.72	1.70	1.62
	.025	2.51	2.31	2.20	2.12	2.07	1.97	1.91	1.88	1.79
	.010	2.98	2.70	2.55	2.45	2.39	2.25	2.17	2.13	2.01
	.001	4.24	3.75	3.49	3.33	3.22	2.98	2.86	2.79	2.61
40	.100	1.76	1.66	1.61	1.57	1.54	1.48	1.45	1.43	1.38
	.050	2.08	1.92	1.84	1.78	1.74	1.66	1.61	1.59	1.51
	.025	2.39	2.18	2.07	1.99	1.94	1.83	1.77	1.74	1.64
	.010	2.80	2.52	2.37	2.27	2.20	2.06	1.98	1.94	1.81
	.001	3.87	3.40	3.14	2.98	2.87	2.64	2.51	2.44	2.24
50	.100	1.73	1.63	1.57	1.53	1.50	1.44	1.41	1.39	1.33
	.050	2.03	1.87	1.78	1.73	1.69	1.60	1.55	1.52	1.44
	.025	2.32	2.11	1.99	1.92	1.87	1.75	1.69	1.66	1.55
	.010	2.70	2.42	2.27	2.17	2.10	1.95	1.87	1.82	1.68
	.001	3.67	3.20	2.95	2.79	2.68	2.44	2.31	2.25	2.03
60	.100	1.71	1.60	1.54	1.50	1.48	1.41	1.38	1.36	1.29
	.050	1.99	1.84	1.75	1.69	1.65	1.56	1.51	1.48	1.39
	.025	2.27	2.06	1.94	1.87	1.82	1.70	1.63	1.60	1.48
	.010	2.63	2.35	2.20	2.10	2.03	1.88	1.79	1.75	1.60
	.001	3.54	3.08	2.83	2.67	2.55	2.32	2.19	2.12	1.89

v_1 = Degrees of Freedom for Numerator

v_2	α	10	15	20	25	30	50	75	100	∞
80	.100	1.68	1.57	1.51	1.47	1.44	1.38	1.34	1.32	1.24
	.050	1.95	1.79	1.70	1.64	1.60	1.51	1.45	1.43	1.32
	.025	2.21	2.00	1.88	1.81	1.75	1.63	1.56	1.53	1.40
	.010	2.55	2.27	2.12	2.01	1.94	1.79	1.70	1.65	1.49
	.001	3.39	2.93	2.68	2.52	2.41	2.16	2.03	1.96	1.72
90	.100	1.67	1.56	1.50	1.46	1.43	1.36	1.33	1.30	1.23
	.050	1.94	1.78	1.69	1.63	1.59	1.49	1.44	1.41	1.30
	.025	2.19	1.98	1.86	1.79	1.73	1.61	1.54	1.50	1.37
	.010	2.52	2.24	2.09	1.99	1.92	1.76	1.67	1.62	1.46
	.001	3.34	2.88	2.63	2.47	2.36	2.11	1.98	1.91	1.66
100	.100	1.66	1.56	1.49	1.45	1.42	1.35	1.32	1.29	1.21
	.050	1.93	1.77	1.68	1.62	1.57	1.48	1.42	1.39	1.28
	.025	2.18	1.97	1.85	1.77	1.71	1.59	1.52	1.48	1.35
	.010	2.50	2.22	2.07	1.97	1.89	1.74	1.65	1.60	1.43
	.001	3.30	2.84	2.59	2.43	2.32	2.08	1.94	1.87	1.62
∞	.100	1.60	1.49	1.42	1.38	1.34	1.26	1.21	1.18	1.00
	.050	1.83	1.67	1.57	1.51	1.46	1.35	1.28	1.24	1.00
	.025	2.05	1.83	1.71	1.63	1.57	1.43	1.34	1.30	1.00
	.010	2.32	2.04	1.88	1.77	1.70	1.52	1.42	1.36	1.00
	.001	2.98	2.52	2.27	2.11	1.99	1.73	1.58	1.50	1.00

t Statistic Values

Degrees of Freedom ν	α = Right-Hand Tail Area						
	.250	.100	.050	.025	.010	.005	.001
1	1.000	3.078	6.314	12.706	31.821	63.567	318.309
2	.816	1.886	2.920	4.303	6.965	9.925	22.327
3	.765	1.638	2.353	3.182	4.541	5.841	10.215
4	.741	1.533	2.132	2.776	3.747	4.604	7.173
5	.727	1.476	2.015	2.571	3.365	4.032	5.893
6	.718	1.440	1.943	2.447	3.143	3.707	5.208
7	.711	1.415	1.895	2.365	2.998	3.499	4.785
8	.706	1.397	1.860	2.306	2.896	3.355	4.501
9	.703	1.383	1.833	2.262	2.821	3.250	4.297
10	.700	1.372	1.812	2.228	2.764	3.169	4.144
11	.697	1.363	1.796	2.201	2.718	3.106	4.025
12	.695	1.356	1.782	2.179	2.681	3.055	3.930
13	.694	1.350	1.771	2.160	2.650	3.012	3.852
14	.692	1.345	1.761	2.145	2.624	2.977	3.787
15	.691	1.341	1.753	2.131	2.602	2.947	3.733
16	.690	1.337	1.746	2.120	2.583	2.921	3.686
17	.689	1.333	1.740	2.110	2.567	2.898	3.646
18	.688	1.330	1.734	2.101	2.552	2.878	3.610
19	.688	1.328	1.729	2.093	2.539	2.861	3.579
20	.687	1.325	1.725	2.086	2.528	2.845	3.552
21	.686	1.323	1.721	2.080	2.518	2.831	3.527
22	.686	1.321	1.717	2.074	2.508	2.819	3.505
23	.685	1.319	1.714	2.069	2.500	2.807	3.485
24	.685	1.318	1.711	2.064	2.492	2.797	3.467
25	.684	1.316	1.708	2.060	2.485	2.787	3.450
26	.684	1.315	1.706	2.056	2.479	2.779	3.435
27	.684	1.314	1.703	2.052	2.473	2.771	3.421
28	.683	1.313	1.701	2.048	2.467	2.763	3.408
29	.683	1.311	1.699	2.045	2.462	2.756	3.396
30	.683	1.310	1.697	2.042	2.457	2.750	3.385
35	.682	1.306	1.690	2.030	2.438	2.724	3.340
40	.681	1.303	1.684	2.021	2.423	2.704	3.307
50	.679	1.299	1.676	2.009	2.403	2.678	3.261
60	.679	1.296	1.671	2.000	2.390	2.660	3.232
70	.678	1.294	1.667	1.994	2.381	2.648	3.211
80	.678	1.292	1.664	1.990	2.374	2.639	3.195
90	.677	1.291	1.662	1.987	2.368	2.632	3.183
100	.677	1.290	1.660	1.984	2.364	2.626	3.174
120	.677	1.289	1.658	1.980	2.358	2.617	3.160
∞	.674	1.282	1.645	1.960	2.326	2.576	3.090

Source: J. Banks and R. G. Heikes, *Handbook of Tables and Graphs for the Industrial Engineer and Manager,* ©1984, p. 57 Reprinted by permission of Prentice-Hall, Englewood Cliffs, New Jersey.

━━━━━ Appendix C
SENSITIVITY ANALYSIS

The repeated operation of a model with successive alterations of its state (the variables, their values, or interrelationships) for the purpose of observing the dependence of its behavior on state factors is called *sensitivity analysis*. There are two basic motivations for such analysis: (1) determination of the dependence of model output on assumptions and on the accuracy of input data and (2) the identification of leverage points on which to base strategies for modification of system behavior.

Sensitivity can be computed as the ratio between the fractional change in a parameter that serves as a basis for decision to the fractional change in the variable being tested. The procedure requires recomputing the value of the decision parameter $F(x)$ using the changed value of the variable x:

$$\text{Sensitivity} = \frac{\dfrac{\Delta F(x)}{F(x)}}{\dfrac{\Delta x}{x}} = \frac{\dfrac{F(x_1) - F(x_0)}{F(x_0)}}{\dfrac{x_1 - x_0}{x_0}}$$

where

x_0 = base value of the variable being tested
x_1 = new value of x
$F(x_0)$ = value of decision parameter when the base value of the variable is used
$F(x_1)$ = value of decision parameter at new value of the variable being tested

Variables having sensitivities equal to or greater than unity are the most sensitive. Variables that are sensitive need to be carefully estimated; those that are insensitive do not require such care. In complex models, it is necessary to select the most pivotal and uncertain variables for analysis. There is no foolproof way to make this selection. However, if the behavior of a particular parameter is crucial to the assessment sponsor, the public, or the conclusions of the study, then variables directly affecting that parameter are prime candidates for analysis.

BIBLIOGRAPHY

Aaker, D. A. (1988). *Developing Business Strategies*. New York: John Wiley.

Abrams, M. and Bernstein, H. (1989). *Future Stuff*. New York: Viking.

Adams, H. K. (1974). *Conceptual Blockbusting: A Guide to Better Ideas*. New York: Addison-Wesley.

Adams, J. L. (1986). *The Care and Feeding of Ideas: A Guide to Encouraging Creativity*. New York: Addison-Wesley.

Adler, J. and Leonard, E. A. (1989). "Anniversary Guide for the Decade Ahead," *Newsweek* vol. 114, 78–82.

Allen, T. J. (1977). *Managing the Flow of Technology: Technology Transfer and the Dissemination of Technological Information Within the R&D Organization*. Cambridge, MA: MIT Press.

Allio, R. J. and Pennington, M. (eds.) (1979). *Corporate Planning: Techniques and Applications*. New York: AMACOM.

Ansoff, H. I. (1987). "Strategic Management of Technology," *Journal of Business Strategy,* vol. 7, no. 3, 28–39.

Appelbaum, R. P. (1970). *Theories of Social Change*. Chicago: Markham.

Argenti, J. (1988). "Power to the Planner," *Accountancy,* vol. 101, 119–120.

Argote, L. and Epple, D. (1990). "Learning Curves in Manufacturing," *Science,* vol. 247, 920–924.

Armstrong, J. E. and Harman, W.W. (1977). "Strategies for Conducting Technology Assessment," Report to the National Science Foundation. Stanford, CA: Stanford University, Department of Engineering—Economic Systems.

Arnstein, S. R. (1975). "A Working Model for Public Participation," *Public Administration Review,* vol. 35, 70–73.

Arnstein, S. R. and Christakis, A. (1975). *Perspectives on Technology Assessment*. Jerusalem: Science and Technology Publishers.

Arrow, K. J. (1963). *Social Choice and Individual Values*. New York: Wiley.

Ascher, W. (1978). *Forecasting: An Appraisal for Policy Makers and Planners*. Baltimore, MD: The Johns Hopkins University Press.

Bailetti, A. J., Callahan, J. R., and Elliott, J. (1988). "The Potential Consequences of High Temperature Superconductivity for Canadian Industry," in Khalil, T. M., Bayraktar, B. A., and Edosomwan, J. A. (eds.), *Management of Technology I.* (pp. 496–503). Geneva: Interscience Enterprises Ltd.

Bakus, G. J., Stillwell, W. G., Latter, S. M., and Wallerstein, M.C. (1982). "Decision Making: With Applications for Environmental Management," *Journal of Environmental Management,* vol. 6, no. 6, 493–504.

Banks, J. and Heikes, R. G. (1984). *Handbook of Tables and Graphs for the Industrial Engineer and Manager.* Englewood Cliffs, NJ; Prentice-Hall (Reston).

Becker. H. A. (1988). "Social Impact Assessment by Scenario Projects Combining Quantitative Analysis," *Impact Assessment Bulletin,* vol. 6, 89–102.

Becker, H. A., Klaassen-van den Berg Jeths, A., Kraan-Jetten, A., and van Rijsselt, R. (1986). "Contextual Scenarios on the Elderly and Their Health in The Netherlands," *Impact Assessment Bulletin,* vol. 4, no. 3/4, 15–48.

Becker, H. S. (1983). "Scenarios: A Tool of Growing Importance to Policy Analysts in Government and Industry," *Technological Forecasting and Social Change,* vol. 23, 95–120.

Becker, R. H. and Speltz, L. M. (1986). "Making More Explicit Forecasts," *Research Management,* vol. 29 (July/August), 21–23.

Bell, F. C. (1984/85). "Practical Predictive Models for Quantitative Impact Analyses," *Impact Assessment Bulletin,* vol. 3, no. 3, 24–30.

Berg, M. R. (1975). "Methodology," in Arnstein, S. R., and Christakis, A. (eds.), *Perspectives on Technology Assessment.* (pp. 63–72). Jerusalem: Science and Technology Publishers.

Bezold, C. (1990). "A Highly Visible Source of Alternative Scenarios," *The Futurist,* vol. 24 (March/April), 24–26.

Bhargava, S. C. (1989). "Generalized Lotka-Volterra Equations and the Mechanism of Technological Substitution," *Technological Forecasting and Social Change,* vol. 35, 319–326.

Bierman, H., Jr. and Hass, J. E. (1973). *An Introduction to Managerial Finance.* New York: Norton.

Branch, K., Hooper, D. A., Thompson, J., and Creighton, J. (1984). *Guide to Social Assessment: A Framework for Assessing Social Change.* Boulder, CO: Westview Press.

Brauer, R. L. (1986). *Facilities Planning: The User Requirements Method.* New York: AMACOM.

Bregha, F. (1989). "The Integration of Environmental Considerations in Government Policy," paper presented at the Annual Conference of the International Association for Impact Assessment, Montreal (June).

Bremer, S. A. (1989). "Computer Modeling in Global and International Relations: The State of the Art," *Social Science Computer Review,* vol. 7, no. 4, 459–478.

Bright, J. R. (1972). *A Brief Introduction to Technology Forecasting: Concepts and Exercises.* Austin, TX: The Pemaquid Press (c/o Technology Futures, Inc.).

Bright, J. R. (1978). *Practical Technology Forecasting: Concepts and Exercises.* Austin TX: The Industrial Management Center.

Brockhaus, W. L. (1975). "A Quantitative Analytical Methodology for Judgmental and Policy Decisions," *Technological Forecasting and Social Change,* vol. 7, 127–137.

Brown, J. K. (1980). *This Business of Issues: Coping with the Company's Environment.* New York: The Conference Board.

Buckle, L. G. and Thomas-Buckle, S. R. (1986). "Placing Environmental Mediation in Context: Lessons from 'Failed' Mediations," *Environmental Impact Assessment Review,* vol. 6, 55–70.

Buckley, M. R. and Reed, R. (1988). "Strategy in Action—Techniques for Implementing Strategy," *Long Range Planning,* vol. 21, 67–74.

Burdge, R. J. (1987). "The Social Impact Assessment Model and the Planning Process," *Environmental Impact Assessment Review,* vol. 7, 141–150.

Burdge, R. J., and Opryszek, P. (1981). *Coping with Change: An Interdisciplinary Assessment of the Lake Shelbyville Reservoir.* Champaign-Urbana, IL: University of Illinois.

Business Week (1989). "Innovation in America," special issue, (June).

Buss, D. M., Craik, K. H., and Dake, K. M. (1986). "Contemporary Worldviews and Perception of the Technological System," in Covello, V.T., Menkes, J., and Mumpower, J. (eds.), *Risk Evaluation and Management.* (pp. 93–130). New York: Plenum Press.

Canan, P. and Hennessy, M. (1982). "Community Values as the Context for Interpreting Social Impacts," *Environmental Impact Assessment Review,* vol. 3, no. 4, 351–365.

Cassels, J. R. T. and Johnstone, H. H. (1979). "Do You Still Beat Your Wife and Other Interesting Questions," *Management Accounting* (February), 48–50.

Cetron, M. J. and Bartocha, B. (1973). *Technology Assessment in a Dynamic Environment.* New York: Gordon and Breach.

Cetron, M. J. and Monahan, T. I. (1968). "An Evaluation and Appraisal of Various Approaches to Technological Forecasting," in Bright, J. R. (ed.), *Technological Forecasting for Industry and Government.* (pp. 144–179). Englewood Cliffs, NJ: Prentice-Hall.

Cetron, M. J. and Ralph, C. A. (1971). *Industrial Applications of Technological Forecasting.* New York: Wiley-Interscience.

Chaffin, W. W. and Talley, W. K. (1980). "Individual Stability in Delphi Studies," *Technological Forecasting and Social Change.* vol. 16, 67–73.

Changnon, S. A., Jr., Davis, R. J., Farhar, B., Haas, J. E., Ivens, J. L., Jones, M. V., Klein, D. A., Mann, D., Morgan, G. M., Jr., Sonka, S. T., Swanson, E. R., Taylor, C. R., and Van Blokland, J. (1977). *Hail Suppresion: Impacts and Issues.* Urbana IL: Illinois State Water Survey.

Chankong, V. and Haimes, Y. Y. (1983). *Multiobjective Decision Making.* New York: North Holland.

Chisholm, R. K. and Whitaker, G. R., Jr. (1971). *Forecasting Methods.* Homewood, IL: Richard D. Irwin.

Churchman, C. W. (1971). *The Design of Inquiring Systems.* New York: Basic Books.

Clark, C. H. (1980). *Idea Management: How to Motivate Creativity and Innovation.* New York: American Management Association.

Clarke, A. C. (1984). *Profiles of the Future: An Inquiry into the Limits of the Possible.* New York: Holt, Rinehart, and Winston.

Clarke, K., Ford, D., and Saren, M. (1989). "Company Technology Strategy," *R&D Management,* vol. 19, 215–229.

Clippinger, J. H. (1979). *Assessing Sociotechnological Change: A Case Study and Methods.* Cambridge, MA: Kalba-Bowen Associates.

Coates, J. F. (1971). "Technology Assessment: The Benefits... The Costs.... The Consequences," *The Futurist,* vol. 5, 225–231.

Coates, J. F. (1976). "Technology Assessment: A Took Kit," *Chemtech,* vol. 6, 372–383.

Coates, J. F., Coates, V. T., Jarratt, J., and Heinz, L. (1986). *Issues Management.* Mt. Airy, MD: Lomond.

Coates, V. T. and Finn, B. (1979). *A Retrospective Technology Assessment: Submarine Telegraphy.* San Francisco: San Francisco Press.

Cocks, K. D. and Ive, J. R. (1988). "Evaluating a Computer Package for Planning Public Lands in New South Wales," *Journal of Environmental Management,* vol. 26, 249–260.

Cohen, B. L. (1980). "Society's Valuation of Life Saving in Radiation Protection and Other Contexts," *Health Physics,* vol. 36, 707–722.

Cohen, L. E. and Land, K. C. (1987). "Age Structure and Crime: Symmetry Versus Asymmetry and the Projection of Crime Rates through the 1990s," *American Sociological Review,* vol. 52, 170–183.

Cohen, S., Teece, D., Tyson, L. D., and Zysman, J. (1984). *Competitiveness.* Berkeley, CA: Berkeley Roundtable on the International Economy (BRIE).

Commission for Computer Integration (1989). *Final Report—Computer Integration Commission.* Terre Haute, IN: Rose-Hulman Institute of Technology.

Comte, A. (1875). *System of Positive Policy.* London: Longmans Green.

Conlon, E. J. (1983). "Managing Organizational Change," in Connolly, T. (ed.), *Scientists, Engineers, and Organizations.* (pp. 363–378). Monterey CA: Brooks/Cole.

Consumer Reports (1990). "How to Drive a Hard Bargain in a Soft Market," vol. 55, 213–218 (April).

Cook, T. (1976). "Federal Agencies Impact Assessment Guidelines — Present and Future." In *Environmental Impact Analysis.* (pp. 85–92). Urbana, IL: University of Illinois, Department of Architecture Monograph Series.

Cook, T.D. (1985). "Postpositivist Critical Multiplism," in Shotland, R. L. and Mark, M. M. (eds.), *Social Science and Social Policy.* (pp. 21–62). Beverly Hills, CA: Sage.

Cook, T. D. and Campbell, D. T. (1979). *The Design and Conduct of Quasi-Experiments and True Experiments in Field Settings.* New York: Random House.

Cope, R. G. (1987). *Opportunity from Strength: Strategic Planning Clarified with Case Examples.* Washington, DC: Association for the Study of Higher Education (ASHE-ERIC Higher Education Report No. 8).

Covello, V. T., Menkes, J., and Mumpower, J. (eds.) (1986). *Risk Evaluation and Management.* New York: Plenum Press.

Cowley, G. (1989). "The Electronic Goddess," *Newsweek,* vol. 113, 50 (March 6).

Crews, J. E. and Johnson, G. P. (1975). "A Methodology for Trade-Off Analysis in Water Resources Planning," *ISTA Journal,* vol. 1, (June), 31–35.

Crookall, D., Greenblatt, C. S., Coote, A., Klabber, J. H. G., and Watson, D. R. (eds.) (1987). *Simulation-Gaming in the Late 1980's,* Proceedings of the International Simulation and Gaming Association's 17th Annual Conference, Toulon, France, (July 1–4), New York: Pergamon Press.

Crosby, P. B. (1979). *Quality is Free.* New York: Signet-New American Library.

Culhane, P. J., Firesema, H. P., and Beecher, J. A. (1987). *Forecasts and Environmental Decisionmaking.* Boulder, CO: Westview Press.

Cunningham, S. (1989). "Computer Systems," Norcross, GA: Search Technology Inc. Report (July 26).

Cunningham, S. (1990). *Applications of the Lotka-Volterra Equations to Technology Forecasting*. Atlanta: Technology Policy and Assessment Center, Georgia Institute of Technology.

Dahlman, C. J. and Westphal, L. E. (1981). "The Meaning of Technological Mastery in Relation to Transfer of Technology," *Annals of the American Association of Political and Social Sciences,* no. 458 (November), 12–26.

Dajani, J. S., Sincoff, M. Z., and Talley, W. K. (1979). "Stability and Agreement Criteria for the Termination of Delphi Studies," *Technological Forecasting and Social Change,* vol. 13, 83–90.

Davis, R. (1973). "Organizing and Conducting Technological Forecasting in a Consumer Goods Firm," in Bright, J. R. and Schoeman, M. E. F. (eds.), *A Practical Guide to Technological Forecasting.* (pp. 601–618). Englewood Cliffs, NJ: Prentice-Hall.

Deason, J. P. (1983). *A Multiobjective Decision Support System for Water Project Portfolio Selections.* Ph.D. dissertation, University of Virginia, Charlottesville.

de Bono, E. (1970). *Lateral Thinking: Creativity Step-by-Step.* New York: Harper & Row.

de Smidt, J. T., Barendregt, A., and Wassen, M. J. (1986). "Impact Assessment of Inversion of Groundwater Flow in Wetland Ecosystems," in Becker, H.A. and Porter, A.L. (eds.), *Impact Assessment Today.* (pp. 281–292). Utrecht, The Netherlands: Jan van Arkel.

Delbecq, A. L. (1974). "How To Do a TA for Less Than $5000," in Coates, V. T. and Mock, J. E. (eds.), *Summary of the Southern Regional Conference on Technology Assessment.* Washington, DC: National Science Foundation, Office of Intergovernmental Science and Research Utilization.

Delbecq, A. L. and Van de Ven, A. H. (1971). "A Group Process Model for Problem Indentification and Program Planning," *Journal of Applied Behavioral Science,* vol. 7, no. 4, 466–498.

Delbecq, A. L., Van de Ven, A. H., and Gustafson, D. H. (1975). *Group Techniques for Program Planning: A Guide to Nominal Group and Delphi Processes.* Glenview, IL: Scott Foresman and Company.

Dertouzos, M. L., Lester, R. K., and Solow, R. M. (1989). *Made in America: Regaining the Competitive Edge.* Cambridge, MA: MIT Press.

Dickson, P. R., and Giglierano, J. J. (1986). "Missing the Boat and Sinking the Boat: A Conceptual Model of Entrepreneurial Risk," *Journal of Marketing,* vol. 50 (July), 58–70.

Disanto, J. E. and Frideres, J. S. (eds.) (1986). "Issues of Impact Assessment: Development of Natural Resources," *Impact Assessment Bulletin,* vol. 4, no. 1–2.

Dornbush, R. and Fischer, S. (1987). *Macroeconomics.* New York: McGraw-Hill.

Douglas, M. and Wildavsky, A. (1982). *Risk and Culture, An Essay on the Selection of Technical and Environmental Dangers.* Berkeley, CA: University of California Press.

Doyle, L. B. (1972). "How To Plot A Breakthrough," in Martino, J. P. (ed.), *An Introduction to Technological Forecasting.* (pp. 31–37). New York: Gordon and Breach Science Publishers.

Drucker, P. F. (1974). *Management: Tasks, Responsibilities, Practices.* New York: Harper & Row.

Drucker, P. F. (1985). *Innovation and Entrepreneurship: Practice and Principles.* New York: Harper & Row.

Edelson, E. and Olsen, M. (1983). "Community Leaders' Perceptions of the Impacts of Energy Conservation Measures," in Rossini, F. A. and Porter, A. L. (eds.), *Integrated Impact Assessment.* (pp. 187–201). Boulder, CO: Westview Press.

Edelstein, M. R. (1988). *Contaminated Communities.* Boulder, CO: Westview Press.

Elgin, D. and Mitchell, A. (1977). "Voluntary Simplicity," *The Co-Evolution Quarterly,* vol. 14 (Summer), 4–19.

Emory, C. W. (1976). *Business Research Methods.* Homewood, IL: Richard D. Irwin, Inc.

Enk, G. A. and Hornick, W. F. (1983). "Human Values and Impact Assessment," in Rossini, F. A. and Porter, A. L. (eds.), *Integrated Impact Assessment.* (pp. 56–71). Boulder, CO: Westview Press.

Environment (1989). "Global Change and Our Common Future," vol. 31 (June), 16–24.

Enzer, S. and Leschinsky, D. (1986). "XimpacT User's Guide and Manual, Release 1.0," c/o University of Southern California, Los Angeles.

Ferriss, A. L. (1988). "The Uses of Social Indicators," *Social Forces,* vol. 66, no.3, 601–617.

Fey, W. R. (1971). "The Dynamics of Educational Institutions," *Proceedings of Summer Computer Simulation Conference, Boston,* July 19, Board of Simulation Conferences, Denver: 1110–1119.

Fey, W. R. (1980). "The Philosophy of the System Dynamics Method," *Proceedings of the International Conference on Cybernetics and Society,* Cambridge (October). New York: Institute of Electrical and Electronics Engineers, New York, 689–694.

Finkel, A. M. (1989). "Has Risk Assessment Become Too 'Conservative'?" *Resources,* Summer, no. 96, 11–13.

Fisher, J. C. and Pry, R. H. (1971). "A Simple Substitution Model of Technological Change," *Technological Forecasting and Social Change,* vol. 3, 75–88.

Ford, D. (1988). "Develop Your Technology Strategy," *Long Range Planning,* vol. 21, no. 5, 85–95.

Ford, D. A. (1975). "Shang Inquiry as an Alternative to Delphi: Some Experimental Findings," *Technological Forecasting and Social Change,* vol. 7, 139–164.

Forrester, J. W. (1961). *Industrial Dynamics,* Cambridge, MA: The MIT Press.

Fowles, J. (1978). "The Problem of Values in Futures Research," in Fowles, J. (ed.), *Handbook of Futures Research.* (pp.125–140). Westport, CT: Greenwood Press.

Fransman, M. (1985). "Conceptualizing Technical Change in the Third World in the 1980s: An Interpretive Survey," *The Journal of Development Studies,* vol. 21, 572–652.

Freeman, A., III and Portnoy, R. (1989). "Economics Clarifies Choices about Managing Risk," *Resources for the Future,* Spring, no. 95, 1–4.

Freeman, D. M., Frey, R. S., and Quint, J. M. (1982). "Assessing Resource Management Policies: A Social Well-Being Framework with a National Level Application," *Environmental Impact Assessment Review,* vol. 3, 59–73.

French, S., Harley, R., Thomas, L. C., and White, D. J. (1983). *Multiple Objective Decision Making.* New York: Academic Press.

Fried, J., Armijo, R., and Trejo, M. (1984). "Assessing Techno-Economic Alternatives for Rural Development in Northeast Mexico," *Technological Forecasting and Social Change,* vol. 25, 61–81.

Futures Group, The (1975). *Technology Assessment of Geothermal Energy Resources Development.* Glastonbury, CT: The Futures Group.

Galanc, T. and Mikus, J. (1986). "The Choice of an Optimum Group of Experts," *Technological Forecasting and Social Change,* vol. 30, 245–250.

Gastil, R. D. (1977). *Social Humanities.* San Francisco: Jossey-Bass.

Gibbs, G. I. (1974). *Handbook of Games and Simulation Exercises.* Beverly Hills, CA: Sage.

Giroult, E. (1984). "The Health Component of Environmental Impact Assessment," *International Seminar on Environmental Impact Assessment,* University of Aberdeen, Scotland (July).

Glickman, T. S. and Gough, M. (eds.) (1990). *Readings in Risk.* Washington, DC: Resources for the Future.

Goicoechea, A., Hansen, D. R., and Duckstein, L. (1982). *Multiobjective Decision Analysis with Engineering and Business Applications.* New York: Wiley.

Goldberger, A. S. (1964). *Econometric Theory.* New York: Wiley.

Gondolf, E. W. (1983). "A Rapid Assessment of Community Education in India: Constraints, Criteria, and Method," *Impact Assessment Bulletin,* vol. 3, no. 4, 32–46.

Gordon, W. J. J. (1961). *Synectics.* New York: Harper & Row.

Gore, A., Jr. (1990). "Futurizing the United States Government," *The Futurist,* vol. 24 (March/August), 22–25.

Granger, C. W. J. (1980). *Forecasting in Business and Economics.* New York: Academic Press.

Gray, J. L. and Starke, F. A. (1988). *Organizational Behavior: Concepts and Applications* (4th ed.). Columbus, OH: Merill.

Greenwald, P. and Sondik, E. J. (eds.) (1986). *Cancer Control Objectives for the Nation: 1985-2000.* NCI Monographs No. 2, U.S. Department of Health and Human Services, Public Health Service, National Institutes of Health.

Guilford, J. P. (1959). "Traits of Creativity," in Anderson, H. M. (ed.), *Interdisciplinary Symposia on Creativity (1957–58).* New York: Harper & Row.

Gujarati, D. (1978). *Basic Econometrics.* New York: McGraw-Hill.

Gustafson, D. H., Shukla, R. K., Delbecq, A. L., and Walster, G. S. (1973). "A Comparative Study of Differences in Subjective Likelihood Estimates Made by Individuals, Interacting Groups, Delphi Groups, and Nominal Groups," *Organizational Behavior and Human Performance,* vol. 9, 280–291.

Habermas, J. (1971). *Knowledge and Human Interests.* Boston: Beacon.

Hacke, J. E., Jr. (1972). "Methodological Preface to Technological Forecasting," in Martino, J. P. (ed.), *An Introduction to Technological Forecasting.* (pp. 1–12). New York: Gordon and Breach Science Publishers.

Halverson T., Swain J. J., and Porter A. L. (1989). "Analysis of the Cross-Impact Model," Atlanta: Technology Policy and Assessment Center Technical Report, Georgia Institute of Technology.

Hamblin, R. L., Jacobsen, R. B., and Miller, J. L. (1973). *A Mathematical Theory of Social Change.* New York: John Wiley.

Hammond, K. R., and Adelman, L. (1976). "Science, Values, and Human Judgment," *Science,* vol. 194 (22 October), 389–396.

Hansen, P. (ed.) (1983). *Essays and Surveys on Multiple Criteria Decision Making.* New York, Springer-Verlag.

Harman, W. W. (1983). "Integrated Impact Assessment: The Impossible Dream?" in Rossini, F. A. and Porter, A. L. (eds.), *Integrated Impact Assessment.* (pp.19–28). Boulder, CO: Westview Press.

Hax, A. C. and Majluf, N. S. (1988). "The Concept of Strategy and the Strategy Formation Process," *Interfaces,* vol. 18 (May–June), 99–109.

Heath, R. L. (1988). *Strategic Issues Management: How Organizations Influence and Respond to Public Interests and Policies.* San Francisco: Jossey-Bass.

Helmer, O. (1983). *Looking Forward—A Guide to Futures Research.* Beverly Hills, CA: Sage.

Hilbrink, J. O. (1989). "Economic Impact and Technical Change," *IEEE Transactions on Engineering Management,* vol. 36, no. 1, 37–46.

Hohenemser, C., Goble, R., Kasperson, J. X., Kasperson, R. E., Kates, R. W., Collins, P., Goldman, A., Slovic, P., Fischhoff, B., Lichtenstein, S., and Layman, M. (1986). "Methods for Analyzing and Comparing Technological Hazards," in Covello, V. T., Menkes, J. and Mumpower, J. (eds.), *Risk Evaluation and Management.* (pp. 249–274). New York: Plenum Press.

Holt, K. (1987). *Innovation: A Challenge to the Engineer.* New York: Elsevier.

Hoos, I. R. (1978). "Methodological Shortcomings in Futures Research," in Fowles, J. (ed.), *Handbook of Futures Research.* (pp. 54–66). Westport, CT: Greenwood Press.

Hunsaker, C. T., Hunsaker, D. B., Jr., Hashimoto, H. H., Simpson, M. M., Perrine, R. L., and Lindberg, R. G. (1983). "Assessing the Environmental Impacts of Solar/Fossil Power Plants," *Environmental Impact Assessment Review,* vol. 4, 242–248.

Hwang, C. L. and Masud, A. S. (1979). *Multiple Objective Decision Making—Methods and Applications.* New York: Springer-Verlag.

Idso, S. B. (1986). "Review of 'Changing Climate' and 'Can We Delay a Greenhouse Warming?'" *Environmental Impact Assessment Review,* vol. 6, 95–98.

Isard, W. (1972). *Ecologic-Economic Analysis for Regional Development.* New York: Free Press.

Jenster, P. V. (1987). "Using Critical Success Factors in Planning," *Long Range Planning,* vol. 20 (August), 102–109.

Johnston, J. (1972). *Econometric Methods.* New York: McGraw-Hill.

Jolliffe, F. R. (1986). *Survey Design and Analysis.* New York: Ellis Horwood Limited (Halsted Press).

Kahn, H. (1965). *On Escalation: Metaphors and Scenarios.* Poughkeepsie, NY: Hudson Institute.

Kahn, S. (1986). "Economic Estimates of the Value of Life," *IEEE Technology and Society Magazine,* vol. 5, no. 2, 24–31.

Kane, J. (1972). "A Primer for a New Cross-impact Language—KSIM," *Technological Forecasting and Social Change,* vol. 4, 129–142.

Kash, D. E. (1977). "Observations on Interdisciplinary Studies and Government Roles," in Scribner, R. and Chalk, R. (eds.), *Adapting Science to Social Needs.* (pp. 147–178). Washington, DC: American Association for the Advancement of Science.

Kash, D. E., White, I. L., Bergey, K. H., Chartock, M. A., Devine, M. D., Leonard, R. L., Salomon, S. N., and Young, H. W. (1973). *Energy Under the Oceans: A Technology Assessment of Outer Continental Shelf Oil and Gas Operations.* Norman, OK: University of Oklahoma Press.

Kasperson, R. E., Renn, O., Slovic, P., Brown, H. S., Emel, R., Goble, R., Kasperson, J. X., and Ratick, S. (1988). "The Social Amplification of Risk: A Conceptual Framework," *Risk Analysis,* vol. 8, 177–187.

Kates, R. W., Clark, W. C., Fischhoff, B., Kasperson, R. E., Lichtenstein, S., and Slovic, P. (1977). *Managing Technological Hazard: Research Needs and Opportunities.* Boulder, CO: Institute for Behavioral Science, University of Colorado.

Katz, A. (1958). "An Operational Analysis of an Electronic Systems Firm," Master's thesis, MIT, Cambridge, MA.

Keeney, R. L. (1986). "The Analysis of Failures of Risks of Fatalities," in Covello, V. T., Menkes, J., and Mumpower, J. (eds.), *Risk Evaluation and Management.* (pp. 233–248). New York: Plenum Press.

Keeney, R. L. and Raiffa, H. (1976). *Decisions with Multiple Objectives: Preferences and Value Trade-offs.* New York: Wiley.

Keilman, N. (1986). "The Unpredictability of Population Trends," *Impact Assessment Bulletin,* vol. 4, no. 3/4, 49–80.

Kelly, D., Cote, R. P., Nicholls, B., and Ricketts, P. (1987). "Developing a Strategic Assessment and Planning Framework for the Marine Environment," *Journal of Environmental Management,* vol. 25, 219–230.

Kelly, P., Kranzberg, M., Rossini, F. A., Baker, N. R., Tarpley, F. A., and Mitzner, M. (1978). *Technological Innovation.* San Francisco: San Francisco Press.

Keyworth, G. A., II (1990). *Goodbye Central: Telecommunications and Computing in the 1990s.* Briefing Paper No. 117. Indianapolis, IN: Hudson Institute.

Khakee, A. (1985). "Futures-Oriented Municipal Planning," *Technological Forecasting and Social Change,* vol. 28, 63–83.

Kidder, T. (1981). *The Soul of a New Machine.* Boston: Little, Brown.

Klabber, J. H. G., Scheper, W. S., Takkenberg, C. A. T., and Crookal, D. (eds.) (1989). *Simulation-Gaming: On the Improvement of Competence in Dealing with Complexity, Uncertainty, and Value Conflicts.* Proceedings of the International Simulation and Gaming Association's 19th Annual Conference, Utrecht, The Netherlands, (August 16–19). New York: Pergamon Press.

Klein, J. A. (1989). "The Human Costs of Manufacturing Reform," *Harvard Business Review,* vol. 67 (Mar/Apr), 60–66.

Kline, S. J. and Rosenberg, N. (1986), "An Overview of Innovation," in Landau, R. and Rosenberg, N. (eds.), *The Positive Sum Strategy.* Washington, DC: National Academy Press.

Knight, K. E. (1985). "A Functional and Structural Measurement of Technology," *Technological Forecasting and Social Change,* vol. 27, 107–127.

Koberg, D. and Bagnall, J. (1974). *The Universal Traveler.* Los Altos, CA: William Kaufman, Inc.

Kuhn, T. S. (1970). *The Structure of Scientific Revolutions* (2nd ed.). Chicago: University of Chicago Press.

Landau, R. and Rosenberg, N. (1986). *The Positive Sum Strategy.* Washington, DC: National Academy Press.

Leistritz, F. L. and Ekstrom, B. L. (1988). "Integrating Social Impact Assessment into the Planning Process," *Impact Assessment Bulletin,* vol. 6, no. 2, 17–19.

Leistritz, F. L. and Murdock, S. H. (1981). *The Socioeconomic Impact of Resource Development: Methods for Assessment.* Boulder, CO: Westview Press.

Lenz, R. C. (1985). "Rates of Adoption/Substitution in Technological Change." Austin, TX: Technology Futures, Inc.

Leontief, W. (1985). "The Choice of Technology," *Scientific American,* vol. 252, no. 6, 37–45.

Leopold, L. B., Clarke, F. E., Henshaw, B. B., and Balsley, J. R. (1971). *A Procedure for Evaluating Environmental Impact.* Washington, DC: U.S. Geological Survey (Circ. 645).

Linstone, H. A. (1989). "Multiple Perspective: Concept, Applications and User Guidelines," *Systems Practice,* vol. 2, no. 3 307–331.

Linstone, H. A. and Turoff, M. (eds.) (1975). *The Delphi Method: Techniques and Applications.* Reading, MA: Addison-Wesley.

Lipinski, A. and Loveridge, D. (1982). "Institute for the Future's Study of the UK, 1978–95," *Futures,* vol. 14, 205–239.

Little, A. D. (1975). *The Consequences of Electronic Funds Transfer.* Cambridge, MA: Arthur D. Little, Inc.

Little, R. L. and Krannich, R. S. (1988). "A Model for Assessing the Social Impact of Natural Resource Utilization on Resource-Dependent Communities," *Impact Assessment Bulletin,* vol. 6, no. 2, 21–35.

Lough, W. T. and White, K. P., Jr. (1988). "A Technology Assessment of Nuclear Power Plant Decommissioning," *Impact Assessment Bulletin,* vol. 6, no. 1, 71–88.

Lovejoy, S. B. (1983). "Employment Predictions in Social Impact Assessment: An Analysis of Some Unexplored Variables," *Socio-Economic Planning Sciences,* vol. 17, 87–93.

Luchsinger, V. P. and Van Blois, J. (1988). "Spinoffs from Military Technology: Past and Future," in Khalil, T. M., Bayraktar, B. A., and Edosomwan, J. A. (eds.), *Management of Technology I.* (pp. 283–290). Geneva: Interscience Enterprises Ltd.

Machlis, G. E. and Rosa, E. A. (1990). "Desired Risk: Broadening the Social Amplificaiton of Risk Framework," *Risk Analysis,* vol. 10, 161–168.

Maddox, N., Anthony, W. P., and Wheatley, W., Jr. (1987). "Creative Strategic Planning Using Imagery," *Long Range Planning,* vol. 20, no. 5, 118–124.

Malinowski, B. (1944). *A Scientific Theory of Culture.* Chapel Hill, NC: University of North Carolina Press.

Mansfield, E. (1975). *Microeconomics Theory and Applications.* New York: W. W. Norton.

Mansfield, E. (1989). *Economics: Principles, Problems and Decisions.* New York: W. W. Norton.

Marchetti, C. (1983). "The Automobile in a System Context: The Past 80 Years and the Next 20 Years," *Technological Forecasting and Social Change,* vol. 23, 3–23.

Marchetti, C. (1987). "The Future of Natural Gas: A Darwinian Analysis," *Technological Forecasting and Social Change,* vol. 31, 155–171.

Marchetti, C. (1988). "Darwin and the Future of ISDN," in Bonatti, M. and Decina, M. (eds.). *Traffic Engineering for ISDN Design and Planning.* (pp. 77–96). New York: North-Holland.

Marchetti, C. and Nakicenovic, N. (1979). *The Dynamics of Energy Systems and the Logistic Substitution Model.* (Report RR-79-13). Laxenburg, Austria: International Institute for Applied Systems Analysis.

Margoluis, R. (1988). "The Application of Rapid Qualitative Evaluation Methods to Environmental and Disaster Impact Assessment," *Proceedings of the International Workshop* on *Impact Assessment for International Development,* Barbados, West Indies, International Association for Impact Assessment, (May 31–June 4, 1987), Belhaven, NC: available through the Federal Environmental Assessment Review Office (FEARO) 750-510 Cambie St., Vancouver BC V6B 2P2, Canada.

Markley, O. W. and Hurley, T. J., III (1983). "A Brief Technology Assessment of the Carbon Dioxide Effect," *Technological Forecasting and Social Change,* vol. 23, 185–202.

Marks, J. V. (1986). "Public Involvement in Forward Planning for River Basin Management: The Case of the South Saskatchewan River Basin Planning Program," in Becker, H. A. and Porter, A. L. (eds.), *Impact Assessment Today.* (pp. 227-249). Utrecht, The Netherlands: Jan van Arkel.

Martin, J. E. (1986). "Environmental Health Impact Assessment: Methods and Sources," *Environmental Impact Assessment Review,* vol. 6, no. 1, 7–48.

Martin, M. and Trumbly, J. E. (1987). "A Project Accountability Chart (PAC)," *Journal of Systems Management,* March, vol. 38, 21–24.

Martino, J. P. (1972). *Technological Forecasting for Decision Making.* New York: American Elsevier.

Martino, J. P. (1983). *Technological Forecasting for Decision Making* (2nd ed.). New York: North-Holland.

Martino, J. P. (1987). *Using Precursors as Leading Indicators of Technological Change.* Report UDR-TR-87-109. Dayton, OH: University of Dayton.

Martino, J. P. (pending). *Technological Forecasting for Decision Making* (3rd ed.).

Marx, K. (1967). *Capital.* New York: International Publishers.

Mason, T. W. (1986). "Entrepreneurs, High Technology and Economic Development: New Uses for Assessment Tools," *Impact Assessment Bulletin,* vol. 3, no. 4, 47–55.

Mason, T. W. (1988). "Economics and Impact Assessment: Ceteris Paribus or Mutatis Mutandis," *Impact Assessment Bulletin,* vol. 6, no. 3–4, 165–171.

Mathisen, D. A. (1977). "The Energy Age," *New Engineer,* vol. 6, 17–27.

McConnell, S. W. and Khalil, T. M. (1988). "Evaluation of New Technology: A Methodology and Case Study," in Khalil, T. M., Bayraktar, B. A., and Edosomwan, J. A. (eds.), *Management of Technology I.* (pp.727–736). Geneva: Interscience Enterprises Ltd.

McIntyre, S. H. (1988). "Market Adaptation as a Process in the Product Life Cycle of Radical Innovations and High Technology Products," *Journal of Product Innovation Management,* vol. 5, no. 2, 140.

Meadows, D. H., Meadows, D. L., Randers, J., and Behrens, W. W., III (1972). *The Limits to Growth.* New York: Universe Books.

Mihai, V. (1984-1985). "Some Contributions to the Estimation of Effects in Impact Studies," *Impact Assessment Bulletin,* vol. 3, no. 3, 18–23.

Miller, J. G., and Vollmann, T. E. (1985). "The Hidden Factory," *Harvard Business Review,* vol. 63, (September/October), 142–150.

Mishan, E. J. (1973). *Economics for Social Decisions: Elements of Cost Benefit Analysis.* New York: Praeger.

Mishan, E. J. (1976). *Cost Benefit Analysis.* New York: Praeger.

Mitchell, A., Dodge, B. H., Kruzic, P. G., Miller, D. C., Schwartz, P., and Suta, B. E. (1975). "Handbook of Forecasting Techniques." Stanford Research Institute Report to

U.S. Army Corps of Engineers Institute for Water Resources, AD-A019 280. (available from NTIS, Springfield VA.)

Mitroff, I. I. and Turoff, M. (1973). "The Whys Behind the Hows," *IEEE Spectrum*, vol. 10, no. 3, 62–71.

Moore, K. D. and Gilmore, J. S. (1983). "Socioeconomic Impacts of Western Energy Resource Development: State of the Art and Limitations," in Rossini, F. A. and Porter, A. L. (eds.), *Integrated Impact Assessment*. (pp. 219–236). Boulder, CO: Westview Press.

Morecroft, J. D. (1986). "The Dynamics of a Fledgling High Technology Growth Market: Understanding and Managing Growth Cycles," *System Dynamics Review*, vol. 2, no. 1, 36–61.

Mouly, G. J. (1963). *The Science of Educational Research*. New York: American Book Company.

Mumpower, J. L., Livingston, S., and Lee, T. J. (1987). "Expert Judgments of Political Riskiness," *Journal of Forecasting*, vol. 6, 51–65.

Murdock, S. H., Leistritz, F. L., and Hamm, R. R. (1986). "The State of Socioeconomic Impact Analysis in the United States: An Examination of Existing Evidence, Limitations and Opportunities for Alternative Futures," *Impact Assessment Bulletin*, vol. 4, no. 3/4, 101–132.

Murdock, S. H., Leistritz, F. L., Hamm, R. R., and Hwang, S-S. (1982). "An Assessment of Socioeconomic Assessments: Utility, Accuracy, and Policy Considerations," *Environmental Impact Assessment Review*, vol. 3, 333–350.

Nadler, G. (ed.) (1987). *1987 International Congress on Planning and Design Theory*. New York: American Society of Mechanical Engineers.

Naisbitt, J. (1982). *Megatrends*. New York: Morrow.

Naisbitt, J. and Aburdene, P. (1989). *Megatrends 2000*. New York: Morrow.

National Academy of Engineering (1987). *Technology and Global Industry: Companies and Nations in the World Economy*. Washington, DC: National Academy Press.

National Bureau of Standards (1987). *NBS Research Reports* (August).

Nelms, K. R. and Porter, A. L. (1985). "EFTE: An Interactive Delphi Method," *Technological Forecasting and Social Change*, vol. 28, 43–61.

Neste, L. (1988). "Tech Forecasting System Implementation Plan," DA Memo 0291, Norcross, GA: Search Technology.

Nijkamp, P. (1986). "Multiple Criteria Analysis and Integrated Impact Analysis," *Impact Assessment Bulletin*, vol. 4, no. 3/4, 226–261.

Osborn, A. F. (1953). *Applied Imagination: Principles and Procedures for Creative Problem Solving*. New York: Charles Scribner & Sons.

Papp, R. (1986). "The Alternative Fuel Cycle Evaluation of the Federal Republic of Germany," in Becker, H. A. and Porter, A. L. (eds.), *Impact Assessment Today*. (pp. 351–365). Utrecht, The Netherlands: Jan van Arkel.

Pareto, V. (1935). *The Mind and Society*. New York: Harcourt, Brace.

Parsons, T. (1951). *The Social System*. New York: Free Press.

Paustenbach, D. J. and Keenan, R. E. (1988). "Health Risk Assessment in the 1990's," *Hazmat World*, vol. 1, no. 3, 48–56.

Pearson, G. L., Pomerov, W. L., Sherwood, G. A., and Winder, J. S., Jr. (1975). "A Scientific and Policy Review of the Final Environmental Impact Statement for the Initial Stage, Garrison Diversion Unit (North Dakota)." Indianapolis, IN: The Institute of Ecology.

Peck, M. J. (1961). *Competition in the Aluminum Industry 1945–58*. Cambridge: Harvard University Press.

Peters, H. P. (1986). "Social Impact Analysis of Four Energy Scenarios," *Impact Assessment Bulletin*, vol. 4, no. 3/4, 149–167.

Porter, A. L. (1987a). "A Two-factor Model of the Effects of Office Automation on Employment," *Office, Technology and People*, vol. 3, 57–76.

Porter, A. L. (1987b). "International Impacts of Technology," *Impact Assessment Bulletin*, vol. 5, no. 3, 21–24.

Porter, A. L. (1988). "Technology Feasibility Assessment System," DA Memo # 0219, Norcross, GA: Search Technology, Inc.

Porter, A. L. (1989). *PC 2000: A Forecast of Microcomputer Developments for the Year 2000*. Atlanta: Technology Policy and Assessment Center, Georgia Institute of Technology.

Porter, A. L., Connolly, T., Heikes, R. G., and Park, C. Y. (1981). "Misleading Indicators: The Limitations of Multiple Linear Regression in Formulation of Policy Recommendations," *Policy Sciences*, vol. 13, 397–418.

Porter, A. L., Roessner, J. D., and Kuehn, T. J. (1987). *High Tech Performance/Potential*. Chicago: National Science Foundation Workshop on Indicators of International Technology Transfer.

Porter, A. L. and Rossini, F. A. (1977). "Evaluation Designs for Technology Assessments and Forecasts," *Technological Forecasting and Social Change*, vol. 10, 369–380.

Porter, A. L. and Rossini, F. A. (1984). "Interdisciplinary Research Redefined: Multi-Skill Problem-Focused Research In the STRAP Framework," *R&D Management*, vol. 14, 105–111.

Porter, A. L. and Rossini, F. A. (1986). "Multiskill Research," *Knowledge: Creation, Diffusion, Utilization*, vol. 7, no.3, 219–246.

Porter, A. L. and Rossini, F. A. (1987). "Technological Forecasting," in Singh, M. G. (ed.) *Encyclopedia of Systems and Control*. (pp. 4823–4828). Oxford: Pergamon.

Porter, A. L. and Rossini, F. A. (1990). "Technological Forecasting," in Sage, A. P. (ed.), *Concise Encyclopedia of Information Processing in Systems and Organizations*. (pp. 511–517). Oxford: Pergamon Press.

Porter, A. L., Rossini, F. A., Carpenter, S. R., and Roper, A. T. (1980). *A Guidebook for Technology Assessment and Impact Analysis*. New York: North-Holland.

Porter, M. E. (1990). "The Competitive Advantage of Nations," *Harvard Business Review* (March/April), 73–93.

Porter, M. E. and Millar, V. E. (1985). "How Information Gives You Competitive Advantage," *Harvard Business Review*, vol. 63 (July/August), 149–160.

Prehoda, R. W. (1972). "Technological Forecasting and Space Exploration," in Martino, J. P. (ed.), *An Introduction to Technological Forecasting*. (pp. 39–51). New York: Gordon and Breach Science Publishers.

Quinn, J. B., and Gagnon, `. E. (1986). "Will Services Follow Manufacturing Into Decline?" *Harvard Business Review*, vol. 64 (November/December), 95–103.

Raj, D. (1972). *The Design of Sample Surveys*. New York: McGraw-Hill.

Redelfs, M., and Stanke, M. (1988). "Citizen Participation in Technology Assessment: Practice at the Congressional Office of Technology Assessment," *Impact Assessment Bulletin*, vol. 6, no. 1, 55–70.

Reich, R. (1989). "The Quiet Path to Technological Preeminence," *Scientific American,* vol. 261 (October), 41–47.

Reitman, W., Weischedel, K. R., Boff, M. E., Jones, M. E., and Martino, J. P. (1985). "Automated Information Management Technology (AIMTECH): Consideration for a Technology Investment Strategy," Air Force Aerospace Medical Research Laboratory, Wright-Patterson Air Force Base, Ohio, (AFAMRL-TR-042).

Rescher, N. (1969). "What is Value Change? A Framework for Research," in Baier, K. and Rescher, N. (eds.), *Values and the Future.* (pp. 68–109). New York: The Free Press.

Richardson, G. P. and Pugh, A. L., III (1981). *Introduction to System Dynamics Modeling with DYNAMO.* Cambridge, MA: MIT Press.

Rieger, W. G. (1986). "Directions in Delphi Developments: Dissertations and Their Quality," *Technological Forecasting and Social Change,* vol. 29, 195–204.

Roberts, E. B. (1964). *The Dynamics of Research and Development.* New York: Harper & Row.

Roe, N. A. (1988). "Reconciling EIA with Variable Time Constraints: Experience and a Solution from Different Types of Development," *Proceedings of the International Workshop on Impact Assessment for International Development,* Barbados, West Indies (May 31–June 4, 1987). International Association for Impact Assessment, 415–427. Available through FEARO, 750-510 Cambie St. Vancouver, BC V6B 2P2, Canada.

Roessner, J. D., Mason, R. M., Porter, A. L., Rossini, F. A., Schwartz, A. P., and Nelms, K. R. (1985). *The Impact of Office Automation on Clerical Employment, 1985–2000.* Westport, CT: Quorum Books.

Roessner, J. D. and Porter, A. L. (1990). "High Technology Capacity and Competition," in Chatterji. M. (ed.), *Technology Transfer in the Developing Countries.* (pp.94–103). London: Macmillan.

Roessner, J. D., Porter, A. L., and Fouts, S. (1988). "Technology Absorption, Institutionalization, and International Competitiveness in High Technology Industries," in Khalil, T. M., Bayraktar, B. A., and Edosomwan, J. A. (eds.), *Management of Technology I.* (pp. 779–790). Geneva: Interscience Enterprises Ltd.

Rogers, M. (1989). "Marvels of the Future," *Newsweek* (December 25), 77–78.

Roper, A. T. (1986). "Mapping Project Scope with Stakeholder Input." Paper presented at the 1986 Annual Conference of the International Association for Impact Assessment, Philadelphia, (May 25–30).

Roper, A. T. (1988). "A Technique for the Early Stages of an Assessment," *Proceedings of the International Workshop on Impact Assessment for International Development,* Barbados, West Indies (May 31–June 4, 1987). International Association for Impact Assessment, available through FEARO 750-510 Cambie St., Vancouver, BC V6B 2P2, Canada.

Roper, A. T. and Mason, T. W. (1983). "Consultant Report: General Housewares Corp. Process Evaluation Function." Terre Haute, IN: Center for Technology and Policy Studies.

Roper's (1987). *The Public Pulse,* vol. 2, no. 1. New York: Roper's.

Ross, W. A. (1987). "Evaluating Environmental Impact Statements," *Journal of Environmental Management,* vol. 25, 137–147.

Rossini, F. A. and Porter, A. L. (1981). "Interdisciplinary Research: Performance and Policy Issues," *SRA Journal,* vol. 13 (Fall), 8–24.

Rossini, F. A. and Porter, A. L. (1982). "Forecasting the Social and Institutional Context of Technological Developments," *Proceedings of the International Conference on Cybernetics and Society.* (pp. 486–490). New York, IEEE.

Rossini, F. A. and Porter, A. L. (eds.) (1983). *Integrated Impact Assessment.* Boulder, CO: Westview Press. 154–167.

Rossini, F. A., Porter, A. L., Kelly, P., and Chubin, D. E. (1978). *Frameworks and Factors Affecting Integration Within Technology Assessments.* Report to the National Science Foundation. Atlanta, GA: Georgia Institute of Technology.

Rossini, F. A., Porter, A. L., and Zucker, E. (1976). "Multiple Technology Assessments," *Journal of the International Society for Technology Assessment,* vol 2, 21–28.

Rouse, W. B. and Rogers, D. M. A. (1990). "Technological Innovation: What's Wrong? What's Right? What's Next?" *Industrial Engineering* (April), 43–50.

Rowe, W. D. (1989). "Alternative Risk Evaluation Paradigms," in Haimes, Y. Y., and Stakhiv, E. Z. (eds.), *Risk Analysis and Management of Natural and Man-Made Hazards.* (pp. 1–21). New York: American Society of Civil Engineers.

Rumberger, R. W. and Levin, H. M. (1985). "Forecasting the Impact of New Technologies on the Future Job Market," *Technological Forecasting and Social Change,* vol. 27, 399–417.

Saaty, T. L. (1980). *The Analytic Hierarchy Process: Planning, Priority Setting, Resource Allocation.* New York: McGraw-Hill.

Saaty, T. L. (1982). *Decision Making for Leaders: The Analytical Hierarchy Process for Decisions in a Complex World.* Belmont, CA: Lifetime Learning Publications.

Saaty, T. L. and Kearns, K. P. (1985). *Analytical Planning: The Organization of Systems.* Oxford: Pergamon Press.

Sachs, A. and Clark, P. B. (1980). "Scoping the Content of Environmental Impact Statements: An Evaluation of Agencies' Experience." Draft report for the Council on Environmental Quality and the Geological Survey, U.S. Department of the Interior.

Sackman, H. (1974). "Delphi Assessment, Expert Opinion, Forecasting, and Group Process," Rand Corporation Report R-1283-pl.

Sadler, B. (1983). "Project Justification in Environmental Assessment of Major Developments." Paper presented at International Workshop on Environmental Planning for Large-Scale Development Projects, Whistler, BC (October 2–5).

Sage, A. P. (1977). *Methodology for Large Scale Systems.* New York: McGraw-Hill.

Sage, A. P. (1990). "Inference and Impact Analysis," in Sage, A. P. (ed.), *Concise Encyclopedia of Information Processing in Systems & Organizations.* (pp. 259–271). Oxford: Pergamon Press.

Sassone, P. G. and Schaffer, W. A. (1978). *Cost Benefit Analysis: A Handbook.* New York: Academic Press.

Schmidtlein, F. A., and Milton, T. H. (1989). "College and University Planning: Perspectives from a Nationwide Study," *Planning for Higher Education,* vol. 17, no. 3, 1–20.

Schnaars, S. P. (1989). *Megamistakes.* New York: The Free Press.

Schumpeter, J.H. (1934). *The Theory of Economic Development.* Cambridge, MA: Harvard University Press.

Schweiger, D. M., and Sandberg, W. R. (1989). "The Utilization of Individual Capabilities in Group Approaches to Strategic Decision-making," *Strategic Management Journal,* vol, 10, 31–43.

Schweitzer, M. (1981). "The Basic-Questions Approach to Social Impact Assessment," *Environmental Impact Assessment Review,* vol. 2, 294–299.

Schwing, R. C., Evans, L., and Shreck, R. M. (1983). "Uncertainties in Diesel Engine Health Effects," *Risk Analysis,* vol. 3, no. 2, 129–31.

Scott, B. R. and Lodge, G. C. (1985). *U.S. Competitiveness in the World Economy.* Boston, MA: Harvard Business School Press.

Scriven, M. (1967). "The Methodology of Evaluation," in Taylor, R., Gagne, R., and Scriven, M. (eds.), *Perspectives of Curriculum Evaluation.* Chicago: Rand McNally.

Sechrest, L. (1985). "Social Science and Social Policy: Will Our Number Ever Be Good Enough?" in Shotland, R. L. and Mark, M. M. (eds.), *Social Science and Social Policy.* (pp. 63–95). Beverly Hills, CA: Sage.

Segal, A. (1986). "From Technology Transfer to Science and Technology Utilization," in McIntyre, J. R. and Papp, D. S. (eds.), *The Political Economy of International Technology Transfer.* (pp. 95–115). Westport, CT: Quorum Books.

Sharif, M. N. and Kabir, C. (1976). "A Generalized Model for Forecasting Technological Substitution," *Technological Forecasting and Social Change,* vol. 8, no. 4, 353–364.

Sharif, M. N. and Sundararajan, V. (1984). "Assessment of Technological Appropriateness: The Case of Indonesian Rural Development," *Technological Forecasting and Social Change,* vol. 25, 225–237.

Shi, H., Porter, A. L., Rossini, F. A. (1985). "Microcomputers in Developing Countries: Industrialization in the Information Age," *International Journal of Applied Engineering Education,* vol. 1, no. 5, 321–327.

Shields, P. G., Rosenthal, K. M., and Holz, G. K. (1986). "The Use of a Computer in Detailed Land Resource Assessment," *Journal of Environmental Management,* vol. 23, 75–88.

Shillito, M. L. (1973). "Pareto Voting," *Proceedings, Society of American Value Engineers,* vol. 8, 131–135.

Shrader-Frechette, K. S. (1985). *Risk Analysis and Scientific Method.* Dordrecht, The Netherlands: D. Reidel.

Shrader-Frechette, K. (1988). "Producer Risk, Consumer Risk and Assessing Technological Impacts," *Impact Assessment Bulletin,* vol. 6, no. 3-4, 156–164.

Shumacher, E. F. (1973). *Small is Beautiful: Economics as if People Mattered.* New York: Harper & Row.

Shurig, R. (1984). "Morphology: A Tool for Exploring New Technology," *Long Range Planning,* vol. 17, no. 3., 129–140.

Sirinaovakul, B., Czajkiewicz, A., Khalil, T. M., Bayraktar, B. A., and Edosomwan, J. A. (1988). "A Spatial Diffusion Model for Advanced Manufacturing Technology." *Technology Management 1. Proceedings of the First International Conference.* (pp. 291–301). Geneva: Interscience Enterprises.

Slovic, P., Fischhoff, B., and Lichtenstein, S. (1981). "Perceived Risk: Psychological Factors and Social Implications," in Warner, F. and Slater, D. H. (eds.), *The Assessment and Perception of Risk.* London: The Royal Society.

Slovic, P., Fischhoff, B., and Lichtenstein, S. (1986). "The Psychometric Study of Risk Perception," in Covello, V. T., Menkes, J., and Mumpower, J. (eds.), *Risk Evaluation and Management.* (pp. 3–24). New York: Plenum Press.

Smits, S. J., Rossini, F. A., and Davis, L. M. (1987). "Managing the Present from the Future: The Challenge of Preparing for Technology in Rehabilitation," *Journal of Rehabilitation Administration*, vol. 11, no. 4, 121–130.

Solow, R. (1957). "Technical Change in the Aggregate Production Function," *Review of Economics and Statistics*, vol. 39, 312–320.

Sonntag, N. C., Greig, L. A., Everitt, R. R., Sadler, B., and Wiebe, J. D. (1988). "Environmental Management Framework for International Development Projects," *Proceedings of the International Workshop on Impact Assessment for International Development*, Barbados, West Indies (May 31–June 4,1987). International Association for Impact Assessment, 543–558. Belhaven, NC. Available through FEARO 750-510 Cambie St., Vancouver, BC V6B 2P2, Canada.

Sorokin, P. A. (1962). *Social and Cultural Dynamics*. New York: American Book Co. (Bedminster Press).

Spengler, O. (1939). *The Decline of the West*. New York: Knopf.

Stans, J. C. (1986). "Prediction and Assessment of Impacts on the Water Environment," in Becker, H. A. and Porter, A. L. (eds.), *Impact Assessment Today*. (pp. 293–321). Utrecht, The Netherlands: Jan van Arkel.

Starr, C. (1969). "Social Benefit versus Technological Risk," *Science*, vol. 65, no. 3899, 1232–1238.

Starr, C. and Whipple, C. (1984). "A Perspective on Health and Safety Risk Analysis," *Management Science*, vol. 30, no. 4, 452–463.

Stewart, H. B. (1989). *Recollecting the Future*. Homewood, IL: Dow Jones-Irwin.

Stover, J. G. and Gordon, T. J. (1978). "Cross Impact Analysis," in Fowles, J. (ed.), *Handbook of Futures Research*. Westport, CT: Greenwood Press.

Sumanth, D. J. (1988a). "Challenges and Opportunities in Managing 'Technology Discontinuities' on S-Curves," in Khalil, T. M., Bayraktar, B. A., and Edosomwan, J. A. (eds.), *Management of Technology I*. (pp. 271–279). Geneva: Interscience Enterprises Ltd.

Sumanth, D. J. (1988b). "A Total Systems Approach to Technology Management for Inter-Organizational Competitiveness," in Khalil, T. M., Bayraktar, B. A., and Edosomwan, J. A. (eds.), *Management of Technology I*. (pp. 799–808). Geneva: Interscience Enterprises Ltd.

Susskind, L. E. (1983). "The Uses of Negotiation and Mediation in Environmental Impact Assessment," in Rossini, F. A. and Porter, A. L. (eds.), *Integrated Impact Assessment*. (pp. 154–167). Boulder, CO: Westview Press.

Susskind, L., McMahon, G., and Rolley, S. (1987). "Mediating Development Disputes: Some Barriers and Bridges to Successful Negotiation," *Environmental Impact Assessment Review*, vol. 7, 127–138.

Sutherland, J. W. (1988). "Intelligence-Driven Strategic Planning and Positioning," *Technological Forecasting and Social Change*, vol. 34, 279–303.

Swain, J. J., Halverson, T., Rossini, F. A., and Porter, A. L. (1989). "Markov Formulation of Cross-Impact Analysis for Impact Assessment and Forecasting," Technology Policy and Assessment Center Technical Report. Atlanta: Georgia Institute of Technology.

Tarr, J. A. (1977). *Retrospective Technology Assessment—1976*. San Francisco: The San Francisco Press.

Taylor, S. J. (1983). "Institutional Analysis in the Context of Integrated Impact Assessment and Planning," in Rossini, F. A. and Porter, A. L. (eds.), *Integrated Impact Assessment.* (pp. 98–104). Boulder, CO: Westview Press.

Technology Administration (1990). *Emerging Technologies.* Washington, DC: U.S. Department of Commerce.

Teece, D. J. (1988). "Capturing Value from Technological Innovation: Integration, Strategic Partnering, and Licensing Decisions," *Interfaces,* vol. 18, no. 3, 46–61.

Thierauf, R. J. (1987). *A Problem-Finding Approach to Effective Corporate Planning.* Westport, CT: Quorum Books.

Thompson, M. S. (1980). *Benefit Cost Analysis for Program Evaluation.* Beverly Hills, CA: Sage.

Toffler, A. (1969). "Value Impact Forecaster—A Profession of the Future," in Baier, K. and Resher, N. (eds.), *Values and the Future.* (pp. 1–30). New York: The Free Press.

Tonn, B. E. (1986). "Using Possibility Functions for Long-Term Environmental Planning," *Futures,* vol. 18, 795–807.

Tribus, M. (1969). *Rational Descriptions, Decisions, and Designs.* New York: Pergamon Press.

Tushman, M. L. and Anderson, P. (1986). "Technological Discontinuities and Organizational Environments," *Administrative Science Quarterly,* vol. 31, 439–465.

U.S. Army Corps of Engineers (1973). Information Supplement No. 1 to Section 122 Guidelines CER 1105-2-105 (15 December 1972). Washington, DC: U.S. Government Printing Office.

U.S. Council on Environmental Quality (1974). "Guidelines for the Preparation of Environmental Impact Statements," Washington, DC: U.S. Government Printing Office.

U.S. Department of Commerce (monthly). *Business Conditions Digest.* Washington, DC: U.S. Government Printing Office.

U.S. Department of Commerce (monthly). *Survey of Current Business.* Washington, DC: U.S. Government Printing Office.

U.S. Environmental Protection Agency, Region X (1973). "Environmental Impact Statement Guidelines." Seattle, WA: EPA.

U.S. Environmental Protection Agency (1985). "The Use of Risk Assessment in Regional Operations." EPA 90415-85 140 (November). Washington, DC: U.S. Government Printing Office.

U.S. Nuclear Regulatory Commission (1979). *Environmental Standard Review for the Environmental Review of Construction Permit Applications for Nuclear Power Plants.* NUREG-0555. Washington DC: Author.

Valentine, L. M. and Danten, C. A. (1983). *Business Cycles and Forecasting.* Cincinnati: South-Western Publishing Co.

Van de Ven, A. H. (1974). *Group Decision Making and Effectiveness—An Experimental Study.* Kent, OH: Kent State University Press.

Vanston, J. H. (1985). "Technology Forecasting: An Aid to Effective Technology Management." Austin, TX: Technology Futures, Inc.

Vlachos, E. (1981). "The Use of Scenarios for Social Impact Assessement," in Finsterbusch, K. and Wolf, C. P. (eds.), *Methodology of Social Impact Assessment* (2nd ed.). (pp. 162–174). Stroudsburg, PA: Hutchinson Ross Publishing.

Volland, C. S. (1987). "A Comprehensive Theory of Long Wave Cycles," *Technological Forecasting and Social Change*, vol. 32, 123–145.

von Oech, R. (1986). *A Kick In The Seat Of The Pants*. New York: Harper & Row.

Walker, J., Jupp, D. L. B., Penridge, L. K., and Tian, G. (1986). "Interpretation of Vegetation Structure in Landsat MSS Imagery: A Case Study in Disturbed Semi-arid Eucalypt Woodlands. Part 1. Field Data Analysis," *Journal of Environmental Management*, vol. 23, 19–33.

Walker, W. E. (1986). "The Use of Screening in the PAWN Study," *Impact Assessment Bulletin*, vol. 4 no 3/4, 203–225.

Wallender, H.W., III (1979). *Technology Transfer and Management in the Developing Countries*. Cambridge, MA: Ballinger.

Warner, M. L., Moore, J. L., Chatterjee, S., Cooper, D. C., Ifeadi, C., Lawhon, W. T., and Reimers, R. S. (1974). *An Assessment Methodology for the Envrironmental Impact of Water Resource Projects*. EPA-600/5-74-016. Washington, DC: Office of Research and Development, U.S. EPA.

Watson, R. H. (1978). "Interpretive Structural Modeling: A Useful Tool for Technology Assessment," *Technological Forecasting and Social Change*, vol. 11, 165–185.

Weiss, C. H. (1972). *Evaluation Research*. Englewood Cliffs, NJ: Prentice Hall.

Wenk, E., Jr., and Kuehn, T. J. (1977). "Interinstitutional Networks in Technological Delivery Systems," in Haberer, J. (ed.), *Science and Technology Policy*. (pp. 153–175). Lexington, MA: Lexington Books.

Wheeler, D. R. and Shelley, C. J. (1987). "Toward More Realistic Forecasts for High-Technology Products," *Journal of Business and Industrial Marketing*, vol. 2, no. 3, 36–44.

Wheelwright, S. C. and Hayes, R. H. (1985). "Competing Through Manufacturing," *Harvard Business Review*, vol. 63 (January/February), 99–109.

Wheelwright, S. C. and Makridakis, S. (1980). *Forecasting Methods for Management*. New York: Wiley.

White, K. P., Jr., Pikey, W. D., Gabler, N. C., and Hollowell, T. (1986). "Simulation Optimization of the Crashworthiness of a Passenger Vehicle in Frontal Collisions Using Response Surface Methodology," *1985 Proceedings of the SAE*. Warrendale, PA: Society of Automotive Engineers.

White, K. P., Jr., Sage, A. P., Rodammer, F. A., and Peters, C. T., Jr. (1985). "The Environmental Advisory Service (EASe): A Decision Support System for Comprehensive Screening of Local Land-use Development Proposals and Comparative Evaluation of Proposed Land-use Plans," *Environment and Planning B*, vol. 12, 221–234.

White, K. P., Jr. and White, D. J. (1988). "Framing U.S. Automobile Safety Standards: Can Multiple-objective Methods be Justified in Regulatory Policy Making?" Working paper, Charlottesville: Department of Systems Engineering, University of Virginia.

Wiederholt, B. J., Weaver, M. D., Coberly, V. J., and Porter, A. L. (1989). "Intelligent Information Retrieval: Prospects for the Future Based on Expert Opinion." DA Memo, Norcross, GA: Search Technology, Inc.

Williams, R. M., Jr. (1967). "Individual and Group Values," *The Annals of the American Academy of Political and Social Sciences*, vol. 371, 20–37.

Willyard, C. H. and McClees, C. W. (1987). "Motorola's Technology Roadmap Process," *Research Management*, vol. 30, no. 5, 13–19.

Wilmoth, G. H., Jarboe, K. P., and Sashkin, M. (1984–85). "A Simple Model for Assessing Technological Impact," *Impact Assessment Bulletin*, vol. 3, no. 3, 38–41.

Winner, L. (1977). *Autonomous Technology.* Cambridge, MA: MIT Press.

Wise, G. (1976). "The Accuracy of Technological Forecasts: 1890- 1940," *Futures*, vol. 8, 411–419.

Woelfel, J. and Fink, E. (1980). *The Measurement of Communication Processes: Galileo Theory and Method.* New York: Academic Press.

Wolf, C. P. (1983). "Societal Impact Assessment of Nuclear and Alternative Energy Systems," in Rossini, F. A. and Porter, A. L. (eds.), *Integrated Impact Assessment.* (pp. 255–274). Boulder, CO: Westview Press.

Wolfe, L. D. D. (1987). "Methods for Scoping Environmental Impact Assessments: A Review of Literature and Experience." Prepared for the Federal Environmental Assessment Review Office, Vancouver, BC by Larry Wolfe Associates, 661 Barnham Roade, West Vancouver, BC, Canada V7S 1T6.

Wonnacott, R. J. and Wonnacott, J. H. (1979). *Econometrics.* New York: Wiley.

World Bank (1982). *The Environment, Public Health, and Human Ecology: Considerations for Economic Development.* Washington, DC: author.

World Health Organization (1982). *Rapid Assessment of Sources of Air, Water and Land Pollution.* WHO Offset Publication No. 62. Geneva: author.

Xu, H., Porter, A. L., and Roessner, J. D. (1990). "Indicators of High Technology Competitiveness," Paper presented at the Institute of Management Sciences/Operations Research Society of America, TIMS/ORSA Meeting, Las Vegas (May).

Yin, R. K. (1989). *Case Study Research* (2nd ed.). Newbury Park, CA: Sage.

Zeckhauser, R. J. and Viscusi, W. K. (1990). "Risk Within Reason," *Science*, vol. 248 (4 May), 559–564.

Zwicky, F. (1962). "Morphology of Propulsive Power," Pasedena, CA: Society for Morphological Research.

Zwicky, F. (1969). *Discovery, Invention, Research: Through Morphological Approach.* Toronto: Macmillan Co.

INDEX

A

Accuracy, *see* Evaluation; Validity
Adaptive weighting (see Sec. 9.4)
ADVANTIG, **256–257**
AHP, *see* Analytic hierarchy process
AIMTECH, **234–238**, **377–378**
Air Force (U.S.), 41, 115–116; *see also* AIMTECH
Air quality, 302, 320
Alternative futures, *see* Scenarios
Aluminum industry, 276
Analogies, 104
Analytic Hierarchy Process (see Sec. 18.5), 401
Annual fractional growth, 199
Appropriate technology, *see* Technology, appropriate
Army Corps of Engineers (U.S.), 335
Artificial intelligence, 67
Asimov, 260
Assumptions
 core, 57
 drag, 63
Attitudes, 313
Attribute analysis, 103, 106
Autocorrelation, 167
Automated storage and retrieval systems (AS/RS), 252–253

B

Backward planning, 368
Bayes rule, 226, 239
Bellamy, Edward, 380
Benefit/cost analysis, 368
 societal (see Sec. 17.4)
 within firm, 324–327
Biases, 56–57, 167, 265
Black & Decker, Inc., 43
Bounding, 296–297
Brainstorming, 108–110, 209
Brundtland Commission, 341

C

Cable television, *see* Television, cable
Canada, 305
Case study, 382
CATV, *see* Television, cable
Checklists, 104–105
China, 117, 121
Churchman, 89
CI (*see* Cross-impact)
Clean Air Act Amendments (U.S.), 320, 321
Coefficient of determination, *see* R^2
Commerce, (U.S.) Department of, 15

Commission for Computer Integration (Rose-Hulman Institute of Technology), 75, 202, 208, 211–**212**, **214–216**
Committees, 207–209
Communicating; *see also* Project Management, communications
  among experts, 211
  to users, 97–98, 129–130
Compaq Computers, 41
Competitive advantage (see Sec. 1.2)
Competitiveness
  technological, **7**
  Computers, **6–7**, **24**, 56, 276; *see also* Commission for Computer Integration
  Confidence intervals (C.I.), 153, 159, 166
  Congress; *see also* OTA
  Critical Trends Assessment Act, **262**
Congressional Clearinghouse for the Future, 117
Consistency ratio, 365–367
Consumer surplus, 336
Context
  assumptions for scenarios, 264–265
  change, 307
  mapping, 126, 127
  oversight, 56
Copper mining, **196**
Cost/benefit analysis, *see* benefit/cost analysis
Crawford slip writing, 110
Creativity techniques (see Ch. 7), 401–402
  group (see Sec. 7.3)
Criteria, *see* Evaluation, criteria
Critical success factors, 41
Critiquing, *see* Evaluation; Validity
Cross-impact (CI) (see Sec. 12.2), 65, 239–241, 402; *see also* KSIM, Ximpact
  conditional probabilities, 225, 228
  dynamic, 241–243; *see also* Cross-impact, time-dependent
  event-on-event (see Sec. 12.2), 239–241
  event-on-trend, 238–239, 241–246
  Monte Carlo simulation, 227–228
  occurrence matrix, **224**

  problems, 239
  procedure, 239–240
  nonoccurrence matrix, **227**
  second order effects, 299
  stochastic, *see* Cross-impact, traditional
  traditional (see Sec. 12.2.1)
  time-dependent (see Sec. 12.2.2), 240–241
  trend-on-trend, 241–246
Cyclic models, 18; *see also* Long wave models
  business, 277

D

Data, 27, 90–91, 150; *see also* Information
  bases, 48
  plotting, *see* Plotting
  secondary, 310
Defense, (U.S.) Department of, 67
Delphi, 91, **214–217**
  variants, 218–219
Demand modeling, 265
Demographic models, 188, 312
Department of Commerce (U.S.)
  Business Conditions Digest, 278
  Survey of Current Business, 279–280, 283
Development; *see also* R&D, Technology growth
  sustainable, 341
Dose-response, 344
Dummy variables, 166
DYNAMO, 248, 250, 258

E

Economic development
  firms' suitability, 296
Economics (see Ch. 14)
  and technological change, 273–276
  indicators, **278**
  output model, 277
Ecosystem, 319
Educational system, **251–252**
EFTE procedure, **205**, **218–219**
EIA, *see* Environmental impact assessment

Employment
  impacts, 308–309
  projections, 312
Energy development, 313
Energy forecast, 51, **192–194**, 265–266,
    297–298; *see also* Power plants,
    nuclear
Envelope curves, 61, 170
Environmental health impact assessment,
    322
Environmental impact assessment (see
    Sec. 16.8), 39
  methods, 318–319
  modeling, 319
  U.S. requirements, 290
Equity, 356
Evaluation (see Ch. 18); *see also* Biases;
    Forecasting, errors; Utility; Validity
  alternatives, 358–359
  criteria, 354, 354–**355**
  criteria weighting, **362**
  formative, 383
  impact assessment, 291–292
  interrogation model, 52, **53–54**, 355
  of forecasts and assessments, 383–386
  summative, 384
Expected value, **329–331**
EXPERT CHOICE, 364
Expert opinion methods (see Ch. 11),
    94–95, 310
  and scenarios, 265
  expert selection (see Sec. 11.2)
  group input techniques, 207–209
  individual input techniques, 206–207
  judgment, 300
  technique selection (see Sec. 11.3)
  zeitgeist, 186
Exponential growth, 61, 164, 170,
    191–192, 404–405
Exponential smoothing, **146–147**
Exposure assessment, 344
Expressed preferences, 346
Externalities, **340–341**
Extrapolative future, 386

F

F statistic, **App. B.1**, **160–162**, 166
Fiber optics, **6–7**
Fisher-Pry model (see Sec. 10.3.1),
    59–60, 155–156, 164, 188,

191–192, 195, 404–405; *see also*
    S-shaped curves
  vs. Gompertz (see Sec. 10.3.3)
Forecast
  accuracy, *see* Evaluation
  for 1990s, **55–56**
Forecasting (see Ch. 4); *see also* Methods
  commandments, **385–386**
  economic (see Ch. 14)
  errors, 56–57, 186, 384–385; *see also*
    Evaluation
  extrapolative, 50, 72
  long range, 188, **192–194**
  normative, 33, 50, 72
  social, *see* Social forecasting
  technique families, 64
  technique selection (see Sec. 6.3.1)
  upper bounds, 187
Fractionation, 103
Friedman, Milton, 279
Future
  extrapolative, 386
  normative, 386
  perspective, 381
  uncertainty, 12–13, 49–**50**, 386
Futures foregone index, 360

G

Galileo method, 313
Gaming (see Sec. 12.5)
Gantt chart (see Sec. 5.3.2), 403–404
Gatekeepers, 132
Gaussian distribution, 122
GE, **12**
Goal setting (see Sec. 3.2.4)
Gompertz model (see Sec. 10.3.2), 59,
    60, 164, 182–183, 191–192,
    404–405; *see also* mortality model
    and S-shaped curves
  vs. Fisher-Pry (see Sec. 10.3.3)
Gore, Albert, 262, 317
Graphing, *see* Plotting
Growth models (see Sec. 4.2.1), 59–61, 65

H

Hazard
  energy, 347
  identification, 343
  materials, 347

Health
    care scenarios in the Netherlands,
        268–269
    impacts, 321–322
    risk assessment, 342–345
High temperature superconductivity,
    304–305
Huxley, Aldous, 260–261
Hyperbolic functions, 164

I

IBM, 276
Impact and probability of occurrence
    matrix, **127**
Impact analysis (see Ch. 16)
Impact assessment (see Sec. 3.2.2; Ch.
    15), 22; *see also* Environmental
    impact assessment, Social impact
    assessment, Risk assessment,
    Technology assessment
    brief, 292–294
    criteria, 291–292
    evaluation of, 381–383
    quality, 293–294
    scales of effort, **294**
    steps, **292–293**
Impact identification (see Sec. 15.3)
Impact likelihood versus desirability
    matrix, 360–361
Impacts
    behavioral, 315
    cultural, 313–315
    fiscal, 312
    health, 321–322
    horizontal, 305–307
    integrative, 307
    institutional, 308–309
    international, 317–318
    legal, 316–317
    matrix, 299
    narrowing the set, 300
    organizational, 308
    political, 315–316
    psychological, 313–314
    vertical, 304–305
Information
    electronic databases, 133
    requirements, 74
Input-output analysis (see Sec. 14.5)

Inquiring systems (see Sec. 6.1)
Innovation
    processes (see Sec. 2.4), 273–276
Internal rate of return, 327
Interindustry transaction table, **280**, 285
Interpretive structural modeling (ISM),
    362
Interviews, 206–207, 298
Issues management, 115

K

KSIM (Kane's simulation) (see Sec. 12.3),
    403

L

Land (use) impacts, 302, 319–320
Lateral thinking, 101–102
Law of diminishing returns, 273
Leading indicators (lead-lag), 64, 157,
    278
Learning curve, 173–**174**
Least squares calculations, *see* Regression
    analyses, least squares calculations
Legal analysis, 316–317
Leontief, Wassily, 279
Life, value of, **350**
Limits to Growth, The, 384
Linear model, 191, 404–405; *see also*
    Regression, linear
Lock, the, **52**
Logistic curves, *see* S-shaped curves
Long wave models, 194; *see also* cyclic
    models
Lotka-Volterra equation (see Sec. 10.4)

M

Maddox Planning Framework, **40–41**
Managing
    from the future (see Ch. 19), 12
    with uncertainty, 13
Manufacturing, 10–12, **256–257**
Mapping, **127**, 305
Marchetti's multiple substitution analyses,
    192–197
Market-driven, 11
Market economy, 31
Market penetration, 185, 194
Market potential, 190

Markov process, 230–238
MAUT, 369–374
Mediation, 375–376
Measures (see Sec. 18.4), 55, 140–141,
 354
 levels, 359–361
Megamistakes, **384–385**
Megatrends, 94, 381
Metaphors, *see* Analogies
Methods
 correlative, 64–**65**
 direct, 64–**65**
 structural, **65**–66
Mission statements (see Sec. 3.2.4)
Models, 90, **96**; *see also* Simulation,
 Technological growth
 naive (see Sec. 9.3), 147
Monitoring (see Ch. 8), 9, 93–**94**, 379
 contextual, 115, 119
 dispersed, **132–133**
 system, **120**
MOOT, 367–372
Morphology (morphological analysis),
 66, 105–106, 266
Mortality model, 182–183, *see also*
 Gompertz model
Motorola, Inc., 44
Multiattribute utility theory, *see* MAUT
Multiple objective decision-aiding
 techniques, **372**
Multiple objective optimization theory,
 see MOOT
Multiple perspectives, 385
Multipliers, **311–312**
Multinational, **4–5**
Myth of abundance, 349

N

Nanotechnology, 220–221
National Environmental Policy Act (U.S.),
 39, 290
NEPA, *see* National Environmental Policy
 Act
Net present value, *see* Present value
Newly Industrialized Countries (NICs), 8
National Institute for Standards and
 Technology, *see* NIST
NGP, *see* Nominal Group Process
NIMBY (not in my backyard), 356

NIST, 121, 125
Noise, 321
Nominal group process, 111, **209–212**,
 297
Nonmonetary effects, 339
Normative future, 386
Nuclear Regulatory Commission (U.S.),
 267
Nuclear power, *see* Power plants, nuclear

O

Observation methods, 310
Opportunity costs, 324–325
Organizational change, 308
Orwell, 261
Oscillatory models, **196**
OTA (Office of Technology Assessment),
 290

P

Pairwise comparison, 361–363
Panels, 91
Parabolic functions, 164
Paradigm changes, 92
Pareto optimality, 336
Participation, 374–375
 within organization, 35
Patent analysis, **306**
Pearl curve, 164–165, 176, 191–192; *see*
 also Fisher-Pry model
PERT (see Sec. 5.3.1), 404
Planning (see Ch. 3)
 backward, 33
 business, 35
 central, 31
 strategic, 32
 technology (see Sec. 3.1.2)
Plotting, 150, 163, 173
Policy capture, **355–356**
Population forecasts, *see* Demographic
 models
Posture technique, 219–220
Power functions, 164, 173
Power plants, 271, **301–302**
 nuclear, 75, 267, 298–299, 308–309,
 317
Precursors of technological change,
 121–122

Preference structures, 374; *see also*
    Revealed preferences; Risk
    perception
Present value, **324–328**
Problem formulation, 292–294
Producers' surplus, 337
Productivity, 8, 307
Project accountability chart (PAC) (see
    Sec. 5.3.3)
Project evaluation and review technique;
    see PERT
Project management (see Ch. 5)
    communications, 84–85
    multiskill teams, 80–84
Project scheduling (see Sec. 5.3),
    403–404
Public goods, **340–341**

R

R^2, 160, 166
Radiation, 321
R&D (Research and Development), 9,
    74, 132, 170–172, 274
Random words, *see* Trigger words
Ranking, 359–360
Rating, 359–360
Regression analyses (see Sec. 9.5),
    66
    and causality, **148–149**
    approximation method, **154–156**
    least squares calculations (see Sec.
        9.5.2), **165**
    linear (see Sec. 9.5)
    multiple, 165
    seasonal adjustment, 166
Relative importance scale, **364**
Relevance trees, 299
Replacement, 186
Requirements analysis, 358
Research, *see* R&D
Residuals, 152
Revealed preferences, 346, 350
Reverence, 357
Reversal, 103–104
Risk assessment, 39, 329
    environmental, 342
    health, 343–345
    political, 314
    probabilistic, 342

social (see Sec. 17.5)
    within firm (see Sec. 17.3)
Risk
    characterization, 345
    classes, 347
    perception, 340, **347–349**, 349
    premium, 331
    sequential decisions, **348–349**
Role-playing, 316
Roper Organization, 124
Rose-Hulman Institute of Technology, 75;
    see also Commission for Computer
    Integration

S

Saaty, Thomas, *see* Analytic hierarchy
    process
Sample standard deviation (s_y), 151,
    161
Sample variance (s_y^2), 151, 161
Scanning techniques, 298–299
Scenarios (see Ch. 13), 64, 68, 73,
    96–97, 239, 314, 387–388
    Analytic hierarchy process, 366
    construction checklist, **263**
    descriptive, 261
    health care in the Netherlands,
        268–269
    future histories, 260
    normative, 261
    nuclear plant decommissioning, 267
    risk, 329–331
    snapshots, 261
    users, 264
Screening methods, 318–319
Search Technology, Inc., 41
Secondary data analysis, 310
Sensitivity analysis, 174, 363
Service(s), 11
Shang inquiry, 218
SIA, *see* Social impact assessment
Simulation (see Ch. 12), 66, 68, 248, 257
    gaming, *see* gaming
    Monte Carlo, 227–228
Sketch methods, 319
Social change
    theories (see Sec. 2.1), 386
Social forecasting (see Sec. 4.3), 49, 51,
    291

Social impact assessment, 39, 291–292, 309–310
Social indicators, 68–69
Socioeconomic impact assessment, **310–312**
Sociotechnical system, 20
external influences, **20**
theory, 386
S-shaped curves (see Sec. 10.3), 61, 163, 169, 241; *see also* Fisher-Pry model; Gompertz model
SST (supersonic transport), 139
STELLA, 248, 250
Substitution, *see* Technological substitution
Surveys, 212–214; *see also* Expert opinion methods
Suspended judgment, 102–103
Sustainable development, 341
Synectics, 110–111
System, 189–190; *see also* System dynamics; Technology delivery system
world, 254
System dynamics (see Sec. 12.4)

T

t statistic, **App. B.2**, 160, **162–163**, 166
TDS, *see* Technology delivery system
Technical progress function, *see* Technological progress function
Technological change, *see also* Technology growth
historical lessons, 381–383
propagation, 304–305
Technological competition, 192
Technological development, *see* Technology growth
Technological diffusion, *see* Technology diffusion
Technological discontinuities, **131**, **139**, 305
Technological progress function, 10, 170–**172**
Technological substitution, 73, 164, 186, 191–192; *see also* Fisher-Pry model
-multiple substitution analysis, 192–197

Technology
appropriate, 37, 274
audit, 43–**45**
decomposing, 275
description, 126
-driven, 11
diffusion, 58
forecasting, *see* Forecasting
high, **7**
infrastructure for, 276
replacement, 186
roadmap, **44**
spin-off, 307
successor, 194
Technology assessment, 38, 289
comparing technologies, 289
retrospective, 38, 380
Technology delivery system (see Sec. 2.3), 5, 130, 307, 309, 316
Technology growth, **58–61**; *see also* S-shaped curves
and values, 312
constant growth model, 143–144
constant fractional growth model, 144–145
economics, 273–274
empirical cases, 197
general model, *see* Lotka-Volterra equation
transformations, **164**
Television, 56
cable (CATV), 140, **141–145**, **152–162**, **178–187**, 396
effects of, **38**
high definition (HDTV), 56
Threshold (safety), 344
Three-point method, *see* Regression analyses, approximation method
Time constant, 194
Time series, 167; *see also* Regression analyses, Trend analysis
Time value of money, 325, 339
TOOLKIT (see Appendix A), 15, 76, 105, 106, 158, 167, 199, 228, 257–258, 300, 360, 364, 379
Toxic wastes, 320
Tracing techniques, 299–300
Transcendence, 356–357
Transformations, 163–165
least squares calculations with, **165**

Trigger words, **107**
Trend analysis (see Chapters 9 & 10), **93**,
    404–405
  and scenarios, 265
  steps (see Sec 10.2)
Trend extrapolation, *see* Trend analysis

U

Upper bounds, 187
Utility, 269, 356
Uncertainty, 329, **348–349**,
    386

V

Validity, 148–149, 179–180, 269–270

Value of life, **350**
Values, 313–314
Variables, *see* Measures
Verne, Jules, 260
Vertical thinking, 102
Visions, 387–389

W

Water quality, 301, 320
Whirlpool Corporation, 34, 74, **116–117**,
    289–290
World Health Organization, 322

X

Ximpact, 237–238